Environmental Analysis

HANDBOOK OF ANALYTICAL SEPARATIONS

Series Editor: ROGER M. SMITH

In this series:

Vol. 1: Separation Methods in Drug Synthesis and Purification
Edited by K. Valkó
Vol. 2: Forensic Science
Edited by M.J. Bogusz
Vol. 3: Environmental Analysis
Edited by W. Kleiböhmer

Environmental Analysis

Edited by

WOLFGANG KLEIBÖHMER

Institut für Chemo- und Biosensorik
D-48149 Münster, Germany

2001

ELSEVIER
Amsterdam – London – New York – Oxford – Paris – Shannon – Tokyo

ELSEVIER SCIENCE B.V.
Sara Burgerhartstraat 25
P.O. Box 211, 1000 AE Amsterdam, The Netherlands

First edition 2001

Library of Congress Cataloging-in-Publication Data
Environmental analysis / edited by Wolfgang Kleiböhmer.
      p. cm -- (Handbook of analytical separations; v. 3)
   ISBN 0-444-50021-9
    1. Environmental chemistry--Methodology. 2. Chemistry, Analytic--Methodology. 3.
Pollutants--Analysis. I. Kleiböhmer, Wolfgang. II. Series

QD63.S4 H36 2000 vol. 3
[TD193]
660'.2842 s--dc21
[628.5'028'7]

2001046022

British Library Cataloguing in Publication Data
Environmental analysis. - (Handbook of analytical
  separations; v. 3)
  1. Environmental chemistry 2. Separation (Technology)
  I. Kleiböhmer, Wolfgang
  577.1'4
  ISBN 0444500219

ISBN: 0-444-50021-9

Editor's preface

This book is not supposed to be an addition to the long list of excellent textbooks dealing with analytical separation techniques in environmental analysis. The objective of this work is to give an up-to-date review on both new solutions for well-known, but still lasting, problems and on solutions for new problems/questions in environmental analysis.

Therefore the book covers a critical compilation of analytical methods for the determination of a wide range of environmental priority pollutants, including amines, polycyclic aromatic hydrocarbons, pesticides, phenols, PCBs, and also including organometallic species, polycyclic aromatic sulphur heterocycles and metabolites of polycyclic aromatic hydrocarbons. For all environmental pollutants, an overview and an assessment of value of a broad range of relevant methods including conventional techniques but also of advanced separation and sample preparation techniques are given, with all possible matrices being considered.

The book does not only give up-to-date material on sampling and sample pre-treatment, extraction techniques, clean-up, pre-fractionation, different types of chromatographic methods and quality assurance, but also information on actual and upcoming analytical problems, such as the determination of polycyclic aromatic sulphur heterocycles and of metabolites of PAH. The final chapter is devoted to the important role of analytical separation methods in water quality control.

Experienced and well-known authors have contributed their knowledge and expertise to their individual chapters. I hope this volume will prove helpful, and I would like to thank all colleagues who accepted to be involved in this project for their enthusiasm and dedication as well as for their cooperation. Last, but not least, I thank Reina Bolt at Elsevier Science for her active collaboration and patience during the production of this book.

Wolfgang Kleiböhmer

Series editor's preface

This volume on Environmental Analysis is the part of what is growing to be the Handbook of Analytical Separations. This volume reflects the importance of analytical separation methods in monitoring and identifying the many compounds of environmental importance. It includes chapters on the main groups of analytes of interest from PAHs and PCBs to phenols, sulphur compounds and pesticides. These methods illustrate the wide range of analytical techniques that have been employed in the measurement of environmental constituents and different matrices that have been examined.

The Handbook of Analytical Separations will be a comprehensive work, which is intended to recognise the importance of the wide range of separation methods in analytical chemistry. Since the first report of chromatography almost a 100 years ago, separation methods have expanded considerably, both in the number of techniques and in the breadth of their applications. The objective of the Handbook is to provide a critical and up-to-date survey, rather than a detailed review, of the analytical separation methods and techniques used for the determination of analytes across the whole range of applications. The Handbook will cover the application of analytical separation methods from partitioning in sample preparation through gas, supercritical and liquid chromatography to electrically driven separations. The intention is to provide a work of reference that will provide critical guidance to the different methods that have been applied for particular analytes, their relative value to the user and their advantages and pitfalls. The aim is not to be comprehensive but to ensure a full coverage of the field weighted to reflect the acceptance of each alternative method to the analyst.

The individual self-contained volumes will each encompass a closely related field of applications and will demonstrate those methods which have found the widest applications in the area. The emphasis is expected to be on the comparison of published and established methods which have been employed in the application area rather than the details of experimental and novel methods. The volumes will also identify future trends and the potential impact of new technologies and new separation methods. The volumes will therefore provide up-to-date critical surveys of the roles that analytical separations play now and in the future in research, development and production, across the wide range of the fine and heavy chemical industry, pharmaceuticals, health care, food production and the environment. It will not be a laboratory guide but a source book of established and potential methods based on the literature that can be consulted by the reader.

I am pleased to acknowledge that the value of the Handbook will be dependent on the volume editors and the contributors that they will bring to each topic. It is their experience and expertise that will provide the insights into the present and future development of separation methods.

Roger M. Smith
Series Editor

Contents

List of contributors

JAN T. ANDERSSON — *Department of Analytical Chemistry, University of Münster, Wilhelm-Klemm-Strasse 8, D-48149 Münster, Germany*

JOHN R. DEAN — *School of Applied and Molecular Sciences, University of Northumbria at Newcastle, Newcastle upon Tyne NE1 8ST, UK*

JACOB DE BOER — *Netherlands Institute for Fisheries Research, P.O. Box 68, 1970 AB IJmuiden, The Netherlands*

LISA J. FITZPATRICK — *School of Applied and Molecular Sciences, University of Northumbria at Newcastle, Newcastle upon Tyne NE1 8ST, UK*

FRITZ H. FRIMMEL — *Engler-Bunte-Institut, Chair of Water Chemistry, Universität Karlsruhe (TH), Engler-Bunte-Ring 1, D-76131 Karlsruhe, Germany*

M. TERESA GALCERAN — *Department of Analytical Chemistry, University of Barcelona, Diagonal 647, E-08028 Barcelona, Spain*

OLGA JÁUREGUI — *Department of Analytical Chemistry, University of Barcelona, E-08028 Barcelona, Spain*

ANNE HÖNER — *Technische Universität Berlin, Strasse des 17. Juni 135, D-10623 Berlin, Germany*

HIROYUKI KATAOKA — *Faculty of Pharmaceutical Sciences, Okayama University, Tsushima, Okayama 700-8530, Japan*

HIAN KEE LEE — *Department of Chemistry, National University of Singapore, 3 Science Drive 3, Singapore 117543, Republic of Singapore*

SHIZUO NARIMATSU — *Faculty of Pharmaceutical Sciences, Okayama University, Tsushima, Okayama 700-8530, Japan*

ROLF-DIETER WILKEN — *ESWE-Institute for Water Research and Water Technology, D-65201 Wiesbaden, Germany*

SHIGEO YAMAMOTO — *Faculty of Pharmaceutical Sciences, Okayama University, Tsushima, Okayama 700-8530, Japan*

CHRISTIAN ZWIENER — *Engler-Bunte-Institut, Chair of Water Chemistry, Universität Karlsruhe (TH), Engler-Bunte-Ring 1, D-76131 Karlsruhe, Germany*

CHAPTER 1

Gas chromatographic analysis of environmental amines with selective detectors

Hiroyuki Kataoka, Shigeo Yamamoto and Shizuo Narimatsu

Faculty of Pharmaceutical Sciences, Okayama University, Tsushima, Okayama 700-8530, Japan

1.1 INTRODUCTION

As environmental issues and global environmental change are generating an increasing amount of attention world-wide, the occurrence of hazardous chemicals with significant toxicities, such as carcinogenicity, mutagenicity, teratogenicity and reproductive abnormalities, in the biosphere have received a great deal of attention in recent years. As science and technology progressed and man's appetite for a carefree life expanded, the problems of environmental pollution became more complex and difficult to alleviate [1,2]. The environment contains a variety of naturally occurring and man-made pollutants to which humans are exposed. Every year many new substances are also synthesized that differ radically from the natural products that exist in biosystems. Many of these substances are not biodegradable and will thus progressively pollute the environment. Among the many environmental pollutants, the occurrence and determination of amines have recently received a great deal of attention. These amines occur in a number of ambient environments such as air, water, soil and foods, and become a source of serious social and hygienic problems [3,4]. Therefore, knowledge of the pollutions with environmental amines has become an urgent requirement for the protection of the terrestrial environment.

1.1.1 Source and toxicity of environmental amines

Aliphatic and aromatic mono-, di- and polyamines are naturally occurring compounds formed as metabolic products in microorganisms, plants and animals, in which the principal routes of amine formation include the decarboxylation of amino acids, amination of carbonyl compounds and degradation of nitrogen-containing compounds. These amines are considered to play a significant role as storage sources of nitrogen

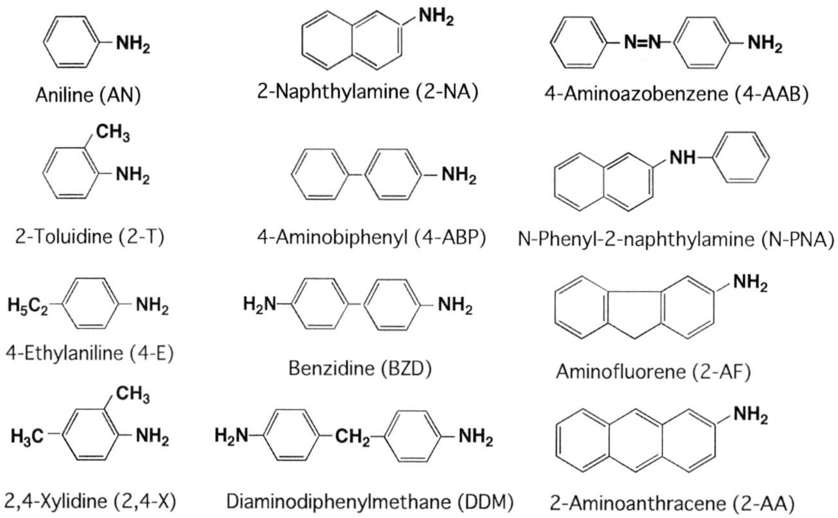

Aniline (AN) 2-Naphthylamine (2-NA) 4-Aminoazobenzene (4-AAB)

2-Toluidine (2-T) 4-Aminobiphenyl (4-ABP) N-Phenyl-2-naphthylamine (N-PNA)

4-Ethylaniline (4-E) Benzidine (BZD) Aminofluorene (2-AF)

2,4-Xylidine (2,4-X) Diaminodiphenylmethane (DDM) 2-Aminoanthracene (2-AA)

Fig. 1.1. Typical toxic aromatic amines in the environment.

and precursor for the synthesis of hormones, alkaloids, nucleic acids and proteins. Amines are also widely used as raw materials or at an intermediate stage in the manufacture of industrial chemicals [3–7]. In particular, aromatic amines such as aniline and its substituted analogues, phenylene diamine and diphenylamine have been widely used as industrial intermediates [3,8–11] in the manufacture of carbamate and urethane pesticides, dyestuffs, cosmetics, pharmaceuticals, photographic developers, shoe polish, resins, varnish and perfumes. These amines have also been employed in the rubber industry as antioxidants and antiozonants [6] and as components in epoxy and polyurethane polymers [5,12]. Typical toxic aromatic amines are shown in Fig. 1.1. An additional source for amines in the environment is the abiotic and biotic degradation of animal waste, domestic waste and industrial products [10,13,14]. Many of these amines have been discharged into the atmosphere [12,15–32] and water [20,33–52] from anthropogenic sources such as cattle feedlot and near livestock buildings [15], waste incineration, sewage treatment, automobile exhaust, cigarette smoke, [13,27,53–59] soil [24,25,40,43,60], foods [61–66] and various industries [3,4,7,8,67]. Thus these amines have also been detected in biological samples [14,37,68–74]. Many of amines have an unpleasant smell and are hazardous to health, i.e. as sensitizers and irritants to the skin, eye, mucous membrane and respiratory tract [75,76]. Acute and chronic exposure to some aromatic amines produces symptoms of headache, dizziness, nausea and methemoglobin [72]. Some of amines are also suspected to be allergic and mutagenic or carcinogenic substances due to their adsorption tendency in tissues [77,78]. In particular, the toxicity of aromatic amines to mammals and fish is well established [78–80], and many of these amines are known to be highly mutagenic and carcinogenic and to form adducts with proteins and DNA [8,77,81–90]. Several polycyclic aromatic amines such as benzidine, 4-aminobiphenyl and 2-naphthylamine, have also been classified by the International Agency for Research on Cancer (IARC) as known human carcinogens.

Volatile N-nitrosamines

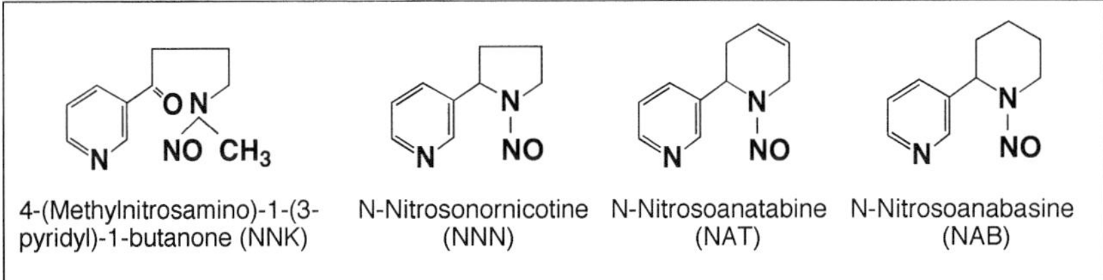

Tobacco-specific N-nitrosamines

Fig. 1.2. Typical toxic *N*-nitrosamines in the environment.

Aliphatic and aromatic amines are not only toxic of themselves but also become toxic *N*-nitrosamines through chemical reactions of a secondary or tertiary amines with nitrosating agents such as nitrite or nitrate [91–95]. The reaction usually takes place in acidic solution, e.g. in the human diets, the environments and in vivo in the stomach or small intestine of experimental animals [96–99]. The reaction of nitrosating agents with primary amines produces short-lived alkylating species that react with other compounds in the matrix to generate products (mainly alcohols) devoid of toxic activity in the relevant concentrations. The nitrosation of secondary amines leads to the formation of stable *N*-nitrosamines while that of tertiary amines slowly produces a range of labile *N*-nitroso products. Although reaction pathways are uncertain, it has also been demonstrated in model experiments that amines react with NOx and OH radicals in air to form nitrosamines and nitramines [67,100]. Moreover, it is considered that volatile *N*-nitrosamines in cigarette smoke are formed from decarboxylation of *N*-nitroso amino acids, pyrolytic nitrosation of secondary amines and concentrated decarboxylation/nitrosation of amino acids during combustion process [94]. Typical *N*-nitrosamines are shown in Fig. 1.2. *N*-Nitrosamines are widely distributed in foodstuffs [66,92,93,101–105] and various human environments such as indoor and outdoor airs [97,106,107], combustion smokes [56,94,108–116], water [117–121], sediment [122], household dishwashing liquids [123], foods [92,101,103–105,124], rubber products [124–132], metalworking fluids [133], drug formulations [134,135] and agricultural chemical formulations [123,136,137]. The presence of carcinogenic tobacco-specific nitrosamines in tobacco products have also been demonstrated and seven nitrosamines have been identified. Furthermore, some *N*-nitrosamines have been detected in human gastric juice, saliva and cervical mucus [138,139]. *N*-Nitrosamines represent a major class of important chemical mutagens, carcinogens, teratogens and immunotoxic agents, which have been described as a serious hazard to human health [77,140–143]. The toxic

Pyrolytic mutagens

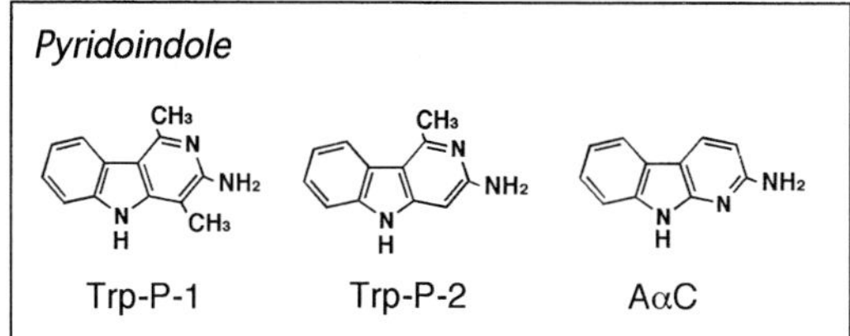

Thermic mutagens

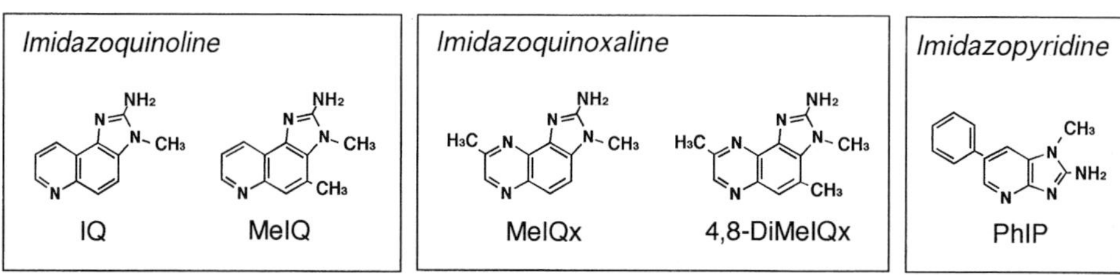

Fig. 1.3. Typical mutagenic and carcinogenic heterocyclic amines.

effects almost always include centritubular necrosis of the liver, and hemorrhaging is also common at high doses. Recent developments in environmental carcinogenesis have demonstrated that *N*-nitrosamines lead to a wide variety of tumors in many animals, which are likely to result in the formation of DNA adducts [77,88,144–148]. The toxicity of *N*-nitrosamines can be manifested even at a μg kg^{-1} level.

On the other hand, a new series of heterocyclic amines formed during heating of amino acids, proteins, creatinine and sugars are potent mutagens in the Ames/*Salmonella* assay [149–156]. Up to the present, 23 heterocyclic amines have been isolated as mutagens, and the structures of two main amine groups are shown in Fig. 1.3. The first group 'pyrolytic mutagens', pyridoindole and pyridoimidazole, are formed by pyrolysing single amino acids and single proteins at temperatures above 300°C. The second group 'thermic mutagens', imidazoquinoline, imidazoquinoxaline and imidazopyridine, are formed from creatine/creatinine, amino acids and sugars at temperatures less than 300°C. Many of these mutagenic heterocyclic amines have been isolated and identified not only from various proteinaceous foods [157–177] including cooked meats and fish, but also from environmental components such as airborne particles and diesel-exhaust particles [178,179], combustion smokes [180–186] indoor air [187–189], cooking fumes [190–192], rain water [187,188,193] and river water [194–196]. Moreover, some mutagenic heterocyclic amines have been detected in biological samples such as urine, plasma, bile and feces [166,197–200]. These facts suggest that heterocyclic amines may be emitted into the atmosphere through combustion of various materials such as foods, wood, grass, garbage and petroleum, and discharged into the water through domestic waste and human waste, although their mechanisms have not been determined. Some heterocyclic amines have much higher mutagenic activity than typical mutagens/carcinogens such as aflatoxin B$_1$, AF-2 and benzo[*a*]pyrene and have been verified to be carcinogenic in rats and mice [149–153,155,201,202]. These toxicities are known to arise from formation of adducts with DNA [203,204]. These compounds induced tumors in the liver, small and large intestine, Zymbal gland, clitoral gland, skin, oral cavity and mammary gland in rats, and the liver, forestomach, lung, hematopoietic system, lymphoid tissue and blood vessels in mice. 2-Amino-3-methylimidazo[4,5-*f*]quinoline (IQ) was also found to be carcinogenic in the monkey, inducing hepatocellular carcinomas [205,206]. Moreover, recent investigations revealed that heterocyclic amines also possess cardiotoxic effect [202] and various pharmaco-toxicological activities such as convulsant activities [207,208] and potent inhibitory effects on platelet function and dopamine metabolism [208–212]. However, significant risks on human health and environment through long-term exposure and bioaccumulation of heterocyclic amines are scientifically unclear.

1.1.2 Selective detection of amines by gas chromatography

Gas chromatography (GC) has been widely used for amine analysis because of its inherent advantages of simplicity, high resolving power, high sensitivity, short analysis time and low cost. However, GC analysis of free amines generally has some inherent problems related to the difficulty in handling low-molecular mass amines because of their high water solubility and high volatility. Therefore, these amines are difficult to extract from water, and not easily chromatographed due to their polarity. Furthermore, amines tend to be adsorbed and decomposed on the columns, and readily give tailed elution peaks, ghosting phenomena and low detector sensitivity. The adsorption tendency in the analytical systems, i.e. in sample vessels, injection system, glass wool and GC column, is in order primary > secondary > tertiary amines, and it is generally more difficult to chromatograph aliphatic than aromatic amines. A common method of overcoming these problems is to convert such polar compounds to relatively non-polar derivatives more suitable for GC analysis. Derivatization methods have been employed to reduce the polarity of the amino group and improve GC properties. Derivatization reactions, often selective for amine type (primary, secondary, tertiary), have also been used to improve detection and separation of these amines. The reaction schemes for typical derivatization of amines are shown in Fig. 1.4. A number of derivatives such as acyl, silyl, dinitrophenyl, permethyl, Shiff base, carbamate, sulfonamide and phosphonamide compounds have been used for this purpose. For example, introducing halogen- and phosphorus-containing groups in the molecule enhances the response of electron capture detector (ECD) and flame photometric detector (FPD), respectively. These derivatization reactions are described in detail in a previous review [4]. For amine analysis, a wide variety of detectors such as thermionic detector, FPD, ECD and chemiluminescence detector (CLD) can be used, and offer increased selectivity for specific amines. Furthermore, the combined technique of GC–mass spectrometry (MS) can provide structure information for the unequivocal identification of amines and these amines can be determined by mass selective detector (MSD) based on the selected ion monitoring (SIM). By using these detectors, sub-nanogram detection limits can be achieved. Although most detectors respond directly to amines, some detectors, such as FPD and ECD, need conversion of amines to suitable derivative prior to detection.

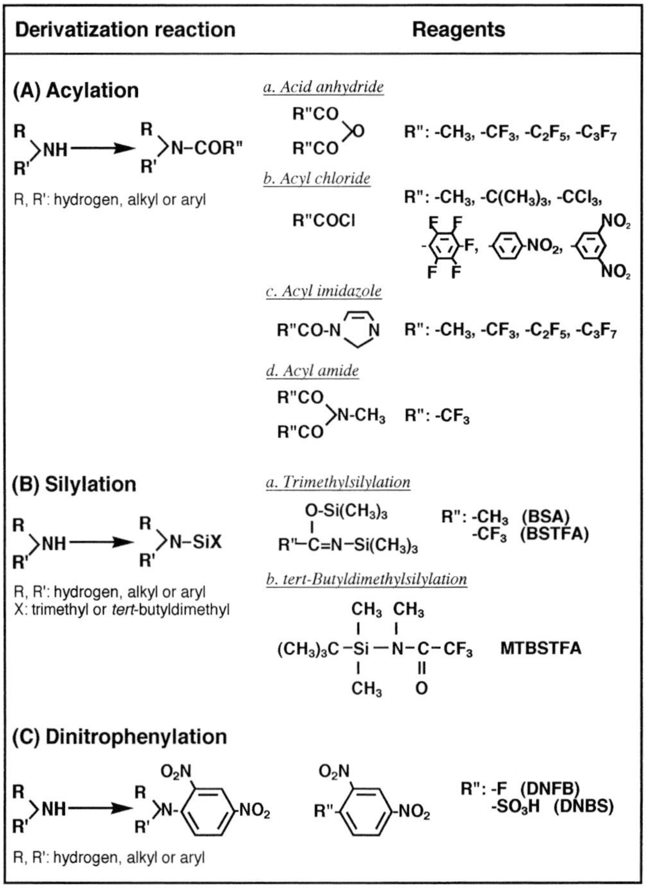

Derivatization reaction	Reagents

(A) Acylation

$R,R'{>}NH \longrightarrow R,R'{>}N{-}COR''$

R, R': hydrogen, alkyl or aryl

a. Acid anhydride

$\begin{matrix}R''CO\\R''CO\end{matrix}{>}O$ R": $-CH_3$, $-CF_3$, $-C_2F_5$, $-C_3F_7$

b. Acyl chloride

$R''COCl$ R": $-CH_3$, $-C(CH_3)_3$, $-CCl_3$,

c. Acyl imidazole

$R''CO-N\underset{\smile}{\frown}N$ R": $-CH_3$, $-CF_3$, $-C_2F_5$, $-C_3F_7$

d. Acyl amide

$\begin{matrix}R''CO\\R''CO\end{matrix}{>}N{-}CH_3$ R": $-CF_3$

(B) Silylation

$R,R'{>}NH \longrightarrow R,R'{>}N{-}SiX$

R, R': hydrogen, alkyl or aryl
X: trimethyl or *tert*-butyldimethyl

a. Trimethylsilylation

$R''{-}C{=}N{-}Si(CH_3)_3$ with $O{-}Si(CH_3)_3$ R": $-CH_3$ (BSA), $-CF_3$ (BSTFA)

b. tert-Butyldimethylsilylation

$(CH_3)_3C{-}Si(CH_3)_2{-}N{-}C(CF_3){=}O$ MTBSTFA

(C) Dinitrophenylation

$R,R'{>}NH \longrightarrow R,R'{>}N{-}(2,4{-}dinitrophenyl)$

R'': $-F$ (DNFB), $-SO_3H$ (DNBS)

R, R': hydrogen, alkyl or aryl

Fig. 1.4. Typical derivatization reaction for amines.

1.1.3 Objective and scope of this review

The presence, identity and quantity of amines in the environment should be established to evaluate possible health hazards. In many cases, environmental amines are present at very low concentration and are often found among a myriad of other compounds from which they must be separated and identified. Continuous intake of amines may induce chronic toxicity, even if they are consumed at a trace level. Toxicity depends on the dose of amines, so that it is very important to know the exact amount of these amines present in the environment to assess their adverse effects on humans. Thus analytical methodology for the isolation and quantification of these amines needs a selective and sensitive detection. The present article is concerned with utilization of the selective GC detectors for amine analysis and their application for the determination of amines in environmental samples. The article consists of following two parts. In the first part (Section 1.2), general aspects of the selective detection of amines by GC are surveyed

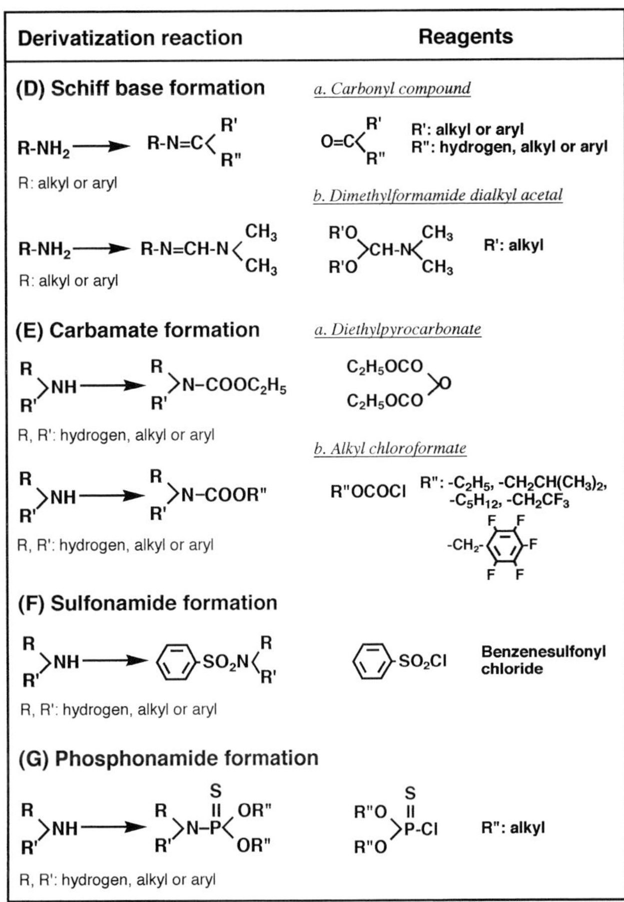

Fig. 1.4 (continued).

according to their detector types. In the second part (Section 1.3), the applications of the selective GC methods in environmental analysis are considered according to the amine types. The article covers not only aliphatic and aromatic primary-, secondary- and tertiary-amines but also *N*-nitrosamines and heterocyclic amines that often occur in various environments as highly toxic compounds. In this article, the references for the past two decades are cited and reviewed. For details of experimental and novel methods, one should refer to the original papers. General aspects of environmental analysis by GC have also been detailed in several books [213–215] and reviews [3,4].

1.2 SELECTIVE DETECTORS FOR GAS CHROMATOGRAPHY OF AMINES

The materials of interest in environmental samples are generally present at very low concentrations and are often found to be among complex matrices containing a number of coexisting substances. Therefore, GC detectors for environmental analysis must be not only sensitive to the minute amounts of analytes, but also selective enough to discriminate against reasonable amounts of coexisting substances. Various selective GC detectors described in Section 1.1.2 have been used for the analysis of environmental amines. The selection of a detector is decided by the instrument availability, the required accuracy and precision, the sensitivity, the sample preparation needed and the cost. In order to function properly, selective detectors must also be operated under optimum conditions. General aspects of gas chromatographic detectors have also been reviewed in several books [215–219].

1.2.1 Thermionic detector

The thermionic detector is selective for N- and P-containing compounds, because a thermionic source of alkali salt efficiently ionizes N- and P-containing organic molecules. There are two types of thermionic detectors, a flame type thermionic detector called flame thermionic detector (FTD) or alkali flame-ionization detector (AFID), and a flameless type detector called thermionic specific detector (TSD) or nitrogen–phosphorus detector (NPD). The flame type can also detect halogen-containing compounds, but the flameless type selectively responses only to N- and P-containing compounds. For N-containing compounds, a pyrolysis takes place, producing intermediate stable CN radicals and then the radicals formed take electrons from the excited atom of the alkali metal, resulting in a cyanide ion. A positive alkali ion migrates to the collector electrode and again liberates an electron. Therefore, the compounds that do not containing a $C-N$ bond, such as nitrogen gas, nitrogen oxides and ammonia cannot respond to NPD. The thermionic detectors require hydrogen and air as with flame ionization detector (FID), but they are at lower flows. Therefore, normal FID-type ionizations are minimal, and the selectivity ratios for detection of N-containing compounds versus carbon detection are $3.5 \times 10^4 : 1$. The sensitivity of thermionic detector is about 0.4–10 pg for N-containing compounds, and is about 50 times more sensitive for nitrogen in comparison with the FID. The main disadvantage of this detector is that its performance deteriorates with time. The alkali salt employed as the bead is continuously lost during the operation of the detector. Furthermore, the thermionic detector is not usable with columns of liquid phases containing halogen, phosphorus, or nitrogen such as XE-60 and OV-210. In addition, the use of column, supports, glass wool treated with phosphoric acid, and leak-detection fluids should be avoided to preserve the lifetime of the active element.

The thermionic detector is one of the most popular selective detectors for GC analysis of amines and is used for the analyses of various amines such as aliphatic [15–20,46,52,61,68,220–225] and aromatic [16–18,21,28,33,53,54,60,61,220,226] amines, N-nitrosamines [101,136] and heterocyclic amines [22,180,227]. Yang et al. [46] reported that 23 low molecular weight aliphatic amines lower than 10 nM can be determined by nitrogen–phosphorus detector (Fig. 1.5A). Aliphatic amines were also detected with TSD after acylation [18,52,221], permethylation [222] and isobutoxycarbonylation [223] (Fig. 1.4). Skarping et al. [28,69,221,222] reported that some aliphatic and aromatic amines can be detected at the fmol level with TSD. Kataoka et al. [54]

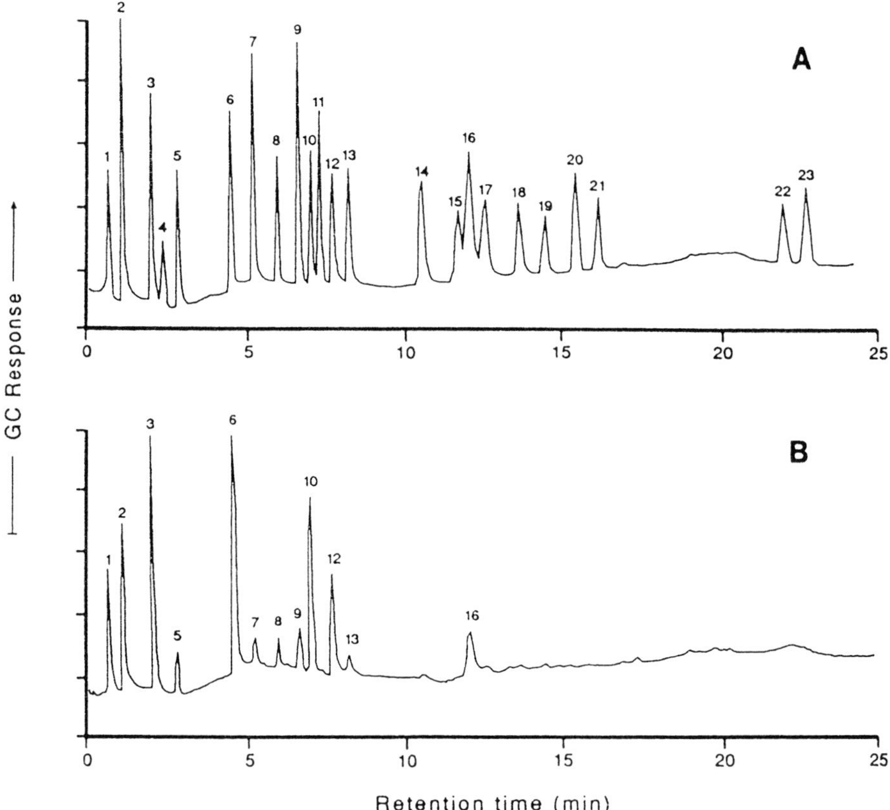

Fig. 1.5. Gas chromatograms of (A) amine standards and (B) a seawater sample. GC conditions: column, 4% Carbowax 20 M and 0.8% KOH on Carbopack B (2 m × 2.5 mm i.d.); column temperature, isothermal at 85°C for 2.5 min, increase to 150°C at 32°C min^{-1} for 6 min and then to 220°C at 10°C min^{-1}; injection temperature, 200°C; detector temperature, 220°C; helium carrier gas flow-rate, 22 ml min^{-1}, detector, NPD. Peaks: 1 = ammonia, 2 = methylamine, 3 = dimethylamine, 4 = ethylamine, 5 = trimethylamine, 6 = isopropylamine, 7 = *n* propylamine, 8 − *tert*-butylamine, 9 − diethylamine, 10 = *sec*-butylamine, 11 = isobutylamine, 12 = pyrrolidine, 13 = *n*-butylamine, 14 = piperidine, 15 = triethylamine, 16 = pyridine, 17 = isoamylamine, 18 = *n*-amylamine, 19 = pyrrole, 20 = dipropylamine, 21 = cyclohexylamine, 22 = tripropylamine, 23 = dibutylamine. (From Yang et al. [46]. Reproduced with permission of the American Chemical Society.)

separated 22 aromatic amines using HP-5 capillary column within 15 min (Fig. 1.6A). This method is based on the derivatization with *n*-propyl chloroformate and subsequent NPD analysis, and can detect the pg level of aromatic amines. Furthermore, Kataoka et al. [180,195,227] developed a simple and rapid derivatization method for GC analysis of heterocyclic amines. Ten heterocyclic amines were determined as their *N*-dimethylaminomethylene derivatives (Fig. 1.4) with NPD. The detection limits of these compounds ranged from 2 to 15 pg.

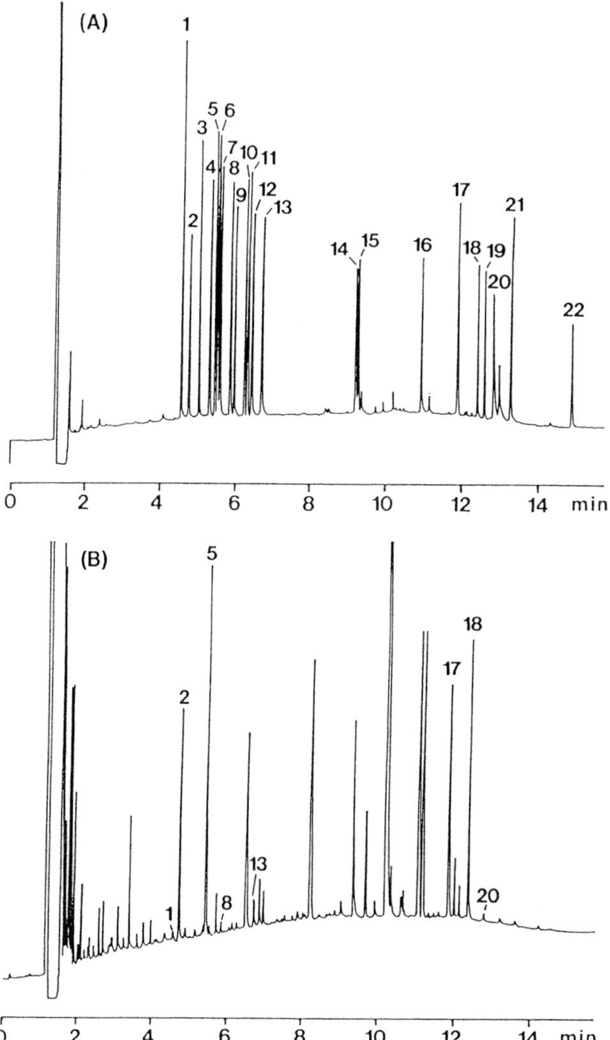

Fig. 1.6. Gas chromatograms obtained from *n*-propoxycarbonyl derivatives of aromatic amines and cigarette smoke sample. (A) Standard amine derivatives (each peak corresponding to 1–5 ng of amine); (B) cigarette smoke sample. GC conditions: column, HP-5 (30 m × 0.32 mm i.d.); column temperature, programmed from 120 to 220°C at 10°C min^{-1} and then programmed from 220 to 320°C at 20°C min^{-1} and isothermal at 320°C for 1 min; injection and detector temperatures, 330°C; carrier gas, helium, programmed from 80 to 120 kPa at 4 kPa min^{-1} and then programmed from 120 to 270 kPa at 30 kPa min^{-1} and held at 270 kPa for 1 min; detector, NPD. Peaks: 1 = aniline, 2 = internal standard (*N*-methyl-*m*-toluidine), 3 = 2-toluidine, 4 = 2,6-xylidine, 5 = 3-toluidine, 6 = 4-toluidine, 7 = 2-ethylaniline, 8 = 2,5-xylidine, 9 = 2,4-xylidine, 10 = 2,3-xylidine, 11 = 3,5-xylidine, 12 = 4-ethylaniline, 13 = 3,4-xylidine, 14 = α-naphthylamine, 15 = *o*-aminobiphenyl, 16 = *N*-phenylnaphthylamine, 17 = internal standard (2,4-diaminotoluene, 18 = 2-aminofluorene, 19 = *p*-aminoazobenzene, 20 = benzidine, 21 = *p,p'*-diaminodiphenylmethane, 22 = 2-aminoanthracene.

1.2.2 Flame photometric detector

FPD measures only specific wavelength with interference filter among the light emitted when a sample is burned in a hydrogen-rich flame, and therefore it gives a high selectivity for P- or S-containing compounds. The emission of S-containing compounds is ranged from 390 to 480 nm and that of P-containing compounds is between 480 and 580 nm. By choosing a filter having a transmission window of wavelength at 390–410 nm for S-containing compounds or at 520–560 nm for P-containing compounds, the maximum emission from S- or P-containing compounds can be sensed by the photomultiplier. On the other hand, the emission of hydrocarbons is alone between 390 and 520 nm, and therefore the emission from the background hydrocarbons will be almost negligible. Sulfur and phosphorus are detected as the S=S and HPO molecules, respectively. These molecules are created in a metastable state and when they decay, they release energy in the form of a photon of a specific wavelength. The light emitted from S-containing compounds is not linear with concentration due to the reactions in the flame, but is approximately proportional to the square of the sulfur atom concentration. The selectivity ratio for detection of S- or P-containing compounds versus carbon detection is $10^4 : 1$. The sensitivity of FPD is about 20 pg for S-containing compounds and 1 pg for P-containing compounds. One serious problem with the FPD is the quenching or re-absorption of the light emitted by the selected species. Hydrocarbon quenching occurs from collisions when the peak of S-containing compounds is co-eluted with another hydrocarbon in relatively high concentration. Self-quenching, that is due to collisional energy absorption, competing chemical reactions or the re-absorption of the photon by an inactivated species, can also occur at high concentrations of the heteroatom species. Precise control of the gas flow to the detector virtually eliminates quenching of the detector response.

Although FPD is widely used for S- and P-containing compounds such as agricultural chemicals and SO_2 in environment, it can also detect amino compounds by conversion into their S- or P-containing derivatives (Fig. 1.4) [55,56,60,63,70,108,228,229]. Hamano et al. [62] and Kataoka et al. [70,228] reported that primary and secondary aliphatic amines could be detected as their *N*-benzenesulfonyl derivatives by FPD with S mode. As shown in Fig. 1.7, 14 primary and 9 secondary amines were separately determined. The detection limits of these amines ranged from 10 to 100 pg. Furthermore, Kataoka et al. reported that aliphatic [63,229] and aromatic [55] amines and *N*-nitrosamines [56,108] could be analyzed as their *N*-dialkylthiophosphoryl derivatives by FPD with P mode. In particular, secondary amines can be selectively converted into their *N*-diethylthiophosphoryl derivatives after treatment with *o*-phthaldialdehyde (OPA), because OPA reacts only with the primary amino group. As shown in Fig. 1.8, primary amines are also derivatized without OPA treatment and detected by FPD, but they are not detected at all with OPA pretreatment. On the other hand, secondary amines are detected irrespective of OPA pretreatment, because these amines do not react with OPA. This technique is also applied to the determination of *N*-nitrosamines, they being denitrosated with hydrobromic acid to produce the corresponding secondary amines and then derivatized.

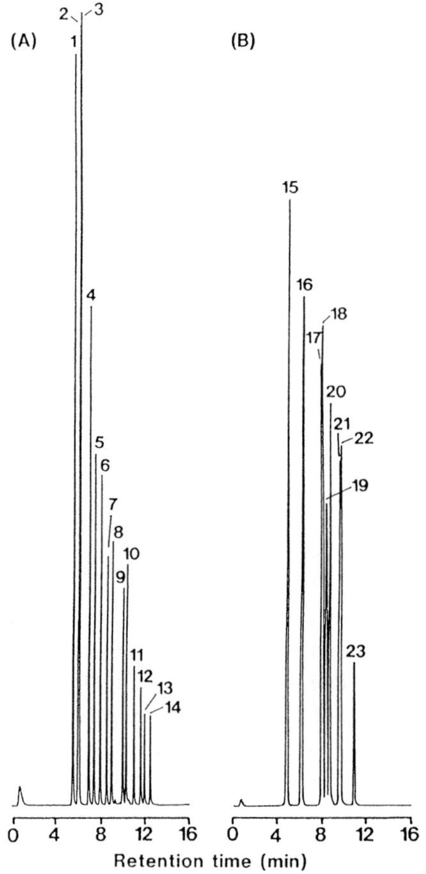

Fig. 1.7. Gas chromatograms obtained from benzenesulfonyl derivatives of (A) primary and (B) secondary amines. GC conditions: column, DB-1 (15 m × 0.53 mm i.d.), column temperature, programmed from 120 to 280°C at 10°C min^{-1}; injection and detector temperatures, 290°C; nitrogen flow-rate, 10 ml min^{-1}; detector, PFD (S mode). Each peak corresponds to 2 ng of amine. Peaks: 1 = methylamine, 2 = ethylamine, 3 = isopropylamine, 4 = *n*-propylamine, 5 = isobutylamine, 6 = *n*-butylamine, 7 = isoamylamine, 8 = *n*-amylamine, 9 = hexylamine, 10 = cyclohexylamine, 11 = heptylamine, 12 = benzylamine, 13 = octylamine, 14 = β-phenylethylamine, 15 = dimethylamine, 16 = diethylamine, 17 = dipropylamine, 18 = pyrrolidine, 19 = morpholine, 20 = piperidine, 21 = dibutylamine, 22 = hexamethyleneimine, 23 = *N*-methylbenzylamine. (From Kataoka [4]. Reproduced with permission of Elsevier Science.)

1.2.3 Electron-capture detector

ECD, that consists of a radioactive source such as Ni^{63}, is selective detector for halogenated or nitrated compounds, organometallics and conjugated carbonyls in which these species are capable of capturing low energy electrons to form negatively charged ions. Electron capture detection involves the following three steps: generation of thermal energy electrons, capture of some of these electrons by electrophilic compound, and

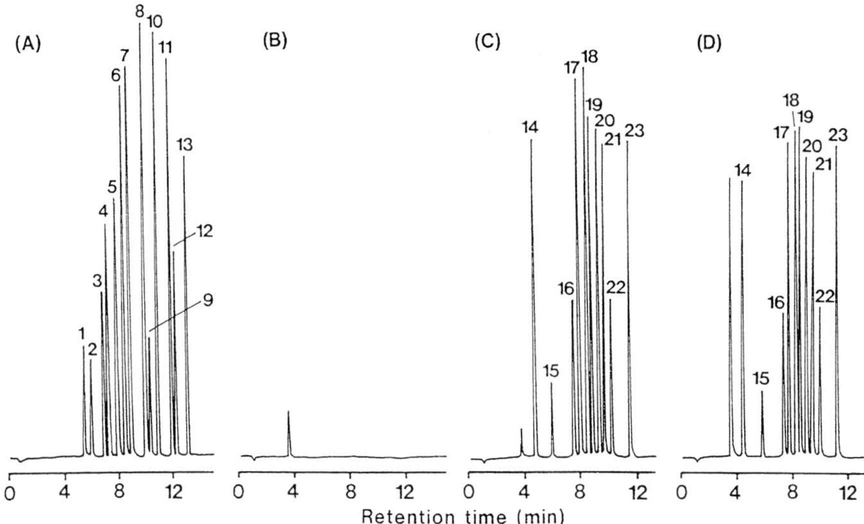

Fig. 1.8. Gas chromatograms obtained from *N*-diethylthiophosphoryl derivatives of primary and secondary amines. (A) Primary amines; (B) primary amines pretreated with OPA; (C) secondary amines; (D) secondary amines pretreated with OPA. GC conditions: column, DB-1701 (15 m × 0.53 mm i.d.), column temperature, programmed from 100 to 260°C at 10°C min^{-1}; injection and detector temperatures, 280°C; nitrogen flow-rate, 10 ml min^{-1}; detector, PFD (P mode). Each peak corresponds to 20 pmol of amine. Peaks: 1 = methylamine, 2 = ethylamine, 3 = isopropylamine, 4 = *n*-propylamine, 5 = isobutylamine, 6 = *n*-butylamine, 7 = isoamylamine, 8 = *n*-amylamine, 9 = hexylamine, 10 = cyclohexylamine, 11 = heptylamine, 12 = benzylamine, 13 = octylamine, 14 = β-phenylethylamine, 15 = dimethylamine, 16 = diethylamine, 17 = dipropylamine, 18 = pyrrolidine, 19 = morpholine, 20 = piperidine, 21 = dibutylamine, 22 = hexamethyleneimine, 23 = *N*-methylbenzylamine. (From Kataoka et al. [229]. Reproduced with permission of Elsevier Science.)

collection and measurement of the unreacted electrons. In order to produce capturable (low energy) thermal electrons, the carrier gas is ionized by β particles from a radioactive source placed in the cell. This electron flow produces a small current, which is collected and measured by a suitable amplifier. When the sample molecule is introduced into the cell, electrons are captured by the sample, resulting in decreased current. Although the selectivity depends on the affinity of compounds for electrode, the selectivity ratio for detection of poly halogen compounds versus carbon detection is $10^5–10^6 : 1$. ECD is extremely sensitive, probably the most sensitive GC detector available. Although the sensitivity of ECD depends on the molecular structure, that of chlorinated compounds is about 0.05–1 pg. For successful electron capture detection, it is important that carrier and purge gases are very clean and dry (99.9995%).

Although ECD is widely used in the detection and analysis of halogenated compounds, in particular, pesticides, it can also detect amino compounds by conversion into their halogen-containing derivatives (Fig. 1.4) [34–36,133,221,230–234]. Miyamoto et al. [230] determined selectively some primary amines as their *N*-benzenesulfonyl-*N*-trifluoroacetyl derivatives by GC–ECD. Skarping et al. [233] reported that aromatic amines could be separated as their perfluoro fatty acid amides by glass capillary column, and determined

Signal Height

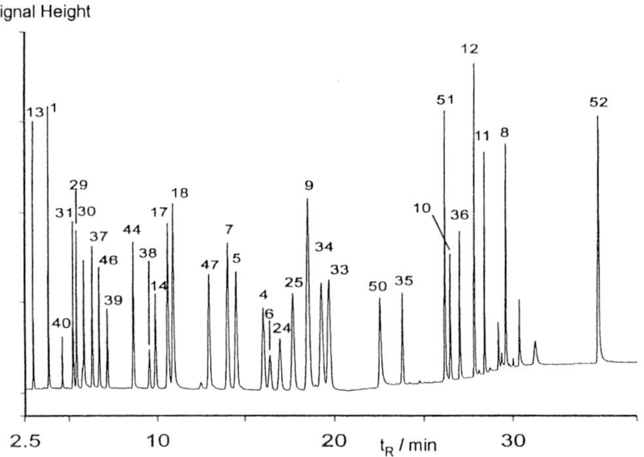

Fig. 1.9. GC separation of 33 iodinated derivatives of aromatic amines after enrichment on HR-P phase. Concentration of aromatic amines, 10–50 ng ml^{-1}. GC conditions: column, 5% phenylmethylpolysiloxane (30 m × 0.25 mm i.d.); column temperature, isothermal at 135°C for 20.5 min, and then programmed from 135 to 235°C at 12.5°C min^{-1} and held at 235°C for 8.5 min; injection and detector temperatures, 250 and 300°C, respectively; carrier gas, nitrogen, 100 kPa; split ratio, 1 : 120; detector, ECD. Peaks: 1 = 4-amino-toluene, 4 = 2,4-diaminotoluene, 5 = 4-amino-2-nitrotoluene, 6 = 2,6-diaminotoluene, 7 = 2-amino-6-nitro-toluene, 8 = 2,6-diamino-4-nitrotoluene, 9 = 2-amino-4-nitrotoluene, 10 = 4-amino-2,6-dinitrotoluene, 11 = 2,4-diamino-6-nitrotoluene, 12 = 2-amino-4,6-dinitrotoluene, 13 = aniline, 14 = 1,3-phenylenediamine, 17 = 3-nitroaniline, 18 = 4-nitroaniline, 24 = 2-amino-3-nitrotoluene, 25 = 2-amino-5-nitrotoluene, 29 = 2,6-dimethylaniline, 30 = 3,4-dimethylaniline, 31 = 3,5-dimethylaniline, 33 = 1-naphthylamine, 34 = 2-naphthylamine, 35 = 2-aminobiphenyl, 36 = 4-aminobiphenyl, 37 = 4-isopropylaniline, 38 = 2,6-di-ethylaniline, 39 = 2-ethyl-6-methylaniline, 40 = 4-chloro-*N*-methylaniline, 44 = 3,4-dichloroaniline, 46 = 4-chloro-2-methylaniline, 47 = 3-chloro-4-methoxyaniline, 50 = 2,6-dinitroaniline, 51 = 3,5-dinitroaniline, 52 = benzidine. (From Schmidt et al. [35]. Reproduced with permission of Elsevier Science.)

using on-column injection and ECD. The detection limits of these amines were ca. 1 pg. Furthermore, Schmidt et al. [35,36] reported the selective and sensitive method for the determination of aromatic amines based on the derivatization to the corresponding iodobenzenes and subsequent GC–ECD analysis. Separation of at least 30 compounds in a single chromatographic run in 30 min is possible (Fig. 1.9), and 52 aromatic amines are investigated. On the other hand, dinitroaniline herbicides could be directly analyzed by GC–ECD, because they possess a nitro group in the molecule [37].

1.2.4 Chemiluminescence detector

CLD such as thermal energy analyzer (TEA), based on the emission spectroscopy, is a selective detector for nitroso and nitroaromatic compounds. The response of TEA to *N*-nitroso compounds is based on the reaction of nitrogen oxide with ozone combined with the preceding pyrolysis of nitroso compounds. The eluate from the GC column enters the pyrolyzer where the selective catalytic decomposition of *N*-nitroso compounds takes place, resulting in increased a nitrosyl radical and an organic radical.

The N−NO bond is the weakest in these compounds. The pyrolyzer eluent expands into the evacuated reaction chamber in which the nitrogen oxide reacts with ozone, resulting in excited nitrogen dioxide. The excited nitrogen dioxide rapidly decays back to its ground state with the characteristic emission of light. Then the emitted radiation is detected by a photomultiplier through a red optical filter. The filter is used to block possible light emission resulting from sources other than NO chemiluminescence. The intensity of the emission is proportional to the nitrosyl radical concentration, and hence to the *N*-nitroso compound concentration. The decomposition of substances with a catalyst at lower temperatures of about 300°C is more advantageous than pyrolysis at elevated temperatures, because the decomposition of the N−NO bond is more selective, even though the detector response to nitrosamines is diminished at lower temperature. The response of the CLD to nitroaromatic compounds increases rapidly with increasing temperature at 800–900°C of the pyrolyzer. Nitrogen-containing compounds are also detected with CLD, based on the catalytic decomposition of these compounds in the presence of oxygen at temperatures of 900–1000°C and on the subsequent ozone oxidation of the nitrogen oxide formed, as with nitroso compounds. The sensitivity of CLD for *N*-nitrosodimethylamine (NDMA) is about 50 pg.

The CLD has been used as a selective detector for *N*-nitrosamines [95,106,107,109–131,134,135]. Billedeau et al. [128] separated volatile *N*-nitrosamines within 13 min (Fig. 1.10). Adams et al. [110] analyzed tobacco-specific *N*-nitrosamines by GC–TEA. On the other hand, the CLD can be used for the selective analysis of not only *N*-nitrosamines but also aliphatic [23–25,64] and aromatic [231] amines by modification of pyrolyzer temperature. Pfundstein et al. [64] reported that primary and secondary amines could be analyzed as their *N*-benzenesulfonyl derivatives by GC–CLD with modified TEA. Thermal pyrolysis of amines at 720°C with a nickel catalyst is used to produce NO radicals.

1.2.5 Mass selective detector

MSD has been recognized as an excellent detector for GC, because the spectral data provides the qualitative information lacking in other GC detectors. When a molecule in the vapor state is bombarded with electrons, bonds are ruptured and the molecule ionizes. The kind and amount of fragments obtained are characteristic of the molecule. The first step in a GC–MS analysis to be qualitative identification, using a complete scan across the entire mass range. Identification can be done with the help of spectral libraries, usually kept in a data base on a computer with the detector-controlling software. The molecular ion and characteristic ions are used for the identification of compounds. The second step is quantitative analysis using SIM. By using a filter to select only a few ions, the sensitivity can be increased. The final step is total ion chromatogram (TIC), based on the plotting of total response (sum of all ions) over time by scanning repeatedly. In quantitative SIM analysis, it is necessary to select compound-specific ions (molecular ion or fragment ion), which give a large signal and have no interferences, are selected for the best chromatography. Next it is necessary to optimize MS response to the monitored ions, calibrate each ion mass, and assign the dwell time for each ion using GC peak width and ion signal ratios.

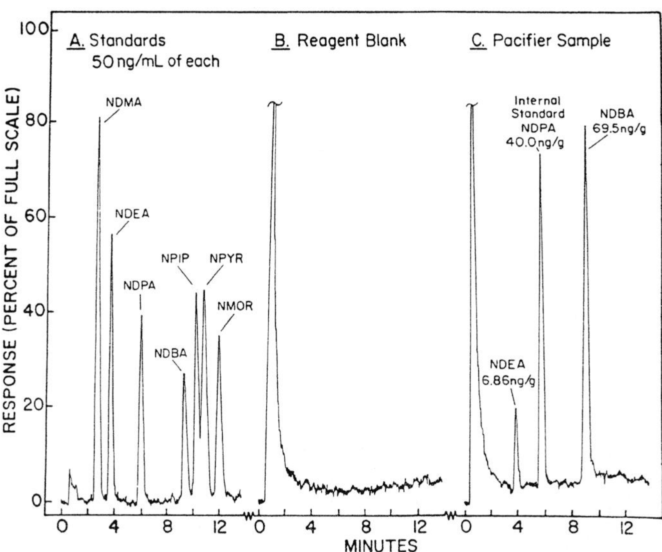

Fig. 1.10. GC–TEA chromatograms of standard *N*-nitrosamines and pacifier sample. (A) standard mixture containing 50 ng ml⁻¹ each of *N*-nitrosamine; (B) reagent blank; (C) pacifier sample. GC conditions: column, 10% Cawbowax/2% KOH on Chromosorb W AW (6 ft × 4 mm i.d.), column temperature, programmed from 150 to 190°C at 4°C min⁻¹; injector temperature, 250°C, carrier gas, nitrogen, 40 ml min⁻¹; detector, TEA; detector furnace, 500°C; cold trap, −150°C. Peaks: NDMA = *N*-nitrosodimethylamine, NDEA = *N*-nitrosodiethylamine, NDPA = *N*-nitrosodipropylamine, NDBA = *N*-nitrosodibutylamine, NPIP = *N*-nitrosopiperidine, NPYR = *N*-nitrosopyrrolidine, NMOR = *N*-nitrosomorpholine. (From Billedeau et al. [128]. Reproduced with permission of AOAC International.)

GC–MS can provide structure information for the unequivocal identification of amines and can be applied to the analyses of various amines, such as aliphatic [38–43,73] and aromatic [9,13–27,47–51,59,64,72,80,234,235] amines, *N*-nitrosamines [115,116,118,137–139] and heterocyclic amines [157–161,190]. Aliphatic amines are analyzed after derivatization with 2,4-dinitrofluorobenzene, benzenesulfonyl chloride and pentafluorobenzaldehyde. Sacher et al. [42] reported that aliphatic primary and secondary amines could be determined as their *N*-benzenesulfonyl and 2,4-dinitrophenyl derivatives, and the method using the former derivatives was more selective and sensitive. On the other hand, Longo and Cavallaro [48] developed a method for the simultaneous identification of 73 primary and secondary aromatic amines as their heptafluorobutyramides (Fig. 1.11). The electrophoretic derivatives were analyzed by GC combined with electron-capture negative-ion CI (EC–NICI) MS. Detection limits of these amines were in the range 0.3–66.3 pg injected in the full-scan mode and 0.01–0.57 pg injected in the SIM mode. Furthermore, Murray et al. [156–159] developed a sensitive and selective method for the determination of several heterocyclic amines as their 3,5-bistrifluoromethylbenzyl derivatives by NICI–MS with SIM mode. The detection limits were about 1 pg. Skog et al. [161] analyzed six non-polar heterocyclic amines by GC–MS without derivatization (Fig. 1.12). The detection limits of these amines by SIM mode detection were in the range 0.1–2 ng per injection.

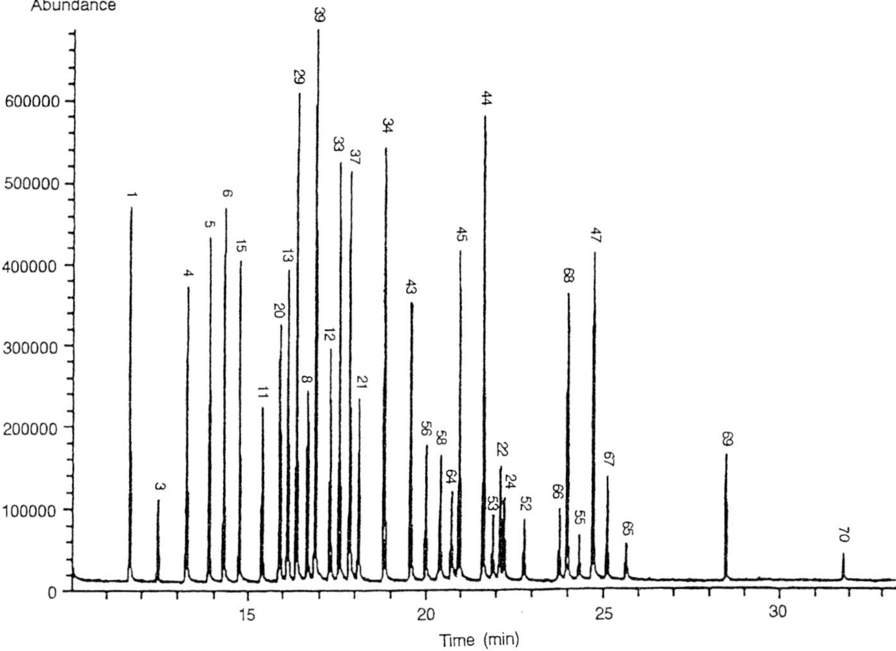

Fig. 1.11. Total ion chromatogram of a mixture of HFBA derivatives under EC-NICI conditions. GCMS conditions: GC column, HP-5 (50 m × 0.2 mm i.d.); column temperature, initially at 50°C for 0.5 min, heated to 110°C at 50°C min^{-1}, to 225°C at 5°C min^{-1}, to 280°C at 20°C min^{-1} and held at 280°C for 15 min; injector and transfer line temperatures, 250 and 280°C, respectively; carrier gas, helium, 250 kPa; ion-source temperature, 200°C; electron energy, 150 eV. Peaks: 1 = aniline, 3 = *N*-methylaniline, 4 = 2-methylaniline, 5 = 3-methylaniline, 6 = 4-methylaniline, 8 = 2,3-dimethylaniline, 11 = 2,6-dimethylaniline, 12 = 3,4-dimethylaniline, 13 = 3,5-dimethylaniline, 15 = 2-ethylaniline, 20 = 2-methoxyaniline, 21 = 4-methoxyaniline, 22 = 2,4-dimethoxyaniline, 24 = 3,4-dimethoxyaniline, 29 = 4-chloroaniline, 33 = 3-chloro-2-methylaniline, 34 = 3-chloro-4-methylaniline, 37 = 5-chloro-2-methylaniline, 39 = 2,4-dichloroaniline, 43 = 3,5-dichloroaniline, 44 = 2,3,4-trichloroaniline, 45 = 2,4,5-trichloroaniline, 47 = 3,4,5-trichloroaniline, 52 = *N*-methyl-4-nitroaniline, 53 = 2-methyl-3-nitroaniline, 55 = 2-methyl-5-nitroaniline, 56 = 2-methyl-6-nitroaniline, 58 = 4-methyl-2-nitroaniline, 64 = 5-chloro-2-nitroaniline, 65 = 4-chloro-3-nitroaniline, 66 = 1-aminonaphthalene, 67 = 2-aminonaphthalene, 68 = 2-aminobiphenyl, 69 = 4-aminobiphenyl, 70 = 4-phenylazoaniline. Each peak corresponds to 50 pg of amine. (From Longo and Cavallaro [48]. Reproduced with permission of Elsevier Science.)

1.3 APPLICATIONS IN ENVIRONMENTAL AMINE ANALYSIS

Amines are present in the environment at low parts per billion or less. When the environmental samples are analyzed by GC with non-selective detector such as FID, many peaks with the same retention times as those of amines are often present in the chromatograms. Although GC analyses with selective detectors can save sample preparation, a certain clean-up procedure for the complex sample matrix is necessary for the reliable and accurate analysis of amines. In order to remove co-eluting interferences and to pre-concentrate amines, the extraction and clean-up of the sample has been performed using

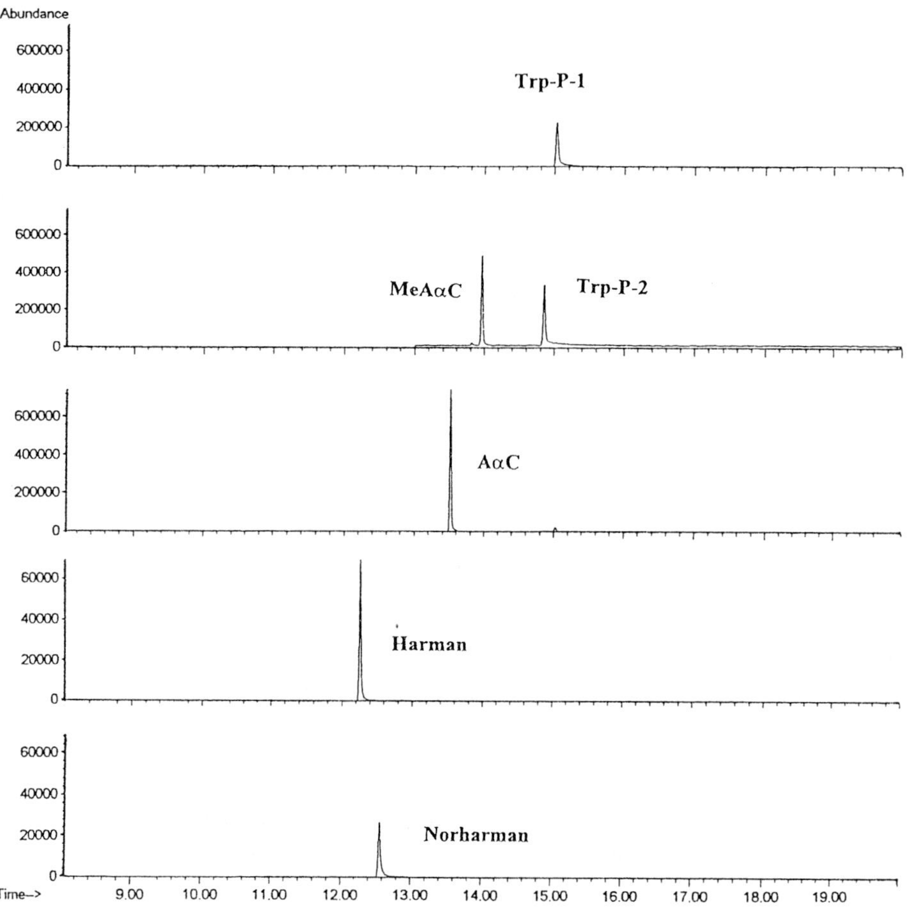

Fig. 1.12. SIM chromatograms of standard heterocyclic amines. GC–MS conditions: GC column, Rtx-50 (50% phenyl–50% methyl polysiloxane, 30 m × 0.32 mm i.d.); Column temperature, initially 100°C for 2 min, increased to 320°C at 20°C min^{-1} and held at 320°C for 7 min; injection temperature, 270°C; separator and ion source temperatures, 250°C; ionization voltage, 70 eV; carrier gas, helium, 1.0 ml min^{-1}. For the reference ions (m/z) the retention times are Trp-P-1 (m/z 211) 15.0 min, Trp-P-2 (m/z 197) 14.9 min, AαC (m/z 183) 13.5 min, MeAαC (m/z 197) 14.0 min, harman (m/z 182) 12.3 min, and norharman (m/z 168) 12.6 min. (From Skog et al. [161]. Reproduced with permission of Elsevier Science.)

a number of different purification techniques, such as distillation [24,25,40,43,48,237], liquid–liquid extraction [27,41,47,53–60,107–115,124–131,134], adsorption with various adsorbent tubes or columns [13,21,23,33,38,136,190–192], preparative high performance chromatography (HPLC) [178,179,184,185,187–189], solid-phase extraction (SPE) with various cartridges [15,35,36] and solid phase microextraction (SPME) [29,37,49,50,73,105,224]. Some amines can be separated by acid–base partition, because these compounds can be extracted with organic solvents not at low pH (pH <1), but at high pH (pH >10). Adsorption with adsorbent tube or column is effective for the separation of amines, but it is time-consuming. Blue-Cotton [183,238], Blue-Rayon

[180,187,188,193,238] or Blue-Chitin [239], cotton, rayon or chitin bearing covalently linked copper phthalocyanine trisulfonate as ligand, can selectively adsorb heterocyclic amines and other mutagens/carcinogens having polycyclic planar molecular structures. Normal or reversed phase HPLC fractionations are also useful for preparation of sample. SPE is simple and rapid, and good recoveries of amines are obtained. On the other hand, SPME recently developed by Pawliszyn and co-workers is an extraction technique using a fused-silica fiber that is coated outside with an appropriate stationary phase, and saves preparation time, solvent purchase and disposal cost, and can improve the detection limits [240–249]. It has been used routinely in combination with GC and GC–MS, and successfully applied to a wide variety of compounds including amines. The environmental amine analyses using various GC detectors are summarized in Table 1.1.

1.3.1 Aliphatic amines

The GC methods for the determinations of aliphatic amines using selective detectors have been applied to various environmental samples such as air [15–20,23–25], cigarette smoke [56], water [20,38–43,46,52] and soil [24,25,40,43]. Many of these methods have been used for free amines [15,17,19,20,23,43,46,224,225,234], and the trace analysis of low-molecular mass aliphatic amines in air has been performed with nitrogen-selective detectors such as NPD, CLD and GC–MS by direct injection or head-space technique. On the other hand, derivatization methods have also been used in water and soil samples [36–38], because these samples cannot be directly analyzed. Kuwata et al. [15] analyzed low molecular weight aliphatic amines in air samples in and around livestock houses by Sep-PAK C_{18} cartridge treatment and subsequent GC with thermionic detection, and detected methylamine (0–12.4 ppb) and trimethylamine (0.28–69.7 ppb). Gronberg et al. [19,20] developed the method for the determination of short chain aliphatic amines in ambient air based on impinger sampling in dilute H_2SO_4, selective enrichment across a PTFE gas membrane and quantification by GC–TSD. The enrichment step was carried out in a flow system directly connected to the GC, methyl-, dimethyl-, trimethyl-, diethyl- and triethylamine were detected at ppt level. Yang et al. [46] analyzed low molecular weight aliphatic amines in sea water samples by circulation diffusion technique and subsequent GC–NPD, and detected methylamine, dimethylamine and trimethylamine at the concentration from <3 to 200 nM (Fig. 1.5B). Abalos et al. [224] used SPME technique for the determination of free volatile C_1–C_6 amines in wastewater and sewage-polluted waters. These amines were directly extracted by headspace SPME and analyzed by GC–NPD. This method is simple, rapid and solvent free. The detection limits and linearity were in the range of 3–56 ng ml^{-1} and 50–600 ng ml^{-1}, respectively. Kataoka et al. [56] reported the determination of secondary amines as their DETP derivatives in cigarette smoke by GC–FPD. This method is selective and sensitive to secondary amines, and the detection limits of amines are 0.05–0.2 pmol. Using this method, it is confirmed that dimethylamine, pyrrolidine, piperidine and morpholine occur in main- and side-stream smokes of cigarettes, and the contents of these amines in side-stream smoke are very high compare with those in main-stream smoke. Kashihira et al. [23] analyzed trimethylamine in freezer air by GC–CLD after

TABLE 1.1

ANALYSIS OF ENVIRONMENTAL AMINES USING SELECTIVE GAS CHROMATOGRAPHIC DETECTORS

Amines	GC detector [a]	Matrices	Reference
Aliphatic amines	NP-FID	Air	Kuwata [15]
	TSD	Working atmosphere	Audunsson and Mathiason [16,17], Skarping et al. [18]
	TSD	Ambient air, rain water	Gronberg et al. [19,20]
	NPD	Sea water	Yang et al. [46]
	NPD	Tap water	Baltussen et al. [52]
	NPD	Waste water, sewage	Abalos et al. [224]
	FPD (P)	Cigarette smoke	Kataoka et al. [56]
	CLD	Atmosphere	Kashihira et al. [23]
	CLD	Marine sediments and atmosphere	Lee and Olson [24,25]
	GC/MS	Airborne	Seeber et al. [13]
	GC/MS	River water, Sediment, Waste water	Koga et al. [38], Avery and Junk [39], Terashi et al. [40], Pietsch et al. [41], Sacher et al. [42]
	GC/MS	Lake water, Sediment	Tsukioka et al. [43]
Aromatic amines	TMD, NPD	Workplace atmosphere	Audunsson et al. [16,17], Skarping et al. [18], Becher [21]
	NPD	Waste water	Riggin et al. [33]
	NPD	Combustion smoke	Dalene and Skarping [53], Kataoka et al. [54]
	NPD, GC/MS	Soil, plant, air	G-Valcarcel et al. [60]
	FPD (P)	Cigarette smoke	Kijima et al. [55]
	ECD	Surface water	Wegman and DeKorte [34]
	ECD	Ground water, waste water, sewage	Schmidt et al. [35,36]
	ECD	River water	Guan et al. [37]
	GC/MS	Air	Meichini et al. [26], Moldovan and Bayona [235]
	GC/MS	Cooking fumes	Chiang et al. [80]
	GC/MS	Combustion smoke, Indoor air	Pieraccini et al. [27,57]
	GC/MS	Cigarette smoke	Forehand et al. [59]
	GC/MS	Ground water, River water, Sediment	Okumura et al. [47], Longo and Cavallaro [48], Muller et al. [49,50]
	GC/MS	Tap water	Vreuls et al. [51]
N-Nitrosamines	TSD, TEA	Herbicide formulation	Scharfe and McLenagham [136]
	FPD (P)	Combustion smoke	Kataoka et al. [56,108]
	ECD	Metalworking fluids	Fadlallah et al. [133]
	TEA	Airborne	Fine et al. [106]
	TEA	Ambient air	Fadlallah et al. [107]
	TEA	Cigarette and tobacco smoke, Tobacco	Brunneman et al. [109–112], Truhaut et al. [113], Tricker et al. [114], Song and Ashley [116]
	TEA	Drinking water, Tap water	Kimoto et al. [117], Sen et al. [118]
	CLD	Drinking water, Ground water	Tomkins et al. [119,120]
	TEA	Household dishwashing liquid	Morrison and Hecht [123]

TABLE 1.1 (CONTINUED)

Amines	GC detector[a]	Matrices	Reference
	TEA	Rubber nipples and pacifiers	Havery and Fazio [125], Osterdahl [126], Thompson [127,128], Gray [129], Sen et al. [130]
	TEA	Elastic rubber netting	Fiddler et al. [124,131]
	CLD	Drug formulation	Dawson et al. [134], Nielsen and Lings [135]
	GC/MS	Cigarette smoke	Arrendale et al. [115], Song and Ashley [116]
	GC/MS	Drinking water	Sen et al. [118]
	GC/MS	Herbicide	Frassanito et al. [137]
Heterocyclic amines	NPD	Air	Andersson [22]
	NPD	Combustion smoke	Kataoka et al. [180]
	NPD	River water	Kataoka et al. [195]
	GC/MS	Cooking fumes	Richlimg et al. [173], Vainiotalo et al. [190]

[a] TSD, thermionic specific detector; NPD, nitrogen–phosphorus detector; FPD, flame photometric detector; ECD, electron-capture detector; CLD, chemiluminescence detector; TEA, thermal energy analyzer; GC/MS, gas chromatography/mass spectrometry.

preconcentration with a Tenax–GC tube, and detected 2.4 ppb of trimethylamine. Koga et al. [38] reported the determination of trace amounts of 12 aliphatic primary and secondary amines as their DNP derivatives in waters from sewage and river by GC–MS. By using this method, low-molecular mass aliphatic amines in water samples can be determined in the range of 1–3 ppb. Avery and Junk [39] reported the determination of trace amounts of seven aliphatic primary amines in waters from tap, river and oil shale process by GC–MS based on the Schiff base formation with PFBA. This method is specific for primary amines and the detection limit is 10 ppb with no sample transfers or manipulations being necessary for sample volumes of 0.5 ml. Terashi et al. [40] reported the determination of eight primary and secondary amines in river water, sea water and sea sediment by GC–MS–SIM based on distillation of the sample under alkaline conditions and subsequent benzenesulfonylation (Fig. 1.13). In this method, the detection limits of amines in water and sediment are 0.02–2 ppm and 0.5–50 ppm, respectively. *n*-Butylamine, di-*n*-propylamine and di-*n*-butylamine were detected in trace amounts in a sea water sample. Pietsch et al. [41] analyzed aliphatic and alicyclic amines in waste water samples by GC–MS after derivatization with trichloroethyl chloroformate and liquid–liquid extraction, and detected morpholine and piperidine. Sacher et al. [42] analyzed primary and secondary amines in industrial waste water and surface water samples by GC–MS as their *N*-benzenesulfonyl derivatives, and detected methylamine, dimethylamine, morpholine and ethanolamine in concentrations up to 100 ng ml^{-1}. Furthermore, ethylamine and diethylamine were found in concentrations up to 30 ng ml^{-1}. Tsukioka et al. [43] reported the determination of lower aliphatic tertiary amines in river water and bottom sediments. These amines were determined by headspace GC–MS with SIM mode after distillation. The detection limits for the

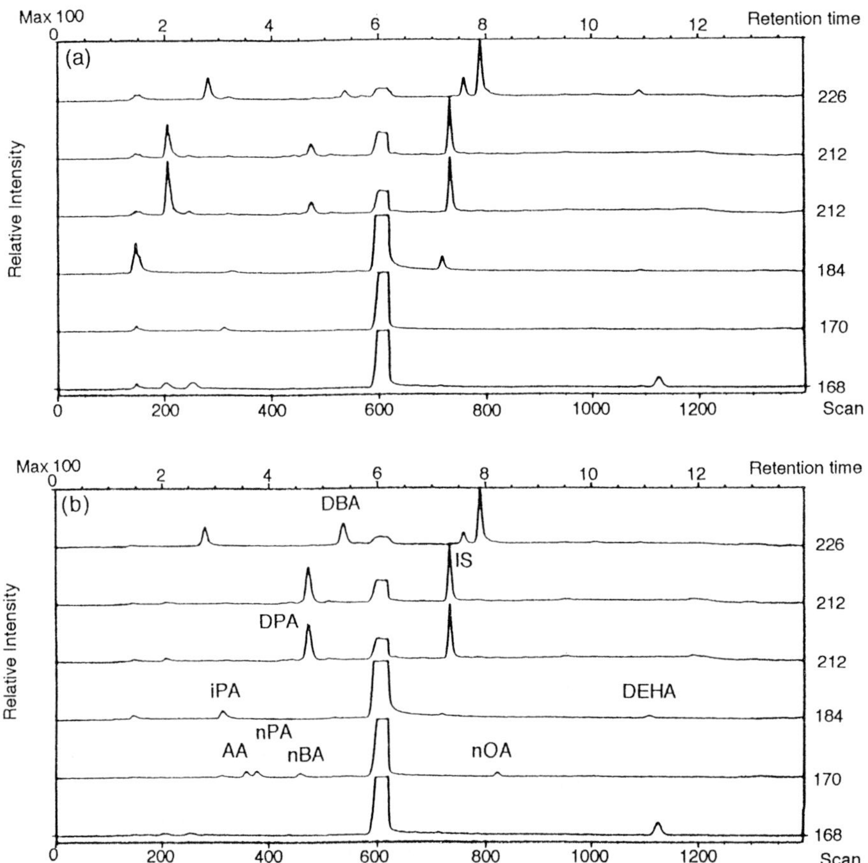

Fig. 1.13. SIM chromatograms of benzenesulfonamide derivatives of amines in a sea water: (a) unspiked; (b) spiked. GC–MS conditions: GC column, SE-54 (25 m × 0.53 mm i.d.); Column temperature, initially 200°C for 1 min, increased to 280°C at 10°C min^{-1} and held at 280°C for 3 min; injection temperature, 280°C; separator and ion source temperatures, 280°C; ionization voltage, 70 eV; carrier gas, helium, 13.0 ml min^{-1}. Peaks, nPA = *n*-propylamine, iPA = isopropylamine, nBA = *n*-butylamine, AA = amylamine, nOA = *n*-octylamine, DPA = di-*n*-propylamine, DBA = diisobutylamine, DEHA = di(2-ethylhexyl)amine, IS = [^{2}H$_{10}$]fluoranthrene. (From Terashi et al. [40]. Reproduced with permission of Elsevier Science.)

final 40-ml samples were 5–50 ng and this method showed excellent sensitivity and selectivity.

1.3.2 Aromatic amines

The GC methods for the determinations of aromatic amines using selective detectors have been applied to various environmental samples such as air [16–18,21,26,27], cooking fumes [80], cigarette smoke [13,27,53–55,57,59], waste water [33–37] and soil [60]. Becher [21] reported the determination of several aromatic amines in workplace

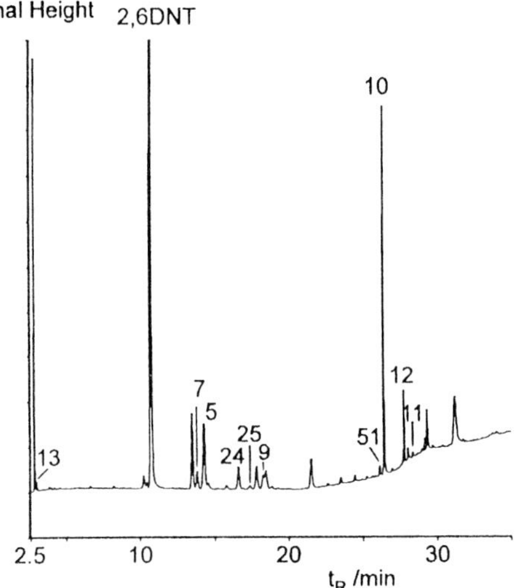

Fig. 1.14. GC–ECD chromatogram of a ground water from Stadtallendorf. GC conditions and peak numbers as in Figs. 1. and 1.5. (From Schmidt et al. [35]. Reproduced with permission of Elsevier Science.)

atmosphere by GC–NPD. Air samples are collected in three-section silica gel tubes and the amines on silica gel are eluted with 2-butanone. G-Valcarcel et al. [60] determined several dinitroaniline herbicides in environmental samples by GC–NPD after solvent extraction. Schmidt et al. [35] reported a method for the selective determination of aromatic amines in water samples. This method is based on the solid-phase extraction at pH 9, subsequent derivatization to the corresponding iodobenzenes and GC–ECD analysis. Aniline and nitroaniline compounds were detected in groundwater samples (Fig. 1.14). Pieraccini and colleagues [27,57] reported the determination of 17 primary aromatic amines as their pentafluoropropionamides in cigarette smoke and indoor air by GC–MS–SIM. The cigarette is smoked in a home-made smoking machine and the amines in the main- and side-stream smokes are trapped in dilute hydrochloric acid. It is confirmed that side-stream smoke contains total levels of aromatic amines about 50–60 times higher than those of main-stream smoke, and some aromatic amines in ambient air such as offices and houses may be derived from a considerable contamination of aromatic amines in side-stream smoke. Kataoka et al. [54] analyzed aromatic amines in the cigarette smoke samples as their *N-n*-propoxycarbonyl derivatives (Fig. 1.7B) and their *N*-dimethylthiophosphoryl derivatives by GC–NPD and GC–FPD (P mode), respectively, after collection of smoke sample by same manner. Forehand et al. [59] also analyzed aromatic amines in particulate phase cigarette smoke as their heptafluorobu-tyryl derivatives by GC–NCI–MS–SIM after simultaneous distillation and extraction as a unique sample clean-up. Okumura et al. [47] determined aniline and related aromatic amines in river water and sediment samples by GC–MS–SIM after liquid–liquid extrac-tion and steam distillation. The detection limits of the anilines in water and sediment samples were 4.2–31 pg ml^{-1} and 1.2–4.0 ng ml^{-1}, respectively. Longo and Cavallaro [48] analyzed aromatic amines in ground water samples as their heptafluorobutyryl

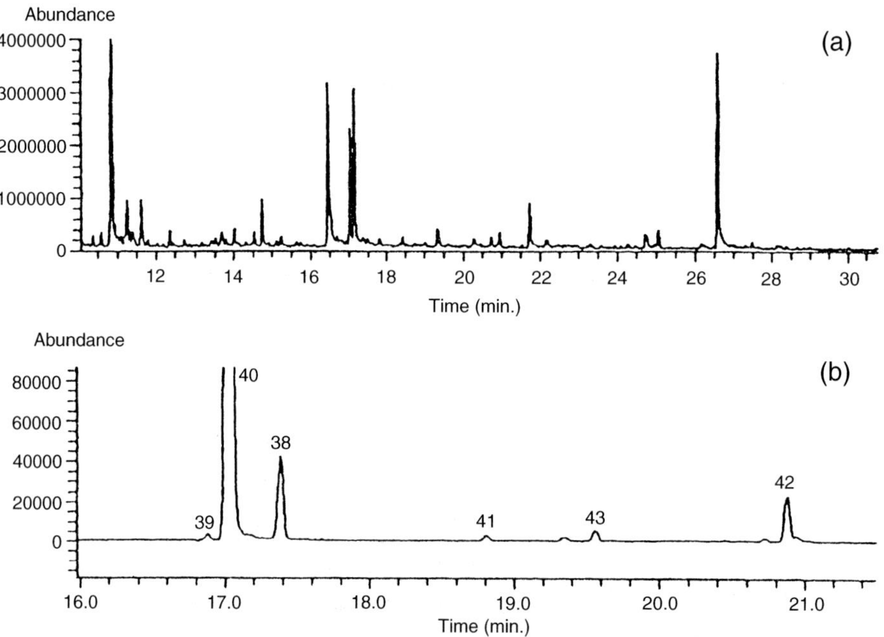

Fig. 1.15. (a) Total ion and (b) SIM (*m/z* 337 ion) chromatograms of a ground water sample. GC conditions and peak numbers as in Fig. 1.7. (From Longo and Cavallaro [48]. Reproduced with permission of Elsevier Science.)

derivatives by GC–EC–NICI–MS after concentration with the Kuderna–Danish evaporator (Fig. 1.15). Muller et al. [49,50] reported the determination of aromatic amines by SPME and GC–MS. This method was simple, rapid, precise and sensitive, various groundwater samples could be directly analyzed. Guan et al. [37] also analyzed nitroaniline herbicides in river water by SPME and GC–ECD.

1.3.3 *N*-Nitrosamines

The GC determinations of *N*-nitrosamines in environmental samples have been carried out in indoor and outdoor airs [95,106,107], combustion smokes [56,94,95,108–116], water [117–120,123], rubber products [124–131], metalworking fluids [133], drug formulations [134,135] and agricultural chemical formulations [123,136,137]. In most of them, *N*-nitrosamines are directly analyzed as free forms by GC–TEA, based on the detection of chemiluminescence emitted from a reaction between releasing NO radicals and ozone after thermal cleavage of N−NO bond in *N*-nitroso compounds. Fadlallah et al. [107] determined volatile nitrosamines in the factory environment by GC–TEA. Air samples were collected by drawing through the Thermosorb/*N*-cartridge, and then the sorbent was eluted with methanol/dichloromethane (1:3). Brunnemann et al. [95,109–112] analyzed volatile and tobacco specific *N*-nitrosamines in main-stream and side-stream smokes of cigarette and tobacco by GC–TEA. The smoke sample was collected in citrate buffer (pH 4.5) containing 20 mM ascorbic acid and then extracted with dichloromethane. NDMA, *N*-nitrosopyrrolidine (NPYR), *N*-nitrosomethyl-ethylamine (NMEA) and some tobacco specific nitrosamines such as *N*-nitrosonornicotine (NNN), 4-(methylnitrosamino)-4-(3-pyridyl)-butanal (NNA) and 4-(methylnitrosamino)-1-(3-pyridyl)-1-butanone (NNK) were detected. The *N*-nitrosamines in environmental tobacco smoke, to which both smokers and non-smokers are exposed, has received a great deal of attention as a source of indoor pollution. Tomkins et al. [119,120] determined *N*-nitrosodimethylamine in drinking and ground waters at ppt levels by GC–CLD (nitrogen mode) (Fig. 1.16). Furthermore, they developed new sample preparation procedure for extraction of *N*-nitrosamine using a C_{18} membrane extraction disk. On the other hand, Thompson and colleagues [127,128] reported the determination of *N*-nitrosamines in rubber nipples and pacifiers at ppb levels by GC–TEA. In these samples, *N*-nitrosodibutylamine was the principal *N*-nitrosamine found, along with trace amounts of NDMA, *N*-nitrosodiethylamine (NDEA) and *N*-nitrosopiperidine (NPIP). The occurrence of *N*-nitrosamines in baby bottle rubber nipples and pacifiers is of special concern because traces of these amines may migrate to infant saliva during sucking, and then be ingested. Although GC–TEA can be used as sensitive and specific method for the analysis of *N*-nitroso compounds, it is too expensive to use in many laboratories. As an alternative method, Kataoka et al. [56,108] reported the determination of seven *N*-nitrosamines by GC–FPD (P mode). The method is based on the denitrosation with hydrobromic acid to produce the corresponding secondary amines and subsequent diethylthiophosphorylation of secondary amines. By using this method, it is confirmed that NDMA, NPYR and NPIP occur in main- and side-stream smoke of cigarettes. In addition, some *N*-nitrosamines in various environmental samples were determined by GC–NPD [136], GC–ECD [133] and GC–MS [115,116,118,137]. Song and Ashley [116] detected tobacco-specific nitrosamines in cigarettes by supercritical fluid extraction and subsequent GC–TEA or GC–MS. As shown in Fig. 1.17, the chromatogram from the TEA detector had a lower background in comparison with that from GC–MS, but signal-to-noise levels were similar for both detections.

1.3.4 Heterocyclic amines

Most heterocyclic amines are polar and less volatile, and tend to elute as broad and tailing peaks due to the strong adsorption to the column and injector during GC analysis, when they are analyzed without derivatization. Therefore, the analyses of these amines have been generally carried out by HPLC [241]. For the GC analysis of heterocyclic amines, several derivatizations using acetic, trifluoroacetic anhydride, heptafluorobutyric anhydride, pentafluorobenzyl bromide, 3,5-bistrifluoromethylbenzyl bromide and 3,5-bistrifluoromethylbenzoyl chloride have been tested [157–160,162, 172,190,197,200]. However, acylation with acid anhydrides yielded derivatives with very poor GC properties. The alkylation products with pentafluorobenzyl bromide, 3,5-bistrifluoromethylbenzyl bromide and 3,5-bistrifluoromethylbenzoyl chloride had good GC properties for some heterocyclic amines. However, these methods gave a

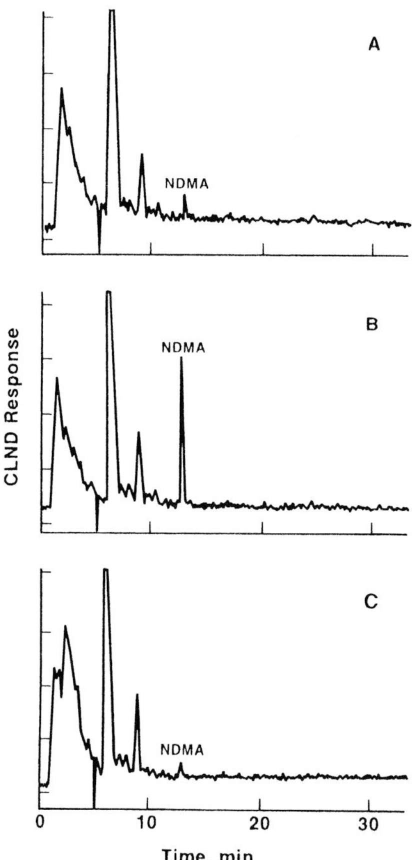

Fig. 1.16. Typical gas chromatograms of samples analyzed for *N*-nitrosodimethylamine using carbon–base membrane disk extractions. (A) Synthetic groundwater blank. (B) Synthetic groundwater sample fortified to 10 ng L^{-1}. (C) Authentic drinking water sample. GC conditions: column, Rtx-200 (crossbond trifluoropropylmethyl, 30 m × 0.53 mm i.d.); column temperature, initially isothermal at 35°C for 5 min, programmed from 35 to 175°C at 6°C min^{-1} and held at 175°C for 5 min; injector temperature, 150°C, carrier gas, helium, 4 ml min^{-1}; detector, CLND (nitrosamine selective mode); detector furnace, 250°C. NDMA = *N*-nitrosodimethylamine. (From Tomkins and Griest [120]. Reproduced with permission of the American Chemical Society.)

mixture of mono- and di-alkylated forms and were used for GC–MS analysis but not for GC analysis. Recently, Kataoka and Kijima [227] developed a simple and rapid derivatization method for GC analysis of mutagenic heterocyclic amines. Ten heterocyclic amines were converted into their *N*-dimethylaminomethylene derivatives (Fig. 1.4) with *N*,*N*-dimethylformamide dimethylacetal and measured by GC–NPD. As shown in Fig. 1.18A, these heterocyclic amines were separated within 7 min, although 2-amino-9*H*-pyrido[2,3-*b*]indole (AαC) and 2-aminodipyrido[1,2-*a* : 3′2′-*d*]imidazole (Glu-P-2) coeluted. The detection limits of these compounds were ranged from 2 to 15 pg. By using this method, AαC, 3-amino-1,4-dimethyl-5*H*-pyrido[3,4-*b*]indole

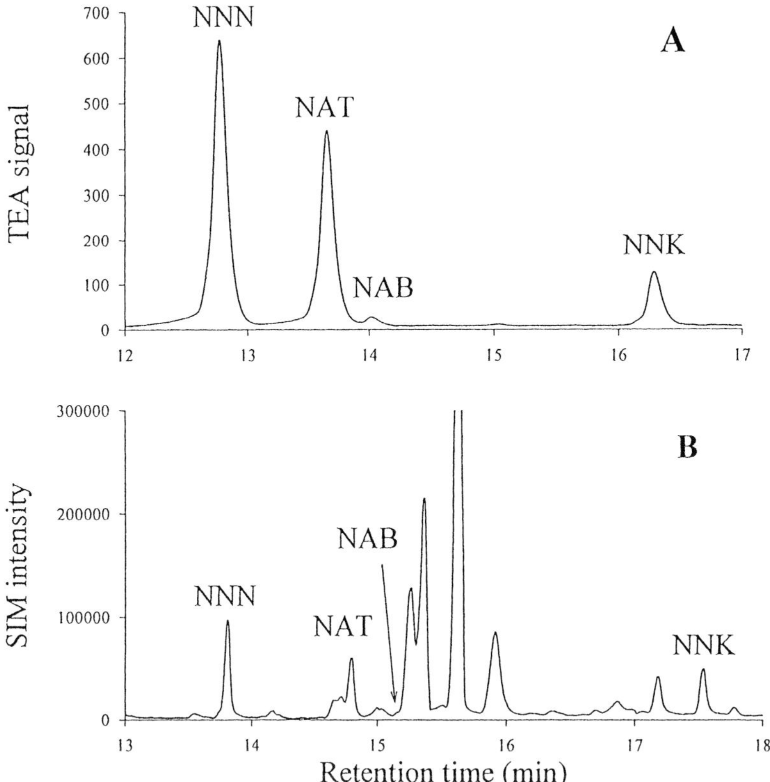

Fig. 1.17. Gas chromatograms of a cigarette tobacco extract solution obtained by (A) GC–TEA and (B) GC–MS using SIM mode. GC conditions: columns, HP-5 (30 m × 0.32 mm i.d. for GC–TEA) and HP-5ms (30 m × 0.25 mm i.d. for GC–MS); column temperature, initially isothermal at 40°C for 1 min, programmed from 40 to 160°C at 20°C min^{-1}, held at 160°C for 1 min, programmed from 160 to 200°C at 4°C min^{-1}, held at 200°C for 1 min, programmed from 200 to 260°C at 15°C min^{-1} and held at 260°C for 1 min; injector temperature, 260°C, carrier gas, helium, 40 cm s^{-1}; TEA, pyrolyzer temperature, 500°C; interface temperature, 200°C, GC–MS, ion source temperature, 230°C. Peaks: NNN – *N*-nitrosonornicotine; NNK = 4-(methylnitrosamino)-1-(3-pyridyl)-1-butanone; NAB = *N*-nitrosoanabasine; NAT = *N*-nitrosoanatabine. (From Song and Ashley [116]. Reproduced with permission of the American Chemical Society.)

(Trp-P-1) 3-amino-1-methyl-5*H*-pyrido[3,4-*b*]indole (Trp-P-2), 2-amino-6-methyldipyrido[1,2-*a* : 3′2′-*d*]imidazole (Glu-P-1), IQ, 2-amino-3,4-dimethylimidazo[4,5-*f*]quinoline (MeIQ) and 2-amino-1-methyl-6-phenylimidazo[4,5-*b*]pyridine (PhIP) were determined in several combustion smoke samples [180] (Fig. 1.18B,C). Kataoka et al. [195] also applied this technique to the analysis of river water sample, and identified IQ, Trp-P-1 and AαC in the water of the Danube River. On the other hand, few GC–MS–SIM data using positive ion electron ionization and negative ion chemical ionization are available on the analysis of heterocyclic amines after derivatization. Vainiotalo et al. [190] analyzed 2-amino-3,8-dimethylimidazo[4,5-*f*]quinoxaline (MeIQx) and 2-amino-3,4,8-trimethylimidazo[4,5-*f*]quinoxaline (4,8-DiMeIQx) in cooking fumes as their

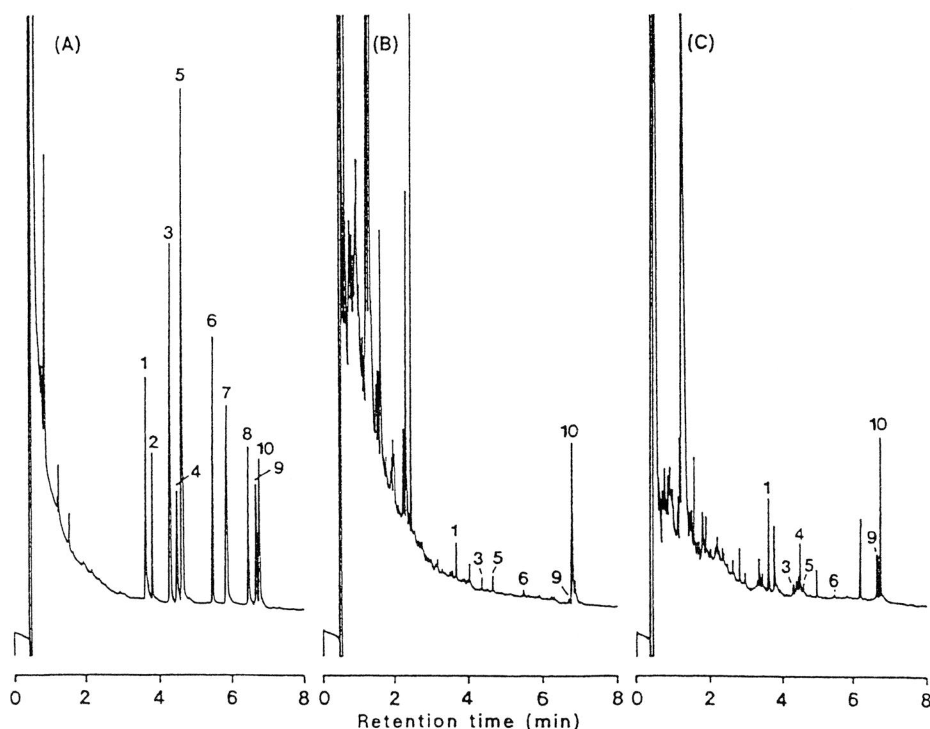

Fig. 1.18. Gas chromatograms obtained from (A) standard heterocyclic amines, (B) cigarette side-stream smoke and (C) combustion smoke of semi-dried fish. GC conditions: column, connected DB-1 (10 m × 0.25 mm i.d.) and DB-17ht (10 m × 0.25 mm i.d.); column temperature, programmed from 230 to 280°C at 10°C min^{-1}, from 280 to 330°C at 25°C min^{-1} and held at 330°C for 1 min; injection and detector temperatures, 340°C; carrier gas, helium, programmed from 180 to 230 kPa at 10 kPa min^{-1}, from 230 to 280 kPa at 25 kPa min^{-1} and held at 280 kPa for 1 min; split ratio, 10:1; detector, NPD. Peaks: 1 = AαC, 2 = Glu-P-1, 3 = Trp-P-1, 4 = Trp-P-2, 5 = IQ, 6 = MeIQ, 7 = MeIQx, 8 = DiMeIQx, 9 = PhIP, 10 = TriMeIQx (internal standard). (From Kataoka et al. [180]. Reproduced with permission of Springer-Verlag.)

3,5-bistrifluoromethylbenzyl derivatives by GC–MS with SIM mode (Fig. 1.19). The cooking fumes were collected through a glass fiber funnel into a sampler which consisted of two glass fiber filters and an XAD-2 sorbent tube. Although these heterocyclic amines give a mixture of mono- and di-bistrifluoromethylbenzyl derivatives, the spectra of the di-bistrifluoromethylbenzyl derivatives possess high mass fragment ions suitable for SIM work. When ions m/z 438 (MeIQx) and m/z 452 (DiMeIQx) were specially monitored, the detection limits of these compounds was 2 pg. The analytical methods for the determination of heterocyclic amines are also described in detail in previous review [250].

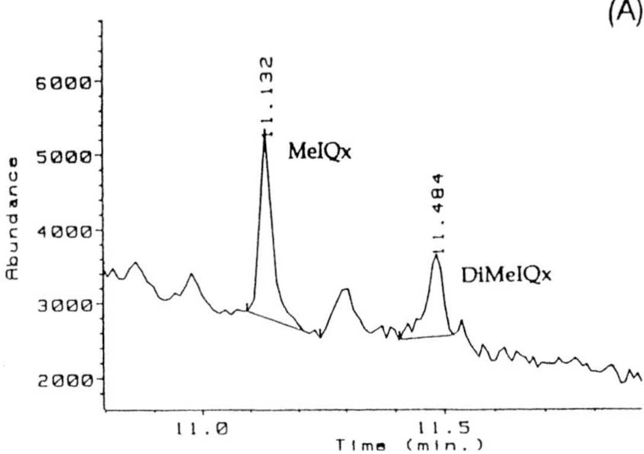

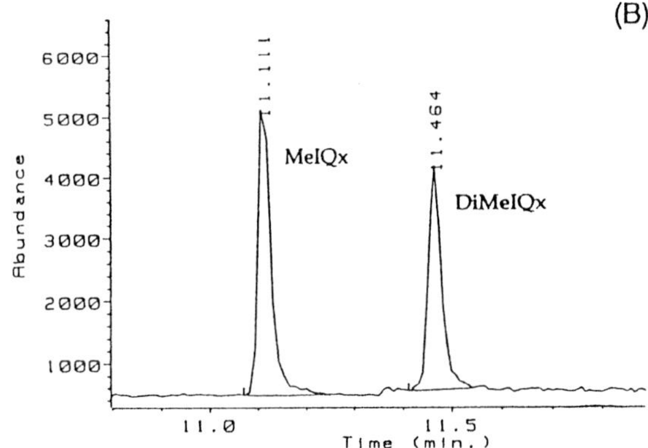

Fig. 1.19. GC-MS (SIM) traces of (A) fumes from fried beef/pork and (B) a standard sample containing 5 ng of MeIQx, 1.5 ng of 4,8-DiMeIQx and 2.5 ng of 7,8-DiMeIQx. The base peaks of the di-bis-TFMB derivatives were monitored (MeIQx: m/z 438. DiMeIQx: m/z 452). GC-MS conditions: column, HP-5 (20 m × 0.2 mm i.d.); column temperature, initially isothermal at 70°C for 1 min, increased to 280°C at 25°C min^{-1} and held at 280°C for 5 min; injector temperature, 250°C; carrier gas, helium, 1 ml min^{-1}; ionization voltage, 70 eV; emission current, 300 μA. (From Vainiotalo et al. [190]. Reproduced with permission of Springer-Verlag.)

1.4 CONCLUSIONS

Environmental pollution has now been recognized as a potentially important problem for public health, because it is very close to daily life of people everywhere in the world. Neglect of the environmental pollution caused widespread contamination of air, soil and water, unhealthy conditions for workers, and the loss of many precious natural resources. Therefore, environmental measurements are essential as a significant index of risk assessment not only in establishing the quantitative relationship between exposure and response against these pollutants, but also in establishing the natural baseline conditions in the environment. Toxic amines, described in this review, are widely distributed in a number of ambient environmental components such as airborne, diesel-exhaust, cigarette smoke, cooking fumes, river water, sewage water, soil and chemical products. The presence of these amines in the environment may be more extensive than previously thought and humans are continually exposed to these compounds in normal daily life. Therefore, monitoring of amines in environments is very important to preserve human health and natural environment, and the practical and reliable methods for determining accurate exposure levels of these amines should be established. For this purpose, various GC methods for the simultaneous determination of amines have been developed. However, the amines in environmental samples are generally present at very low concentrations and are often to be found among complex matrices containing a number of coexisting substances. Therefore, GC detectors for environmental analysis must be not only sensitive to the minute amounts of analytes, but also selective enough to discriminate against reasonable amounts of coexisting substances.

The nitrogen selective thermionic detectors, the nitrosamine specific detector TEA and GC–MS have been widely used for obtaining a high sensitivity and selectivity. In addition, the conversions of amines into fluorinated derivative to introduce ECD response and into sulfur- and phosphorus-derivatives to introduce FPD response have been devised for these purposes. The GC with nitrogen selective thermionic detection is highly sensitive and selective for amine analysis, and available in most laboratories. The selectivity and sensitivity of thermionic detectors depend on the flow rates of hydrogen, air, make-up gas and carrier gas, and a relatively stable response is realized using a flameless type thermionic detector. The GC–MS, capable of measuring simultaneously retention time and molecular mass, are powerful techniques for the identification and quantification of amines in complex matrix samples. The TEA is also selective and sensitive for *N*-nitroso compounds, and its response depends on the pyrolysis temperature. Although the GC–MS–SIM and GC–TEA are most effective for the analysis of a specified amine, they require sophisticated and expensive equipment that is beyond the reach of many laboratories. On the other hand, the GC with FPD and ECD are necessary for derivatization of amines to the corresponding halogen- and sulfur- or phosphorus-containing compounds, respectively, but thereby offer increased selectivity and sensitivity for specific amines. The GC methods with selective detectors, described in this review, have advantages and disadvantages and so far there are no all-powerful analytical methods.

The choice of an analytical method depends on the presence of amines in the environment at low parts per billion or less and the variety and complexity of sample. In order to achieve an efficient isolation and preconcentration of amines, several methods of sample preparation for the analysis of various environmental samples have been developed using a number of different purification techniques, such as distillation, liquid–liquid extraction, column chromatography, solid-phase extraction, extraction with mutagen specific adsorbent and preparative HPLC. The combination of these techniques is used successfully for the extraction and purification of amines. However, many of these methods are time-consuming, laborious and give low recoveries for some specific amines. On the other hand, SPME method is simple and rapid, and is directly applicable to the sample.

Environmental analysis has benefited from the recent advances in GC to include high resolution columns for complex mixtures and improved sensitivity and selectivity in detection. Powerful selective detection tools, coupled with the many technological advances in capillary column, software, and hardware development over the past decade, have added a new dimension. Measurement techniques in environmental analysis have become increasingly more sensitive and selective, in many cases lowering detection and quantitation limits by several orders of magnitude over the past few decades. For this purpose, further developments in the selective GC detector and in the capillary columns, e.g., shorter inactive columns with smaller internal diameters giving ultra-high column efficiency and speed, higher temperature phases and exterior coating for the fused-silica tubing, permit the analysis of both high-temperature and highly volatile amines, are expected. Furthermore, simple, rapid and automated separation analysis of amines in various environmental samples will be achieved by combination with convenient sample preparation technique such as SPME. Last, we hope that this review will serve as a guide to choosing the most effective techniques for the GC analysis of environmental amines.

1.5 REFERENCES

1   T. Colborn, D. Dumanoski and J.P. Myers, Our Stolen Future, Dutton, New York, 1996.

2   J.J. Peirce, R.F. Weiner and P.A. Vesilind, Environmental Pollution and Control, 4th edn., Butterworth-Heinemann, Woburn, MA, 1998.

3   H.A.H. Billiet, J. Chromatogr. Libr., 51B (1992) 583.

4   H. Kataoka, J. Chromatogr. A, 733 (1996) 19.

5   G.D. Clayton and F.E. Clayton (Eds.), Patty's Industrial Hygiene and Toxicology, Vol. 28, Wiley-Interscience, New York, 1981.

6   L. Parmeggiani (Ed.), Encyclopedia of Occupational Health and Safty, Vol. 1, International Labour Office, Geneva, 1983.

7   P.F. Vogt, J.J. Geralis, Ullman's Encyclopedia of Industrial Chemistry, Vol. A2, 5th edn., VCH, Weinheim, 1985.

8   S.E. Manahan, Environmental Chemistry, 4th edn., Lewis, 1990.

9   R.F. Straub, R.D. Voyksner and J.T. Keever, Anal. Chem., 65 (1993) 2131.

10 G.I. Baughman and E.J. Weber, Environ. Sci. Technol., 28 (1994) 267.

11 M.J. Ellenhorn, S. Schonwald, G. Ordog and J. Wasserberger (Eds.), Ellehorn's medical technology, 2nd edn., William and Wilkins, Baltimore, MD, 1997, p. 1496.

12 L. Hansen, B. Aakesson, J. Sollenberg and T. Lundh, Scand. J. Work Environ. Health, 12 (1986) 66.

13 G. Seeber, M.R. Buchmeiser, G.K. Bonn and T. Bertsch, J. Chromatogr. A., 809 (1998) 121.

14 G. Skarping, M. Dalene and P. Lind, J. Chromatogr. A, 663 (1994) 199.

15 K. Kuwata, E. Akiyama, Y. Yamazaki, H. Yamasaki, Y. Kuge and Y. Kiso, Anal. Chem., 55 (1983) 2199.

16 G. Audunsson and L. Mathiasson, J. Chromatogr., 261 (1983) 253.

17 G. Audunsson and L. Mathiasson, J. Chromatogr., 315 (1984) 299.

18 G. Skarping, T. Bellander and L. Mathiasson, J. Chromatogr., 370 (1986) 245.

19 L. Gronberg, P. Loukvist and J.A. Jonsson, Chromatographia, 33 (1992) 77.

20 L. Gronberg, P. Loukvist and J.A. Jonsson, Chemosphere, 24 (1992) 1533.

21 G. Becher, J. Chromatogr., 211 (1981) 103.

22 B. Andersson and K. Andersson, Appl. Occup. Environ. Hyg., 6 (1991) 40.

23 N. Kashihira, K. Makino, K. Kirita and Y. Watanabe, J. Chromatogr., 239 (1982) 617.

24 C. Lee and B.L. Olson, Org. Geochem., 6 (1984) 259.

25 A. Van Nesta, R.A. Duce and C. Lee, Geophys. Res. Lett., 14 (1987) 711.

26 E. Meichini, L. Boniforti and S.D. Marzio, Toxicol. Environ. Chem., 20 (1989) 9.

27   F. Luceri, G. Pieraccini, G. Moreti and P. Dolara, Toxicol. Indust. Health, 9 (1993) 405.
28   S. Skarping, L. Renman, C. Sango, L. Mathiasson and M. Dalene, J. Chromatogr., 346 (1985) 191.
29   L. Pan, J.M. Chong and J. Pawliszyn, J. Chromatogr. A, 773 (1997) 249.
30   P. Simon and C. Lemacon, Anal. Chem., 59 (1987) 480.
31   C.-X. Gao, I.S. Krull and T. Trainor, J. Chromatogr. Sci., 28 (1990) 102.
32   E.D. Zlotorzynska and W. Maruszak, J. Chromatogr. B, 714 (1998) 77.
33   R.M. Riggin, T.F. Cole and S. Billets, Anal. Chem., 55 (1983) 1862.
34   R.C.C. Wegman and G.A.L. De Korte, Int. J. Environ. Anal. Chem., 9 (1981) 1.
35   T.C. Schmidt, M. Less, R. Haas, E. von Low, K. Steinbach and G. Stork, J. Chromatogr. A, 810 (1998) 161.
36   M. Less, T.C. Schmidt, E. von Low and G. Stork, J. Chromatogr. A, 810 (1998) 173.
37   F. Guan, K. Watanabe, A. Ishii, H. Seno, T. Kumazawa, H. Hattori and O. Suzuki, J. Chromatogr. B, 714 (1998) 205.
38   M. Koga, T. Akiyama and R. Shinohara, Bunseki Kagaku, 30 (1981) 745.
39   M.J. Avery and G.A. Junk, Anal. Chem., 57 (1985) 790.
40   A. Terashi, Y. Hanada, A. Kido and R. Shinohara, J. Chromatogr., 503 (1990) 369.
41   J. Pietsch, S. Hampel, W. Schmidt, H.J. Brauch and E. Worch, Fresenius J. Anal. Chem., 355 (1996) 164.
42   F. Sacher, S. Lenz and H.-J. Brauch, J. Chromatogr. A, 764 (1997) 85.
43   T. Tsukioka, H. Ozawa and T. Murakami, J. Chromatogr., 642 (1993) 395.
44   W.S. Gardner and P.A. St John, Anal. Chem., 63 (1991) 537.
45   J. Lehotay, V. Rattay, E. Brandsteterova and D. Oktavec, J. Liq. Chromatogr., 15 (1992) 307.
46   X.-H. Yang, C. Lee and M.I. Scranton, Anal. Chem., 65 (1993) 572.
47   T. Okumura, K. Imamura and Y. Nishikawa, J. Chromatogr. Sci., 34 (1996) 190.
48   M. Longo and A. Cavallaro, J. Chromatogr. A, 753 (1996) 91.
49   L. Muller, E. Fattore and E. Benfinati, J. Chromatogr. A, 791 (1997) 221.
50   E. Fattore, L. Muller, E. Davoli, D. Castelli and E. Benfenati, Chemosphere, 36 (1998) 2007.
51   R.J.J. Vreuls, E. Romijn and U.A.Th. Brinkman, J. Microcol. Sep., 10 (1998) 581.
52   E. Baltussen, F. David, P. Sandra, H.G. Janssen and C. Cramers, J. High Resol. Chromatogr., 21 (1998) 645.
53   M. Dalene and G. Skarping, J. Chromatogr., 331 (1985) 321.
54   H. Kataoka, G. Maruo, K. Kijima, S. Yamamoto and S. Narimatsu, in preparation.
55   K. Kijima, H. Kataoka and M. Makita, J. Chromatogr. A, 738 (1996) 83.
56   H. Kataoka, S. Shindoh and M. Makita, J. Chromatogr. A, 723 (1996) 93.
57   G. Pieraccini, F. Luceri and G. Moneti, Rapid Comm. Mass Spectrom., 6 (1992) 406.
58   G. Grimmer, K.W. Noujack and G. Dettbarn, Toxicol. Lett., 35 (1987) 117.
59   J.B. Forehand, G.L. Dooly and S.C. Moldveanu, J. Chromatogr. A, 898 (2000) 111.
60   A.I.G. Valcarcel, C.S. Brunete, L. Martinez and J.L. Tadeo, J. Chromatogr. A, 719 (1996) 113.
61   R.I.P. Martin, J.M. Franco, P. Molist and J.M. Gallardo, Int. J. Food Sci. Technol., 22 (1987) 509.
62   T. Hamano, Y. Mitsuhashi and Y. Matsuki, Agric. Biol. Chem., 45 (1981) 2237.
63   H. Kataoka, S. Shindoh and M. Makita, J. Chromatogr., 695 (1995) 142.
64   B. Pfundstein, A.R. Tricker and R. Preussmann, J. Chromatogr., 539 (1991) 141.
65   B. Pfundstein, A.R. Tricker, E. Theobald, B. Spiegelhalder and R. Preussmann, Food Chem. Toxicol., 29 (1991) 733.
66   R.A. Scanlan, J.F. Barbour and C.I. Chappel, J. Agric. Food Chem., 38 (1990) 442.
67   B.J. Finlayson-Pitts, J.N. Pitts, Atmospheric Chemistry, Wiley-Interscience, New York, 1986.
68   S. Lowis, M.A. Eastwood and W.G. Brydon, J. Chromatogr., 278 (1983) 139.
69   G. Skarping, M. Dalene, T. Brorson, J.F. Sandstrom, C. Sango and A. Tiljander, J. Chromatogr., 479 (1989) 125.
70   H. Kataoka, S. Ohrui, Y. Miyamoto and M. Makita, Biomed. Chromatogr., 6 (1992) 251.
71   A.Q. Zhang, S.C. Mitchell, R. Ayesh and R.L. Smith, J. Chromatogr., 584 (1992) 141.
72   A. Dasgupta, J. Chromatogr. B, 716 (1998) 354.
73   G.A. Mills, V. Walker and H. Mughal, J. Chromatogr. B, 723 (1999) 281.
74   G. Grimmer, G. Dettbarn, A. Seidel and J. Jacob, Sci. Total Environ., 247 (2000) 81.

75   L. Belin, U. Wass, G. Audunsson and L. Mathiasson, Br. J. Ind. Med., 40 (1983) 251.

76   H. Greim, D. Bury, H.J. Klimisch, M. Oeben-Negele and K. Ziegler-Skylakakis, Chemosphere, 36 (1998) 271.

77   J.S. Wishnok, Anal. Chem., 64 (1992) 1126A.

78   J.M. Sontag (Ed.), Carcinogens in Industry and the Environment, Marcel Dekker, New York, 1981.

79   F.S.H. Abram and I.R. Sims, Water Res., 16 (1982) 1309.

80   T.-A. Chiang, W. Pei-Fen, L.S. Ying, L.-F. Wang and Y.C. Ko, Food Chem. Toxicol., 37 (1999) 125.

81   G. Birner and H.G. Neumann, Arch. Toxicol., 62 (1988) 110.

82   E. Ward, A. Carpenter, S. Markowitz, D. Roberts and W. Halperin, J. Natl. Cancer Inst., 83 (1991) 501.

83   P.L. Skipper, X. Peng, C.K. Soohoo and S.R. Tannenbaum, Drug Metab. Rev., 26 (1994) 111.

84   M. Riffelmann, G. Muller, W. Schmieding, W. Popp and K. Norpoth, Int. Arch. Occup. Environ. Health, 68 (1995) 36.

85   M. Dalene, G. Skarping and P. Brunmark, Int. Arch. Occup. Environ. Health, 67 (1995) 67.

86   E.M. Ward, G. Sabbioni, D.G. DeBord, A.W. Teass, K.K. Brown, G.G. Talaska, D.R. Roberts, A.M. Ruder and R.P. Streicher, J. Natl. Cancer Inst., 88 (1996) 1046.

87   B. Stahlbom, B. Akesson and B. Jonsson, Int. Arch. Occup. Environ. Health, 70 (1997) 393.

88   C. Kutzer, B. Branner, W. Zwickenpflug and E. Richter, J. Chromatogr. Sci., 35 (1997) 1.

89   H.-G. Neumann, I.Z. Baier and C. van Dorp, Arch. Toxicol. Suppl., 20 (1998) 179.

90   G. Sabbioni and A. Beverbach, J. Chromatogr. B Biomed. Sci. Appl., 744 (2000) 377.

91   K.D. Brunnemann, S.S. Hecht and D. Hoffmann, J. Toxicol. Clin. Toxicol., 19 (1982) 661.

92   A.R. Tricker and R. Preussmann, Mutat. Res., 259 (1991) 277.

93   A.R. Tricker and S.S. Kubaki, Food Addit. Contam., 9 (1992) 39.

94   A.R. Tricker and R. Preussmann, Clin. Invest., 70 (1992) 283.

95   K.D. Brunnemann, B. Prokopczyk, M.V. Djordjevic and D. Hoffmann, Crit. Rev. Toxicol., 26 (1996) 121.

96   W. Lijinsky, Oncology, 37 (1980) 223.

97   B. Spiegelhalder and R. Preussmann, Carcinogenesis, 6 (1985) 545.

98   W.R. Licht and M.W. Deen, Carcinogenesis, 9 (1988) 2227.

99   E.W. Van Stee, R.A. Sloane, J.E. Simmons, M.P. Moorman and K.D. Brunnemann, Carcinogenesis, 16 (1995) 89.

100 M.C. Bowman, Handbook of Carcinogens and Hazadous Substances, Marcel Dekker, New York, Basle, 1982.

101 K. Takatsuki and T. Kikuchi, J. Chromatogr., 508 (1990) 357.

102 M.H. Li, C. Ji and S.J. Cheng, Nutr. Cancer, 8 (1986) 63.

103 J.H. Hotchkiss, Cancer Surv., 8 (1989) 295.

104 A.R. Tricker, B. Pfundstein, E. Theobald and R. Preussmann, Food Chem. Toxicol., 29 (1991) 729.

105 N.P. Sen, S.W. Seaman and B.D. Page, J. Chromatogr. A, 788 (1997) 131.

106 D.H. Fine, D.P. Rounbehler and U. Goff, IARC Sci. Publ., 109 (1993) 269.

107 S. Fadlallah, S.F. Cooper, G. Perrault, G. Truchon and J. Lesage, Bull. Environ. Contam. Toxicol., 57 (1996) 867.

108 H. Kataoka, M. Kurisu and S. Shindoh, Bull. Environment. Contam. Toxicol., 59 (1997) 570.

109 K.D. Brunnemann, W. Fink and F. Moser, Oncology, 37 (1980) 217.

110 J.D. Adams, K.D. Brunnemann and D. Hoffman, J. Chromatogr., 256 (1983) 347.

111 K.D. Brunnemann, L. Genoble and D. Hoffman, J. Agric. Food Chem., 33 (1985) 1178.

112 K.D. Brunnemann and D. Hoffmann, Crit. Rev. Toxicol., 21 (1991) 235.

113 R. Truhaut, N.P. Lich, M. Castegnaro, M.C. Bourgade and C. Martin, J. Cancer Res. Oncol., 108 (1984) 157.

114 A.R. Tricker, C. Ditrich and R. Preussmann, Carcinogenesis, 12 (1991) 257.

115 R.F. Arrendale, W.J. Chamberlain, O.T. Chortyk and J.L. Baker, Anal. Chem., 58 (1986) 565.

116 S. Song and D.L. Ashley, Anal. Chem., 71 (1999) 1303.

117 W.I. Kimoto, C.J. Dooley and W. Fiddler, Water Res., 15 (1981) 1099.

118 N.P. Sen, P.A. Baddoo, D. Weber and M. Boyle, Intern. J. Environ. Anal. Chem., 66 (1994) 149.

119 B.A. Tomkins, W.H. Griest and C.E. Higgins, Anal. Chem., 67 (1995) 4387.

120   B.A. Tomkins and W.H. Griest, Anal. Chem., 68 (1996) 2533.

121   J.S. Ho, T.A. Bellar, J.W. Eichelberger and W.L. Budde, Environ. Sci. Technol., 24 (1990) 1748.

122   W.S. Brewer, A.C. Draper and S.S. Wey, Environ. Pollut. Ser. B, Chem. Phys. , 1 (1980) 37.

123   J.B. Morrison and S.S. Hecht, Food Chem. Toxicol., 20 (1982) 583.

124   J.W. Pensabene and W. Fiddler, J. Assoc. Off. Anal. Chem., 77 (1994) 981.

125   D.C. Havery and T. Fazio, Food Chem. Toxicol., 20 (1982) 939.

126   B.-G. Osterdahl, Food Chem. Toxicol., 21 (1983) 755.

127   H.C. Thompson, S.M. Billedeau, B.J. Miller, E.B. Hansen and J.P. Freeman, J. Toxicol. Environ. Health, 13 (1984) 615.

128   S.M. Billedeau, H.C. Thompson and B.J. Miller, J. Assoc. Off. Anal. Chem., 69 (1986) 31.

129   J.I. Gray and M.A. Stachiw, J. Assoc. Off. Anal. Chem., 70 (1987) 64.

130   N.P. Sen, S.W. Seamen and S.C. Kushwaha, J. Assoc. Off. Anal. Chem., 70 (1987) 434.

131   J.W. Rensaber, W. Fiddler and R.A. Gates, J. Agric. Food Chem., 43 (1995) 1919.

132   J.M. Fajen, G.A. Carson, D.P. Roinbehler, T.Y. Fan, R. Vita, U.E. Goff, M.H. Wolf, G.S. Edwards, D.H. Fine, V. Reinhold and K. Biemann, Science, 205 (1979) 1262.

133   S. Fadlallah, S.F. Cooper, M. Fournier, D. Drolet and G. Perrault, J. Chromatogr. Sci., 28 (1990) 517.

134   B.A. Dawson and R.C. Lawrence, J. Assoc. Off. Anal. Chem., 70 (1987) 554.

135   L.B. Niclsen and S. Lings, Med. Hypothesis, 42 (1994) 265.

136   R.R. Scharfe and C.C. McLenagham, J. Assoc. Off. Anal. Chem., 72 (1989) 508.

137   R. Frassanito, E. Benfenati, G. Ciotti and R. Fanelli, Toxicol. Environ. Chem., 45 (1994) 199.

138   B. Prokopczyk, J.E. Cox, D. Hoffmann and S.E. Waggoner, J. Natl. Cancer Inst., 89 (1997) 868.

139   J.W. Dallinga, D.M.F.A. Pachen, A.H.P. Lousberg, J.A.M. van Geel, G.M.P. Houben, R.W. Stockbrugger, J.M.S. van Maanen and J.C.S. Kleinjans, Cancer Lett., 124 (1998) 119.

140   R. Preussmann, B.W. Stewart, N-Nitroso carcinogens in the environment, in: C.E. Searles (Ed.), Chemical Carcinogens, 2nd edn., Amer. Chem. Soc., Washington, DC, 1984.

141   A.R. Tricker, B. Spiegelhalder and R. Preussmann, Cancer Serv., 8 (1989) 251.

142   R. Desjardins, M. Fournier, F. Denizeau and K. Krzystyniak, J. Toxicol. Environ. Health, 39 (1992) 281.

143   E. De Stefani, P. Boffetta, M. Mendilaharsu, J. Carzoglio and H. D-Pellergrini, Nutr. Cancer, 30 (1998) 158.

144   D.E. Shuker, Cancer Surv., 8 (1989) 475.

145   S. Fadlallah, S.F. Cooper, M. Denizeau, S. Mansour, F. Guertin, K. Krzystyniak and M. Fournier, Intern. J. Environ. Anal. Chem., 56 (1994) 165.

146   S.S. Hecht, Mutat. Res., 425 (1999) 127.

147   M. Stiborova, H.H. Schmeiser, M. Wiessler and E. Frei, Cancer Lett., 138 (1999) 61.

148   S.K. Chhabra, L.M. Anderson, C. Perella, D. Dasai, S. Amin, S.A. Kyrtopoulos and V.L. Souliotis, Toxicol. Appl. Pharmacol., 169 (2000) 191.

149   T. Sugimura and K. Wakabayashi, in: M.W. Pariza, H.-U. Aeschbacher, J.S. Felton and S. Sato (Eds.), Mutagens and Carcinogens in the Diet, Wiley-Liss, New York, 1990.

150   J.S. Felton and M.G. Knize, Mutat. Res., 259 (1991) 205.

151   M. Jagerstad, K. Skog, S. Grivas and K. Olsson, Mutat. Res., 259 (1991) 219.

152   K. Wakabayashi, M. Nagao, H. Esumi and T. Sugimura, Cancer Res. Suppl., 52 (1992) 2092s.

153   G. Eisenbrand and W. Tang, Toxicology, 84 (1993) 1.

154   K. Skog, Food Chem. Toxicol., 31 (1993) 655.

155   B. Stavric, Food Chem. Toxicol., 32 (1994) 977.

156   T. Sugimura, M. Nagao and K. Wakabayashi, Environ. Health Perspect., 104 (1996) 429.

157   S. Murray, N.J. Gooderham, V.F. Barner, A.R. Boobis and D.S. Davies, Carcinogenesis, 8 (1987) 937.

158   S. Murray, N.J. Gooderham, A.R. Boobis and D.S. Davies, Carcinogenesis, 9 (1988) 321.

159   S. Murray, A.M. Lynch, M.G. Knize and N.J. Gooderham, J. Chromatogr., 616 (1993) 211.

160   L.M. Tikkanen, T.M. Sauri and K.J. Latva-Kala, Food Chem. Toxicol., 31 (1993) 717.

161   K. Skog, A. Solyakov, P. Arvidsson and M. Jagerstad, J. Chromatogr. A, 803 (1998) 227.

162   M. Takahashi, K. Wakabayashi, M. Nagao, M. Yamamoto, T. Masui, T. Goto, N. Kinae, I. Tomita and T. Sugimura, Carcinogenesis, 6 (1985) 1195.

163 J.S. Felton, M.G. Knize, M. Roper, E. Fultz, N.H. Shen and K.W. Turteltaub, Cancer Res. Suppl., 52 (1992) 2103s.

164 M.G. Knize, J.S. Felton and G.A. Gross, J. Chromatogr., 624 (1992) 253.

165 G.A. Gross and L. Fay, in: R.H. Adamson, J.-A. Gustafsson, N. Ito, M. Nagao, T. Sugimura, K. Wakabayashi and Y. Yamazoe (Eds.), Heterocyclic Amines in Cooked Foods: Possible Human Carcinogens, Princeton Scientific, Princeton, NJ, 1995.

166 K. Wakabayashi, H. Ushiyama, M. Takahashi, H. Nukaya, S.-B. Kim, M. Hirose, M. Ochiai, T. Sugimura and M. Nagao, Environ. Health Perspect., 99 (1993) 129.

167 D.W. Layton, K.T. Bogen, M.G. Knize, F.T. Hatch, V.M. Johnson and J.S. Felton, Carcinogenesis, 16 (1995) 39.

168 S. Robbana-Barnat, M. Rabache, E. Rialland and J. Fradin, Environ. Health Perspect., 104 (1996) 280.

169 N.K.J. Rich, K. Zhao, B.P. Murray, S. Bhadresa, S.J. Crosbie, A.R. Boobis and D.S. Davies, Br. J. Clin. Pharmacol., 42 (1996) 91.

170 M.T. Galceran, P. Pais and L. Puignon, J. Chromatogr. A, 719 (1996) 203.

171 G.A. Gross and A. Gruter, J. Chromatogr., 592 (1992) 271.

172 B. Thomson, Eur. J. Cancer Prev., 8 (1999) 201.

173 E. Richling, M. Kleinschnitz and P. Schreier, Eur. Food Res. Technol., 210 (1999) 68.

174 F. Toribio, E. Moyano, L. Puignou and M.T. Galceran, J. Chromatogr. A, 569 (2000) 307.

175 Z. Balogh, J.I. Gray, E.A. Gomaa and A.M. Booren, Food Chem. Toxicol., 38 (2000) 395.

176 P. Pais and M.G. Knize, J. Chromatogr. B Biomed. Sci. Appl., 747 (2000) 139.

177 P.A. Guy, E. Gremaud, J. Richoz and R.J. Turesky, J. Chromatogr. A, 883 (2000) 89.

178 S. Manabe, N. Kurihara, O. Wada, S. Izumikawa, K. Asakuno and M. Morita, Environ. Pollut., 80 (1993) 281.

179 S. Manabe, S. Izumikawa, K. Asakuno, O. Wada and Y. Kanai, Environ. Pollut., 70 (1991) 255.

180 H. Kataoak, K. Kijima and G. Maruo, Bull. Environ. Contam. Toxicol., 60 (1998) 60.

181 D. Yoshida and T. Matsumoto, Cancer Lett., 10 (1980) 141.

182 T. Matsumoto, D. Yoshida and H. Tomita, Cancer Lett., 12 (1981) 105.

183 M. Yamashita, K. Wakabayashi, M. Nagao, S. Sato, Z. Yamaizumi, M. Takahashi, N. Kinae, I. Tomita and T. Sugimura, Jap. J. Cancer Res., 77 (1986) 419.

184 S. Manabe, O. Wada and Y. Kanai, J. Chromatogr., 529 (1990) 125.

185 Y. Kanai, O. Wada and S. Manabe, Carcinogenesis, 11 (1990) 1001.

186 S. Manabe, K. Tohyama, O. Wada and T. Aramaki, Carcinogenesis, 12 (1991) 1945.

187 S. Manabe, E. Uchino and O. Wada, Mutation Res., 226 (1989) 215.

188 S. Manabe, O. Wada, M. Morita, S. Izumikawa, K. Asakuno and H. Suzuki, Environ. Pollut., 75 (1992) 301.

189 S. Manabe and O. Wada, Environ. Pollut., 64 (1990) 121.

190 S. Vainiotalo, K. Matveinen and A. Reunanen, Fresenius J. Anal. Chem., 345 (1993) 462.

191 H.P. Thiebaud, M.G. Knize, P.A. Kuzmicky, J.S. Felton and D.P. Hsieh, J. Agric. Food Chem., 42 (1994) 1502.

192 H.P. Thiebaud, M.G. Knize, P.A. Kuzmicky, D.P. Hsieh and J.S. Felton, Food Chem. Toxicol., 33 (1995) 821.

193 J. Wu, M.K. Wong, H.K. Lee and C.N. Ong, J. Chromatogr. Sci., 33 (1995) 712.

194 T. Ohe, Mutat. Res., 393 (1997) 73.

195 H. Kataoka, T. Hayatsu, G. Hietsch, H. Steinkeller, S. Nishioka, S. Narimatsu, S. Kunasmuller and H. Hayatsu, Mutat. Res., 466 (2000) 27.

196 Y. Ono, I. Somiya and Y. Oda, Water Res., 34 (2000) 890.

197 S. Murray, N.J. Gooderham, A.R. Boobis and D.S. Davies, Carcinogenesis, 10 (1989) 763.

198 S. Manabe and O. Wada, Environ. Mol. Mutagen., 15 (1990) 229.

199 H. Ushiyama, K. Wakabayashi, M. Hirose, H. Itoh, T. Sugimura and M. Nagao, Carcinogenesis, 12 (1991) 1417.

200 M.D. Friesen, L. Garren, J.-C. Bereziat, F. Kadlubar and D. Lin, Environ. Health Perspect., 99 (1993) 179.

201 H. Ohgaki, S. Takayama and T. Sugimura, Mutat. Res., 259 (1991) 399.

202 R.H. Adamson and U.P. Thorgeirsson, Adv. Exp. Med. Biol., 369 (1995) 211.

203 H.A. Schut and E.G. Snyderwine, Carcinogenesis, 20 (1999) 353.

204 L. Airoldi, R. Pastorelli, C. Magagnotti and R. Fanelli, Adv. Exp. Med. Biol., 231 (1999) 231.

205 R.H. Adamson, U.P. Thorgeirsson, E.G. Snyderwine, S.S. Thorgeirrson, J. Reeves, D.W. Dalgard, S. Takayama and T. Sugimura, Jpn. J. Cancer Res., 81 (1990) 10.

206 R.H. Adamson, U.P. Thorgeirsson and T. Sugimura, Arch. Toxicol. Suppl., 18 (1996) 303.

207 T. Mukai, Y. Yamazoe and R. Kato, Eur. J. Pharmacol., 98 (1984) 35.

208 Y. Kanai, O. Wada and S. Manabe, J. Pharmacol. Exp. Ther., 252 (1990) 1269.

209 S. Manabe, Y. Kanai, S. Ishikawa and O. Wada, J. Clin. Chem. Biochem., 26 (1988) 265.

210 H. Ichinose, N. Ozaki, D. Nakahara, M. Naoi, K. Wakabayashi, T. Sugimura and T. Nagatsu, Biochem. Pharmacol., 37 (1988) 3289.

211 M. Naoi, T. Takahashi, H. Ichinose, K. Wakabayashi, T. Sugimura and T. Nagatsu, Biochem. Biophys. Res. Commun., 157 (1988) 494.

212 R. Hasegawa, H. Tanaka, S. Tamano, T. Shirai, M. Nagao, T. Sugimura and N. Ito, Carcinogenesis, 15 (1994) 2567.

213 F.I. Onuska and F.W. Karasek, Open Tubular Column Gas Chromatography in Environmental Sciences, Plenum Press, New York, 1984.

214 F. Bruner, Gas Chromatographic Environmental Analysis: Principles, Techniques, Instrumentation, VCH, New York, 1993.

215 R.L. Grob and M.A. Kaiser, Environmental problem solving using gas and liquid chromatography, J. Chromatogr. Libr., 21 (1982).

216 M. Dressler, Selective gas chromatographic detectors, J. Chromatogr. Libr., 36 (1986).

217 R. Buffington and M.K. Wilson, Detectors for Gas Chromatography — A Practical Primer, Hewlett Packard, Avondale, PA, 1991.

218 R.P.W. Scott, Chromatographic detectors: design, function and operation, Chromatographic Science Series, Vol. 73, Marcel Dekker, New York, 1996.

219 H. Kataoka, S. Yamamoto and S. Narimatsu, Encyclopedia of Separation Science, Vol. 5, Academic Press, London, UK, 2000, pp. 1982.

220 M. Dalene, L. Mathiasson and J.A. Jonsson, J. Chromatogr., 207 (1981) 37.

221 S. Skarping, B.E.F. Smith and M. Dalene, J. Chromatogr., 303 (1984) 89.

222 M. Dalene, T. Lundh and L. Mathiasson, J. Chromatogr., 322 (1985) 169.

223 T. Lundh and B. Akesson, J. Chromatogr., 617 (1993) 191–196.

224 M. Abalos, J.M. Bayona and F. ventura, Anal. Chem., 71 (1999) 3531.

225 C. Maris, A. Laplanche, J. Morvan and M. Bloquel, J. Chromatogr. A, 846 (1999) 331.

226 S. Skarping, L. Renman and M. Dalene, J. Chromatogr., 270 (1983) 207.

227 H. Kataoka and K. Kijima, J. Chromatogr. A, 767 (1997) 187.

228 S. Ohrui, H. Kataoka, Y. Miyamoto, K. Ohtsuka and M. Makita, Bunseki Kagaku, 40 (1991) 119.

229 H. Kataoka, M. Eda and M. Makita, Biomed. Chromatogr., 7 (1993) 129.

230 Y. Miyamoto, H. Kataoka, S. Ohrui and M. Makita, Bunseki Kagaku, 43 (1994) 1113.

231 G.W. Diachenko, Environ. Sci. Technol., 13 (1979) 329.

232 R.T. Coutts, E.E. Hargesheimer, F.M. Pasutto and G.B. Baker, J. Chromatogr. Sci., 19 (1981) 151.

233 S. Skarping, L. Renman and B.E.F. Smith, J. Chromatogr., 267 (1983) 315.

234 H.-B. Lee, J. Chromatogr., 457 (1988) 267.

235 Z. Moldvan and J.M. Bayona, Rapid Commun. Mass Spectrom., 14 (2000) 379.

236 M. Mohnke, B. Schmidt, R. Schmidt, J.C. Buijten and Ph. Mussche, J. Chromatogr., 667 (1994) 334.

237 J.J. Richard and G.A. Junk, Anal. Chem., 56 (1984) 1625.

238 H. Hayatsu, J. Chromatogr., 597 (1992) 37.

239 H. Hayatsu, T. Hayatsu, S. Arimoto and H. Sakamoto, Anal. Biochem., 235 (1996) 185.

240 E.E. Penton, Adv. Chromatogr., 37 (1997) 205.

241 J. Pawliszyn, Solid Phase Microextraction: Theory and Practice, Wiley-VCH, New York, 1997.

242 J. Pawliszyn (Ed.), Applications of Solid Phase Microextraction, The Royal Society of Chemistry, Cambridge, UK, 1999.

243 H. Kataoka, H.L. Lord and J. Pawliszyn, J. Chromatogr. A, 880 (2000) 35.

244 H.L. Lord and J. Pawliszyn, J. Chromatogr. A, 885 (2000) 153.

245   J. Namiesnik, B. Zygmunt and A. Jastrzebska, J. Chromatogr. A, 885 (2000) 405.
246   N.H. Snow, J. Chromatogr. A, 885 (2000) 445.
247   M.deF. Alpendurada, J. Chromatogr. A, 889 (2000) 3.
248   H.L. Lord and J. Pawliszyn, J. Chromatogr. A, 902 (2000) 17.
249   H. Kataoka, H.L. Lord and J. Pawliszyn, Encyclopedia of Separation Science, Vol. 9, Academic
       Press, London, UK, 2000, p. 4153.
250   H. Kataoka, J. Chromatogr. A, 774 (1997) 121.

W. Kleiböhmer (Ed.), *Environmental Analysis*
Handbook of Analytical Separations, Vol. 3
© 2001 Elsevier Science B.V. All rights reserved

CHAPTER 2

Modern techniques for the analysis of polycyclic aromatic hydrocarbons

Hian Kee Lee

*Department of Chemistry, National University of Singapore, 3 Science Drive 3, Singapore 117543,
Republic of Singapore*

2.1 INTRODUCTION

Studies in the past few decades suggest strongly that while genetic disposition is an important factor, nevertheless the occurrence of cancer in man is also strongly linked to environmental factors. In particular, attention has been focused on the important role played by chemical carcinogens existing in the environment. Polycyclic aromatic hydrocarbons (PAHs) represent one of the largest and most important classes of environmental carcinogens. Fig. 2.1 shows the structures of some of these compounds. (It should be noted that the term PAHs can be taken to mean those benzenoid compounds comprising of carbon and hydrogen atoms only, as well as those substituted ones, with various side chains such as $-NO_2$, $-NH_2$, $-OH$, etc., or those with heteroatoms such as O, S, or N as part of the ring structure.)

PAHs are formed from the combustion of fossil fuels and organic matter and are ubiquitous to the environment. Anthropogenic activities are the major sources of emission of PAHs into the environment [1], although some are released via natural processes. The well-documented carcinogenic and mutagenic properties of many PAHs make them perhaps the most widely studied class of compounds, especially amongst the environmental pollutants. The importance of these compounds is illustrated by the great and continuing interest in developing analytical techniques for them. Characterizing these compounds in environmental samples through a variety of analytical procedures is important to determine their specific emission sources; knowledge of these allows the relevant authorities to strategize their approach to reduce, if not eliminate, PAH emissions into our surroundings.

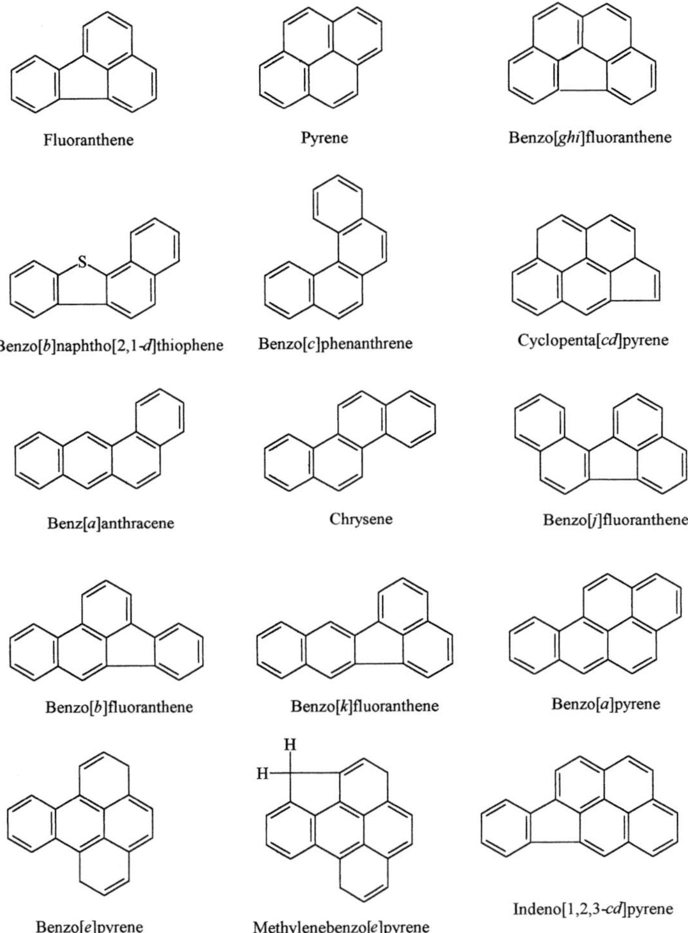

Fluoranthene   Pyrene   Benzo[*ghi*]fluoranthene

Benzo[*b*]naphtho[2,1-*d*]thiophene   Benzo[*c*]phenanthrene   Cyclopenta[*cd*]pyrene

Benz[*a*]anthracene   Chrysene   Benzo[*j*]fluoranthene

Benzo[*b*]fluoranthene   Benzo[*k*]fluoranthene   Benzo[*a*]pyrene

Benzo[*e*]pyrene   Methylenebenzo[*e*]pyrene   Indeno[1,2,3-*cd*]pyrene

Fig. 2.1. Structures of some polycyclic aromatic hydrocarbons.

2.2 DEVELOPMENT OF INSTRUMENTAL TECHNIQUES FOR PAH ANALYSIS

Chromatographic methods are widely employed in the separation and determination of PAHs. Capillary-column gas chromatography (GC) and GC–mass spectrometry (MS) have become the most popular methods for the analysis of complex mixtures of PAHs. The nature of combustion processes and the variety of organic matter and fossil fuels mean that a particular technique needs to be able to handle complex, multi-component mixtures of these compounds. In this respect, GC and/or GC–MS are without doubt the methods of choice. Resolving power, speed of analysis and low detection limits are all significant advantages afforded by GC procedures.

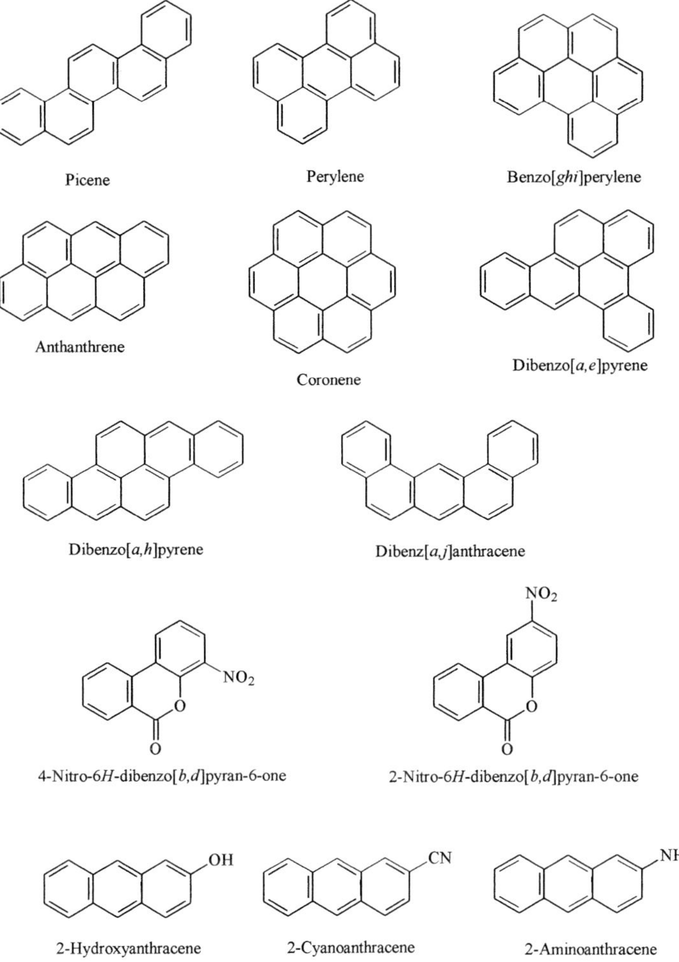

Fig. 2.1 (continued).

Since the 1970s, when researchers such as Lee et al. [2], Bjorseth et al. [3] and Schomburg et al. [4], to name just a few, began using capillary GC to separate PAHs, there have been a vast number of reports on the application of this technique for the analysis of these compounds. As mentioned earlier, one of the advantages of capillary GC is its high resolving power, but some PAHs may be degraded when exposed to high temperatures during separation; in addition, some isomeric PAHs such as benzo[*b*]fluoranthene and benzo[*k*]fluoranthene are very difficult to separate. For these and also the less volatile PAHs, high-performance liquid chromatography (HPLC) is a powerful analytical technique. This is especially so in the past ten years when stationary phases designed to be more compatible to PAH analysis have become commercially available.

Schmit et al. [5] were the first to use a chemically bonded C-18 column to separate PAHs by HPLC. Reversed-phase HPLC based on the use of C-18-packed columns is

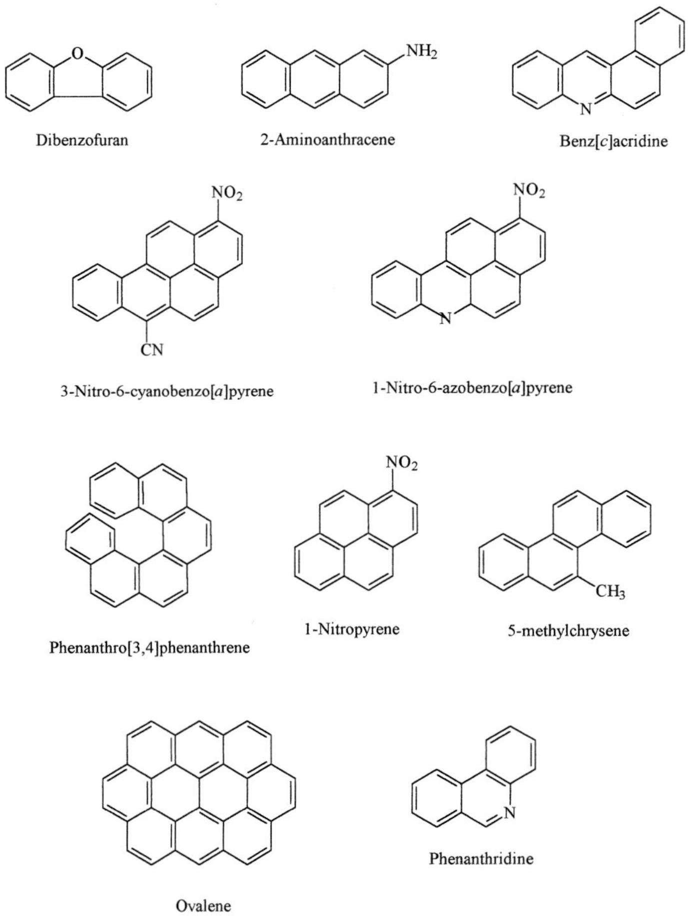

Dibenzofuran

2-Aminoanthracene

Benz[*c*]acridine

3-Nitro-6-cyanobenzo[*a*]pyrene

1-Nitro-6-azobenzo[*a*]pyrene

Phenanthro[3,4]phenanthrene

1-Nitropyrene

5-methylchrysene

Ovalene

Phenanthridine

Fig. 2.1 (continued).

now the most popular LC mode for separation of PAHs [6–15]. Even for the lower-molecular-weight PAHs, whose higher volatilities deem them to be better suited for GC, HPLC provides a strong challenge to the latter in some cases, and is routinely employed.

A rather more recent separation technique that was originally thought to have the potential to replace both GC and HPLC for many analytical applications is supercritical fluid chromatography (SFC). The primary advantage of SFC in harnessing the benefits of both GC and HPLC, has also been exploited for the analysis of many compounds including PAHs. Although the potential and promise of SFC in general have not been completely realized, it should be acknowledged that it does provide a convenient advantage over the other two techniques in respect of the amenability of combining extraction (with supercritical fluids) with separation. As will be discussed below, it is the extraction aspects that supercritical fluids have most to offer in the analytical chemistry of PAHs.

In the 1980s, capillary electrophoresis (CE), which is based on the difference in migration of charged species in electrophoretic media under the influence of an applied electrical field inside small capillaries, was introduced. It is a fast growing technique whose potential has yet to be fully exploited. Not surprisingly, in one of its numerous guises that allows neutral compounds to be separated (micellar electrokinetic chromatography), CE has been applied to PAH analysis. One of the attractive features of CE is the ease of method development. This is due to a combination of speed, high efficiency, and flexibility of manipulating the chemical composition of the system (separation conditions) in a short period of time. Another useful feature is that basic but powerful CE systems can be easily assembled and set up in-house.

2.2.1 Gas chromatography

Since capillary column GC was first used to separate PAHs in 1964 [16], the technique has progressed to the point that it is now the standard method for the determination of these compounds. Amongst the important advantages of GC are resolving power and high detection limits. The beauty of using GC for PAH analysis is that only a slight modification of an existing GC protocol is usually sufficient to meet the requirement of a particular application. Further, by adjusting the carrier gas flow rate, temperature programming, and switching to a similar stationary phase, different types of matrices can be effectively analyzed using GC coupled with different detectors.

The critical PAHs (those biologically active ones that are most likely to be found in the environment and therefore pose the greatest risks to human health) are volatile or semi-volatile, and since the fundamental requirement for a compound to be analyzed by GC is volatility, PAHs containing up to 24 carbon atoms may be analyzed by GC [17]. This general rule has exceptions, of course. For example, a less condensed PAH, even with 30 carbons such as pyranthene, is amenable to GC analysis, whereas its counterpart, naphtha[8,1,2-*abc*]coronene requires high-temperature GC or SFC [17].

Commercially, there are now available many stationary phases from many suppliers designed optimally for separating complex PAH mixtures, including critical ring isomers (see ref. [18] for example). Flame ionization detection is normally adequate for sensitive detection but coupling GC with MS affords greater selectivity through the application of selected ion monitoring.

Bjorseth and Eklund [19] were amongst the first to use simultaneous flame ionization detection and electron-capture detection to determine PAHs (see Fig. 2.2). There is generally a tendency to favor flame ionization detection for the GC detection of these compounds although the use of independent or simultaneous multiple detection obviously offers greater scope for analyte identification purposes.

Typical GC conditions for separating PAHs are [20]: column: HP-5 (a 5%-diphenyl-siloxane, 95%-dimethylsiloxane non-polar phase) (Hewlett-Packard, Palo Alto, CA, USA) or equivalent, 30 m × 0.25 mm I.D., 0.25 μm thickness, or equivalent; initial oven temperature held at 50°C for 12 min, increased linearly at 10°C/min to 280°C which is held for 10 min.

The gas chromatography of PAHs has been extensively reviewed. One of the most

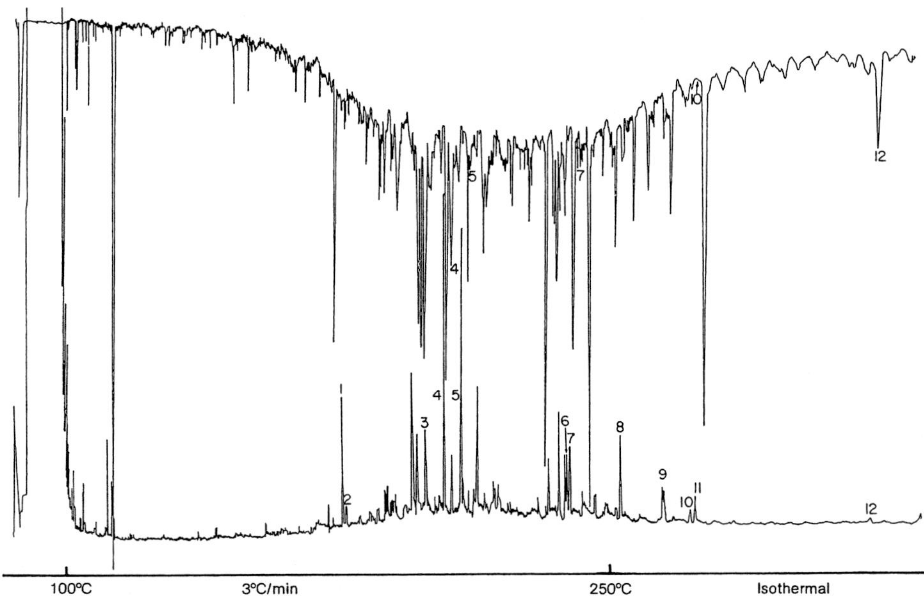

Fig. 2.2. Gas chromatogram of PAHs in urban air particulates detected with simultaneous electron-capture (upper trace) and flame ionization (lower trace) detection. (1) Phenanthrene; (2) anthracene; (3) 3,6-dimethylphenanthrene; (4) fluoranthene; (5) pyrene; (6) benz[*a*]anthracene; (7) chrysene/triphenylene; (8) β,β-binaphthalene; (9) benzo[*b*]- and [*k*]fluoranthene; (10) benzo[*e*]pyrene; (11) benzo[*e*]pyrene; (12) *o*-phenylenepyrene. (Reproduced with permission from ref. [19].)

TABLE 2.1

SOME APPLICATIONS OF GC WITH VARIOUS DETECTION METHODS IN THE ANALYSIS OF PAHs FROM A VARIETY OF MATRICES

Matrix containing PAHs	Detection	Reference
Water samples	GC–electron capture detection (ECD)	[22]
Soil samples	GC–mass spectrometry (MS)	[23]
Air samples	GC–MS	[24]
Marine water	GC–MS	[25]
River sediments	GC–ECD	[26]
Marine sediment	GC–MS	[27]
Eggs of coastal-nesting birds	GC–MS	[28]
Foam	GC–flame ionization detection (FID)	[29]
Marine sediments	GC–MS	[30]
Marine sediments	GC–ECD	[31]
Tar soils	GC–MS	[32]
River sediments	GC–MS	[33]
Soil samples	GC–MS	[34]
Coal	GC–MS	[35]
Marine water	GC–MS	[36]
Diesel particles	GC–MS	[37]
Synthetic mixture of high-molecular-weight PAHs	GC–FID	[38]

recent is that by Wegener et al. [21]. Table 2.1 [21] is a summary of some applications of GC to PAH analysis.

GC will continue to be a popular technique for the analysis of PAHs because of its proven capabilities. The technique is well-established and GC instrumentation is a common feature in routine analytical laboratories. Indeed many of the sample preparation procedures described below involve the use of GC as the main analytical tool.

2.2.2 High-performance liquid chromatography

Since it was developed some twenty-five years ago, HPLC has been widely applied to the separation of PAHs, and is now established as a strong alternative to GC. Some of the reasons given for using GC (volatility of PAHs, speed of analysis, high-resolution separation, etc.), in fact, are slowly becoming non-critical issues. Many analysts are using HPLC as a replacement for GC simply because advances of various aspects of the former technique have allowed them to do so successfully, even if intuitively, GC should be the technique of choice.

Several aspects of HPLC in recent history have provided the impetus for favoring HPLC. Advances in stationary phase chemistries have led to the availability of many packing materials that permit the separation of critical pairs of PAHs. Improvements in traditional HPLC detection systems such as ultraviolet and fluorescence detectors provide high sensitivity and selectivity for PAHs. Additionally, the lower costs of liquid delivery systems have made the more versatile gradient elution mode more affordable in a typical analytical laboratory, although in many cases, isocratic elution works very well for selected groups of PAHs.

Miniaturization of many components of HPLC has been another positive factor; this has allowed the use of small-diameter columns (microbore or even nanobore HPLC) with much higher resolution than conventional 4.6-mm columns. Microscale HPLC has also permitted the improvement in analyte detectability (due to narrower and sharper chromatographic peaks). Advances in detector technology have resulted in increased applicability of techniques such as UV diode-array detection which now provides not only improved analyte identification capabilities but also enhancements in sensitivity (originally its primary disadvantage). The previously mutually exclusiveness of sensitivity and identification specificity is no longer a significant problem.

HPLC is also often used as a clean-up or fractionation procedure for other chromatographic or spectroscopic determination of PAHs. It is therefore feasible for an on-line extraction–separation system to be realized, as in combined HPLC–GC and HPLC–HPLC (see below).

The versatility of using HPLC lies in the flexibility that separation parameters may be varied to afford the best set of conditions for PAH analysis. In GC, only temperature-programming provides a measure of selectivity for a given stationary phase. However, in HPLC, apart from the stationary phase, the mobile phase (either isocratic or gradient elution) and/or column temperature provide an extra dimension. The various detection possibilities also offer another level of versatility that exploits the spectral characteristics of PAHs.

References pp. 69–74

2.2.2.1 Stationary phases

Reversed-phase packing materials (primarily C-18-based) are by far the most popular for the separation of PAHs by HPLC [13]. It should be appreciated that not all commercial C-18-based stationary phases behave similarly, however. Vendors have in recent years made available columns packed with stationary phases specially optimized for separating PAHs. In general, apart from the type of stationary phases used, the particle size, mobile phase flow rate, column length and temperature (if applied) all affect the selectivity of PAH separation.

Some recent applications making use of C-18 reversed-phase HPLC include the determination of PAHs found in meat products [39], indoor and outdoor air [40], natural waters [41,42], edible oils and fats [43], several types of water samples (including river, ground, surface water, etc.) [44], air particulates [45–48], sewage sludge [49]; smoked food [50], smoked fish [51] and seafood [52]. Figs. 2.3 and 2.4 show liquid chromatograms of PAHs extracted from indoor particulates [40] and olive oil [43], respectively.

An interesting approach was undertaken by Kurganov et al. [53] who used a column-switching technique with two columns running with the same eluent while thermostatted at two different temperatures. The columns were Superspher-100 RP-18 (Merck, Darmstadt, Germany). Although the packings were monomeric phases, the observed selectivity was found to be sufficient for isocratic separation of 20-component PAH mixtures in less than 15 min. In most other cases, polymeric phases would be needed for satisfactory separation under gradient elution.

In another unconventional approach, Saito et al. [54] reported the use of C-60 fullerene stationary phase to separate PAHs. The C-60 bonded silica phase had exceptional selectivity for PAHs. Kayali et al. [55] used a micellar sodium dodecylsulfate mobile phase modified with *n*-propanol to separate PAHs on short-chain (C-1 and C-4) columns. Van Stijn et al. [43] reported the use of a Chromspher PI donor–acceptor

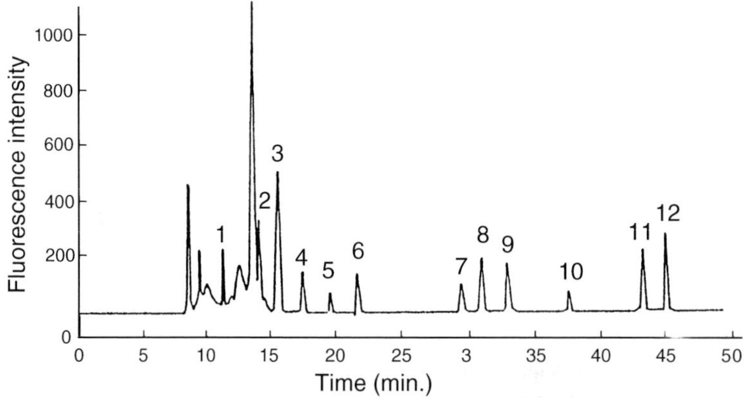

Fig. 2.3. Liquid chromatogram of PAHs in indoor particulates. (1) Naphthalene; (2) acenaphthylene; (3) phenanthrene; (4) anthracene; (5) fluoranthene; (6) pyrene; (7) 1-methylpyrene; (8) benz[*a*]anthracene; (9) chrysene; (10) benzo[*e*]pyrene; (11) benzo[*k*]fluoranthene; (12) benzo[*a*]pyrene. (Reproduced with permission from ref. [40].)

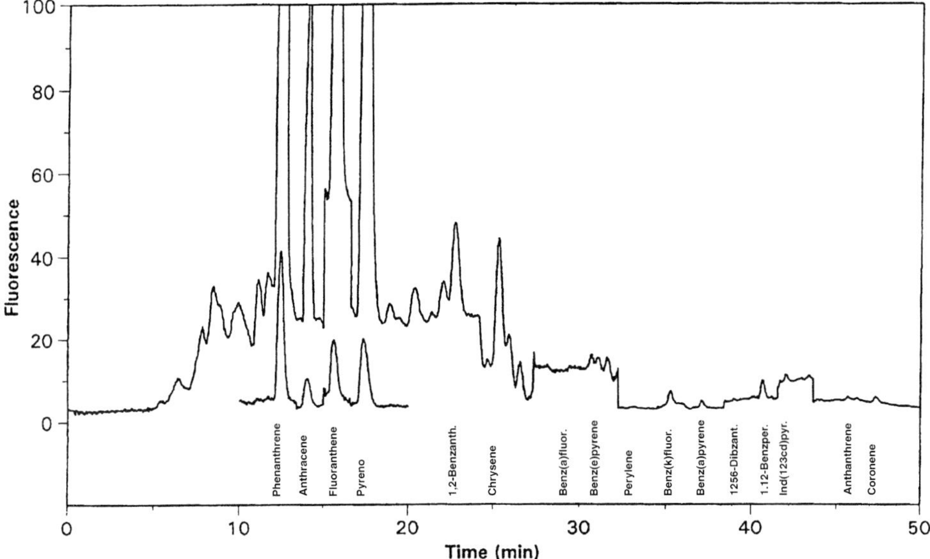

Fig. 2.4. Liquid chromatogram of PAHs in olive oil, including part of a second analysis 15× diluted. (Reproduced with permission from ref. [43].)

complex chromatography column (Chrompack, Bergen op Zoom, The Netherlands) coupled to a 250 cm × 4.6 mm Chromspher 5 PAH analytical column to separate 16 PAHs in fats and oils (see Table 2.2). Polymeric C-18 phases (e.g. silica modified by octadecyltrichlorosilane in the presence of water) [56] and monomeric phases (e.g. silica silanized under dry conditions) are also commonly used as stationary phases [57,58].

2.2.2.2 Mobile phase

An isocratic binary solvent system can be used to separate PAHs, as reported by Lawrence and Weber [59] who used acetonitrile–water (70:30, v/v) to separate 16

TABLE 2.2

PROGRAMMED MOBILE PHASE SYSTEM USED FOR CHROMSPHER5 PAH COLUMN TO SEPARATE SIXTEEN PAHs IN FATS AND OILS [43]

Time (min)	Flow rate (ml/min)	Water (%)	Acetonitrile (%)	Ethyl acetate (%)
0	0.4	15	85	0
2.5	0.4	15	85	0
3.0	1	15	85	0
12	1	6	94	0
20	1	4.5	95.5	0
26	1	4	89	7
39	1	0	30	70
50	1	0	30	70

components. (See also above the description of Kayali et al.'s work [55].) However, the separation time was long (60 min) and the capacity (retention) factor (k') was between 0.45 and 35.64. In addition, some peaks (acenaphthene and fluorene) were not fully resolved. A gradient elution mobile phase system should be considered to obviate such a problem. A typical gradient elution protocol consists of an acetonitrile–water (55 : 45, v/v) mixture, initially maintained for 2 min, then linearly programmed to 100% acetonitrile within 23 min, and maintained for 15 min at a flow rate of 1.25 ml/min [59].

Zhu et al. [40] used methanol–water (50 : 50, v/v, initially), and by increasing the methanol concentration linearly to 96%, was able to separate twelve PAHs, including several important ones such as benz[*a*]anthracene, benzo[*a*]pyrene and benzo[*k*]fluoranthene, over 45 min. Lan et al. [60] established a systematic optimization system (with mobile phase as one parameter to optimize) using a mixed-level orthogonal array design and 16 PAHs were separated successfully in under 20 min.

2.2.2.3 Detection

2.2.2.3.1 UV detection. UV detection is most commonly used for HPLC separations in general. For PAHs, since most of these compounds absorb UV radiation strongly, it is convenient to analyze them using this detection mode [61,62]. There is a range of wavelengths across which PAHs absorb, so the sensitivity and selectivity can be further improved by monitoring at different wavelengths that match the λ_{max} values associated with individual compounds. In this respect, variable-wavelength UV detectors, especially those newer, wavelength-programmable types are most appropriate. Some allow stopped-flow scanning so that spectra of the compounds being detected may be obtained.

Increasingly, to provide a greater degree of confirmation of the identities of PAHs detected, diode-array detection is being used. As stated above, many modern diode-array detectors can now approach the sensitivity of conventional UV detectors while offering additionally the capability of more positive identification of the analytes. Scanning can be effected on-the-fly without having to compromise the chromatography by stopping the flow. The measurement of the absorbance ratio when a peak is being detected simultaneously at two wavelengths also permits the confirmation of peak purity [63].

2.2.2.3.2 Fluorescence detection. Fluorescence detection is another option for HPLC applications. It provides better results than UV detection because of its higher sensitivity and selectivity [10,64–66] for the analytes with the necessary characteristics. It is likely that because of the better selectivity of fluorescence detection, some sample clean-up steps may be avoided. However, difficulties can be encountered with complex matrices (containing fluorescent impurities) and low PAH concentrations. In order to improve the selectivity, fluorescence wavelengths can be programmed to enhance the specificity and the selectivity towards individual PAHs in the mixture so as to minimize interferences from co-eluting species.

A recent development is the use of programmable fluorescence spectrometers [50,56,67–69]. Chiu et al. [50] and Chen et al. [39] used seven programmable wavelength settings to detect sixteen PAHs by fluorescence. Without this programming, acenaphthylene was not detected because of its low fluorescence quantum yield.

TABLE 2.3

EXCITATION (λ_{ex}) AND EMISSION (λ_{em}) WAVELENGTHS USED FOR FLUORESCENCE DETECTION OF PAHs IN SMOKED FOOD [50]

PAHs	λ_{ex} (nm)	λ_{em} (nm)
Naphthalene, acenaphthene, fluorine	270	340
Phenanthrene	254	375
Anthracene, fluoranthene	260	420
Pyrene, benz[*a*]anthracene, chrysene	254	390
Benzo[*b*]fluoranthene, benzo[*k*]fluoranthene, benzo[*a*]pyrene, dibenz[*a,h*]anthracene, benzo[*ghi*]perylene	260	420
Indeno[1,2,3-*cd*]pyrene	293	498
Acenaphthylene	320	533

The observed results indicated that fluorescence detection had about 20–320 times higher sensitivity than UV detection in the analysis of meat products [39]. Table 2.3 shows excitation–emission wavelengths that were used in this work. Similar results of increased sensitivity with fluorescence were observed in other reports [12,13,69].

2.2.2.3.3 Electrochemical detection. Although fluorescence detection is most suitable for the HPLC analysis of many PAHs, it is not universally applicable. Some PAHs, such as acenaphthylene, do not fluoresce (see above) and can be determined only by UV detection with limited sensitivity. Amperometric detection provides an alternative and is universally applicable, since all PAHs can be determined via electrooxidation that gives rise to radical cations [70].

Nirmaier et al. [44] used trichloroacetic acid (TCA) as the supporting electrolyte in a methanol–water eluent to determine eight PAHs. It was found that the sensitivity of the amperometric technique was about 5–10 times better than UV detection. Galceran and Moyano [71] described the applicability of electrochemical detection to the determination of oxy- and nitro-PAHs with reversed-phase HPLC. A voltammetric detector with a 1-µl wall-jet cell with a Ag/AgCl reference electrode and two glassy carbon electrodes were used. The mobile phase used constituted 70:30 to 80:20 acetonitrile–buffer (25 mM acetic acid–sodium acetate). This buffer solution provided the pH, conductivity and ionic strength needed for the electrochemical reactions. The optimum working potential was established at −650 mV. This detector permitted quantification of PAHs with a sensitivity of 3–0.3 ng injected.

2.2.2.3.4 Mass spectrometric detection. One major advance in recent years of liquid chromatographic detection has been the ease with which mass spectrometry (MS) can now be coupled to HPLC systems. The surmounting of problems associated with coupling a liquid-based separation system with a vacuum-based detection technique has been a real boon to the analytical sciences. While HPLC with conventional-bore (ca. 4.6 mm I.D.) columns have previously been coupled with varying degrees of success to mass spectrometry, two major technological advances paved the way for the unprecedented ease and feasibility of operation of HPLC–MS: (a) the advent of microbore or nanobore

LC in which columns of very narrow dimensions allow for eluent flow rates that are more compatible with MS systems; and (b) the introduction of electrospray ionization MS (in which ionization is effected at atmospheric pressure).

Given the foregoing, it is therefore not surprising that even a cursory study of the current literature indicates that two related MS ionization techniques — electrospray ionization (ESI) and atmospheric pressure chemical ionization (APCI) — are beginning to dominate LC–MS applications in the analysis of PAHs. Galceran and Moyano [72] have compared both techniques for the determination of hydroxy-PAHs. They had earlier applied LC–ESI MS in both negative and positive modes to characterize the same compounds; gradient LC with methanol–formic acid/ammonium formate as eluent was used [73].

Applications of LC–APCI or ESI MS to the analysis of PAHs in various samples include the following.

High-molecular-mass (>300 amu) PAHs in mussels, air particulates and coal tar were determined to ascertain the potential sources of pollution in a harbor in Lake Ontario, Canada [74]. In a separate study, bottom sediments, suspended sediments, coal-tar contaminated sediments were determined by the same technique [75]. Thirty-one mono- and polycyclic derivatives with up to five condensed aromatic rings carrying groups such as carboxyl, lactone, hydroxyl, dicarboxylic anhydride and carbonyl, were characterized [76]. The compounds gave interpretable fragmentation patterns. Castillo et al. [77] reported the use of enzyme-linked immunosorbent assay (ELISA) kits followed by LC–APCI MS to determine some carcinogenic PAHs in European industrial effluents. Morikawa et al. [78] determined PAHs by detecting PAH–tropylium complexes using LC–ESI MS. Mansoori [79] used LC–APCI tandem MS to identify and quantify isomeric PAHs in a coal tar reference material. A high-pressure quadrupole collision cell, with which a triple–quadrupole MS was equipped, was used for low-energy collision-induced dissociation studies.

Older MS techniques based on vacuum-ionization processes are still being used, however, particularly particle-beam LC–MS. Bonfanti et al. [80] used normal-phase LC with a particle-beam system to simultaneously identify aliphatic hydrocarbons, PAHs and nitro-PAHs. Pace et al. [81] measured high-molecular-weight PAHs in contaminated soils with reversed-phase LC coupled to a particle-beam MS. There seems little doubt however that the tremendous advances being made in ESI and APCI MS will ensure that they will be the MS techniques of choice for PAH analysis.

2.2.3 Supercritical fluid chromatography [82,83]

Although supercritical fluid chromatography (SFC) was considered to be a major advance in chromatography when it was first introduced, its potential has not been realized subsequently. It would have taken an exceptional separation technique to topple the well-established GC and HPLC, and for a time, SFC was felt to be that technique. Unfortunately for its proponents, it never reached the popularity envisaged for it. Nevertheless, SFC has gained a modicum of acceptance, specifically because supercritical fluids have the ability to dissolve a variety of solutes of high molecular weight and low volatility, allowing its applicability in niche areas of the analytical sciences. Additionally, in most cases, SFC employs non-toxic and cheap carbon dioxide; thus, the problem of extensive organic usage and disposal does not arise. Interest in this technique has continued, although at a lower scale than expected at its advent, in particular, because of the compatibility of the commonly used supercritical fluids with both GC and HPLC detectors and its capabilities in analyzing thermally labile compounds. The greater ease with which both extraction and separation may be integrated without having to switch solvents is another positive factor.

Packed HPLC columns, packed capillary (micro-packed) and open tubular GC columns can all be employed in SFC. The hybrid gas-like and liquid-like properties of supercritical fluids make it easier for GC and HPLC detectors to be interfaced to the technique. For example, the PAHs present in a fraction of an extract from Diesel particulate matter were separated and identified using SFC coupled with an UV multichannel detection system [84]. For the photoionization detection, a high energy UV lamp was used to ionize the organic species present in the column effluent stream. Sim et al. [85] demonstrated sensitivities of the order of picograms for some PAHs using a commercial PID with a packed microbore column. The flame ionization detector is the most widely used detector in SFC. A more selective detector for SFC is the flame photometric detector. Fourier-Transform infrared [86], UV–visible diode-array and fluorescence detection [87] have also been used with SFC. Recently, Moyano et al. [88] interfaced SFC to APCI MS to study hydroxy-PAHs.

Although the early promise of SFC has not been fully realized, the usefulness of supercritical fluids as extraction solvents cannot be denied. Thus, supercritical fluid extraction has become an important sample preparation procedure for PAHs. This is discussed below.

2.2.4 Capillary electrophoresis [89]

Capillary electrophoresis (CE) is a relatively new separation technique in comparison to GC and HPLC. It has developed tremendously since it was introduced in the late 1970s and early 1980s. The amount of literature on it is enormous, and invariably there has been substantial work done on the CE of PAHs. It is interesting to note that PAHs represent a favored group of compounds often used as test materials in the development of new or improved analytical techniques. Just as in the development of GC, HPLC and SFC, multi-component mixtures of PAH that arise from complex combustion processes provide a rigorous test of the applicability and capability of CE. Additionally, the strong hydrophobicities of PAHs serve as a special challenge for CE, a basically aqueous-based technique. While the basic mode of CE, capillary zone electrophoresis, is unsuitable for separating PAHs, which are neutral species, much effort has been expended on exploiting the other CE modes for analyzing complex mixtures of these compounds.

The CE approaches that have been utilized to separate PAHs include: (1) micellar electrokinetic chromatography [90–123]; (2) solvophobic interaction [124–132] or complexation [133–143]; and (3) capillary electrochromatography [144–175]. Some of these applications are highlighted in the following paragraphs.

2.2.4.1 Micellar electrokinetic chromatography

Micellar electrokinetic chromatography (MEKC) has the capability of separating uncharged compounds. In MEKC, charged organized media such as micelles are incorporated into the running buffer. They act as the separation medium for uncharged solutes which partition between the micelle and the buffer.

Because hydrophobic compounds tend to be included in the micelles with high partition ratios, the resolution of such compounds is therefore not successful with simple micellar solutions such as sodium dodecylsulfate (SDS). The addition of urea to a micellar solution was developed for the separation of PAHs [91]. This procedure expands the migration-time window and hence enhances resolution. The addition of organic modifiers has been reported to help the separation of hydrophobic compounds [92]. Dimethylsulfoxide (DMSO) and acetone have been used as such organic modifiers, with SDS as a surfactant.

Cyclodextrin-modified MEKC (CD-MEKC) has also been reported for the separation of highly hydrophobic compounds [94,95] using SDS and various α-, β-, or γ-cyclodextrins, and addition of urea [96]. γ-CD was found to be more effective compared to the other cyclodextrins [95], probably owing to the co-inclusion of a monomeric surfactant molecule together with the analyte molecule. Fig. 2.5 shows the separation of seven PAHs using a buffer containing 2 mM γ-CD [95]. In another report, sixteen PAHs were resolved when using a buffer consisting of 20 mM γ-CD, 5 M urea and 100 mM SDS in 100 mM sodium borate, pH 9.0 [97].

Bile salts are more polar than SDS, and their inclusion in the separation medium leads to a general reduction of k' in MEKC, which is particularly advantageous in dealing with hydrophobic compounds. In addition, the bile salt micelle resembles a reversed or inverted micelle, so it can tolerate high concentration of organic solvents [100,101].

El Rassi and co-workers [102–104] introduced several types of in situ charged micelles. In situ charged micelles refer to dynamically charged entities that are formed via the complexation of borate with surfactants having sugar head groups. These dynamically charged surfactants yield micelles with adjustable surface charge densities that can be conveniently manipulated by changing the borate concentration and the pH of the running electrolyte. The four surfactants, namely octanoylsucrose (OS), octyl-β-D-glucopyranoside (OG), octyl-β-D-maltopyranoside (OM), and manoyl-N-methylglucamide (MEGA9), in the presence of alkaline borate yield micelles characterized by migration-time windows of varying width.

Palmer and co-workers [105] developed a monomolecular pseudo-stationary phase, sodium 10-undecylenate (SUA) oligomer for MEKC. SUA was oligomerized to form a micelle-like structure. The oligomer offers high stability in the presence of organic media and nine PAHs were separated using a buffer containing 30% (v/v) acetonitrile in 20 mM borate buffer pH 8.2 with 5 mM SUA. However, there are some limitations in the use of SUA caused by the solubility of the oligomer. Following the procedure employed for SUA, sodium 10-undecenyl sulfate (SUS) was synthesized and polymerized [106]. As with other molecular surfactants, the critical micelle concentration is zero. The new phase has higher electrophoretic mobility and can be used with high percentages of organic modifiers. Twelve PAHs were separated when using a buffer containing 60%

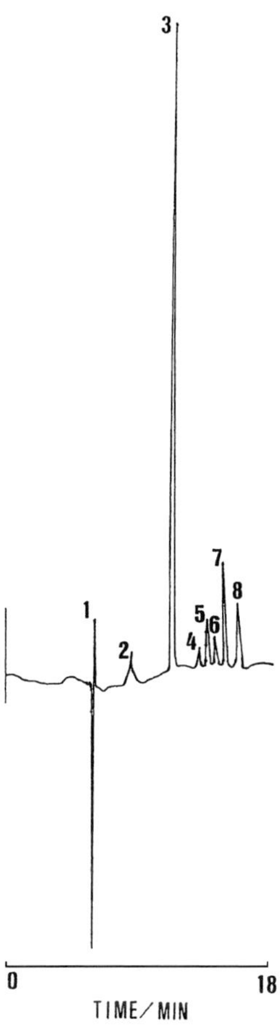

Fig. 2.5. Micellar electrokinetic chromatogram of PAHs. Conditions: 10 mM SDS, 2 mM γ-CD in 0.05 M phosphate–0.1 M borate buffer (pH 7.0); capillary 50 cm × 0.05 mm I.D.; voltage 15 kV; current 36 μA; UV detection wavelength 254 nm. (1) Methanol; (2) acenaphthalene; (3) phenanthrene; (4) perylene; (5) benzo[*a*]pyrene; (6) chrysene; (7) benz[*a*]anthracene; (8) fluoranthene. (Reproduced with permission from ref. [95].)

(v/v) methanol, in borate buffer pH 9.3 with 0.6% (w/v) SUS [107]. However, the use of potassium persulfate as a free radical initiator for the polymerization process of SUS leads to two major limitations: the low synthetic yields and the contamination of the product with sodium sulfate. These problems can be avoided if γ-irradiation is used to initiate polymerization [108] as reported by Shamsi et al. who were able to separate the sixteen PAHs listed by the United States Environmental Protection Agency (USEPA). They showed that the elution order of most of the PAHs generally followed an order relating to the increasing length-to-breadth ratio. Recently, they reported the feasibility of poly-SUS as pseudo-stationary phase for the separation of 12 monomethylbenz[*a*]anthracene [109].

Another approach for the preparation of polymeric pseudo-stationary phase involves

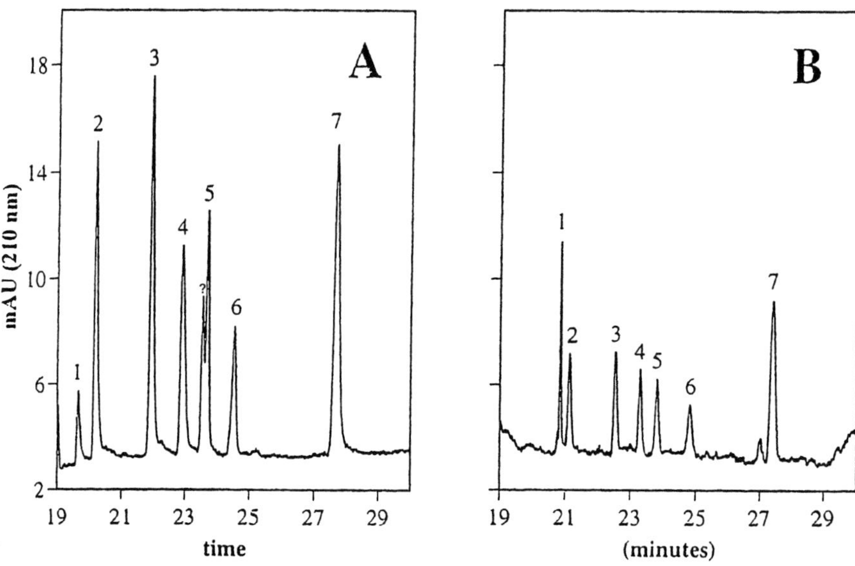

Fig. 2.6. Sweeping MEKC analysis of seven PAHs in the presence of Brij 35 in the sample matrix and separation solutions. Background solution: 50 mM SDS, 0.02 mM Brij 35, and 35% methanol prepared in 100 mM phosphoric acid (pH 1.9); sample solution: PAHs in 0.05 mM Brij 35 in 50 mM phosphoric acid adjusted to the conductivity of the background solution (pH 1.9); injected length of sample solution 1.3 mm (A), 10.3 cm (B); applied voltage −25 kV; concentration of analytes ∼70 ppm (A), ∼0.7 ppm (B); (1) anthracene; (2) phenanthrene; (3) fluorene; (4) acenaphthene; (5) biphenyl; (6) acenaphthylene; (7) naphthalene; capillary 56 cm to the detector (64.5 cm total). (From Quirino and Terabe [116]. Reproduced with permission of the American Chemical Society.)

the attachment of interacting groups, namely alkyl groups, as well as ionic groups to a polymer support such as dendrimers reported by Tanaka et al. [110–112].

The use of polyallylamine (PAA) was been reported [113] as a support to attach alkyl groups of C-8 to C-16 as well as carboxylate groups and they were able to separate the sixteen USEPA-listed PAHs. A double alkyl chain di(2-ethylhehyl)phosphate (DEHP) as a potential anionic micellar pseudophase was reported by Akbay et al. [114].

Yang et al. [115] reported another ionic polymer, poly(methyl methacrylate/ethyl acrylate/methacrylic acid) for the separation of hydrophobic compounds. Recently, in their search for ways to improve the concentration sensitivity for neutral analytes by sample stacking, Quirino and Terabe [116] discovered a new phenomenon now known as sweeping. Sweeping is the picking and accumulation of analyte molecules by the pseudo-stationary phase that penetrates the sample zone. This causes a unique focusing effect, and leads to a significant enhancement of in-capillary enrichment. It was acknowledged, however, that PAHs are recalcitrant analytes to handle for sweeping experiments, specifically relating to adsorption and reproducibility problems [116]. Fig. 2.6 shows chromatograms of PAHs generated from sweeping experiments under the indicated conditions [116].

In an interesting approach, Gottlicher and Bachmann [117] reported the use of a suspension of reversed-phase particles with a diameter of 0.5 μm as a pseudo-stationary phase.

2.2.4.2 Solvophobic interaction

An approach developed for the CE separation of neutral compounds is based on solvophobic interaction between the neutral analytes and ions present in the running buffer. This technique was first reported by Walbroehl and Jorgenson [124]. They used a mixed water–acetonitrile medium to promote solvophobic interaction between a tetrahexylammonium (THA) ion and PAHs, in order to place a charge on an otherwise neutral molecule. The larger, more non-polar solutes, which had the strongest interactions with the THA ion migrated fastest, and they were able to separate four PAHs. Nie et al. [125] used a separation mode based on solvophobic interaction with THA with UV–laser-excited native fluorescence detection for ultrasensitive determination of four PAHs. Shi and Fritz [127] later reported the separation of a mixture of seventeen PAHs using sodium dioctyl sulfosuccinate (DOSS) as the solvophobic additive. Separation was based on the differences in the strength of analyte–DOSS association in acetonitrile–water ($\sim$40%, v/v) solution. The CE system proposed by Shi and Fritz [127] was utilized by Jankowiak et al. [128] to demonstrate, for the first time, that CE can be interfaced with a low temperature fluorescence line-narrowing (FLN) spectroscopy for on-line structural characterization. Detection by laser-induced fluorescence (LIF) spectroscopy, under fluorescence non-line-narrowing and line-narrowing conditions, provided three-dimensional electropherograms and FLN spectra, which led to significantly improved overall resolution and allowed for structural characterization ('fingerprinting') of molecular analytes. Luong and Guo [129] developed a mixed-mode separation using DOSS for the separation of the PAHs listed by the USEPA.

Excellent separations of PAH compounds have been obtained in acetonitrile–water mixture using DOSS or sulfonated lauryl polyoxyethylene sulfate (Brij-S) as an additive and a basic pH. However, these separations require a rather long separation time because the electroosmotic and electrophoretic vectors are in opposite directions (countermigration). Ding and Fritz [131] reported the separation of nineteen PAHs at pH 2.4 with Brij-S as a solvophobic additive.

2.2.4.3 Complexation

The scope of applications of MEKC has been widened by the involvement of organic media or cyclodextrins amongst the many additives being used in this CE mode. Separation of neutral analytes can also be achieved by using only cyclodextrins as the pseudo-stationary phase. Szolar et al. [133] reported a dual-cyclodextrin (CD) phase system for the separation of nonionizable solutes. Two neutral forms, hydroxypropyl-β-cyclodextrin (HPβCD) and methyl-β-cyclodextrin (MβCD), and two negatively charged forms, carboxymethyl-β-cyclodextrin (CMβCD) and sulfobutyl-β-cyclodextrin (SBβCD), were used to demonstrate the principle of separation based on differential partitioning between the cyclodextrins. The technique resembles MEKC in terms of instrumentation and the fundamental relationships for resolution and capacity factor, which are influenced by the existence of a finite elution window. Efficiency is comparable to that achievable by MEKC.

Brown et al. [136] also reported the separation of the EPA-listed PAHs extracted from contaminated soils using 35 mM SBβCD and 15 mM MβCD, with efficiencies for all components greater than 10^5 theoretical plates/meter. LIF detection provided sensitive detection of eleven of the sixteen components, with detection limits measured typically in the low parts per billion range.

2.2.4.4 Capillary electrochromatography

Capillary electrochromatography (CEC) combines the surface-mediated selectivity potential of HPLC and the high efficiency of CE. In contrast to pressurized flow, conventionally used in HPLC, in CEC, electrical flow is employed to drive the bulk flow through the column. Thus, electroosmotic flow resulting from the zeta potential at the column walls and at the packed particles creates the flow of mobile phase. Since this flow is close to plug-like, vs. parabolic in pressurized systems, column efficiencies in CEC are found to be at least an order of magnitude higher than in HPLC. High resolution can be translated ultimately into fast analysis. Both porous and nonporous particles can be used in the packed bed. The diameters of the particles used in CEC are generally similar to those in HPLC (3–5 μm) or sometimes smaller ($\sim$1–2 μm). While such small particle sizes would require high inlet pressures for HPLC, the use of electric field removes this restriction.

Seifar et al. [147] reported the separation of nine PAHs with 1.5 μm octadecylsilane (ODS) modified nonporous silica particles when SDS was added to the mobile phase.

In an interesting approach, capillary columns packed with 3-μm ODS and C-18 bonded silica particles with a supercritical CO_2 carrier and ultrasonication were highly efficient and durable and gave highly symmetrical chromatographic peaks at applied voltages up to 30 kV [149]. The USEPA-listed PAHs were separated and column efficiencies up to 4.8×10^5 theoretical plates/meter were obtained.

Recently, the use of monolithic columns containing a wall-supported continuous porous bed has shown great potential for CEC due to the inherent advantages of high sample capacity, high column bed stability, and absence of end frits.

Three approaches have been employed to prepare monolithic columns. The first approach is to polymerize an organic monomer in the capillary tubing in situ to form a continuous polymeric bed. The second approach is to form a silica-based network using a sol-gel process and then to functionalize the network. The third method consists of fusing the porous particulate packing materials in a capillary in situ using a sintering process.

Tang et al. [155] reported a new method for the preparation of a monolithic capillary column. A fused-silica capillary packed with porous ODS particles using a CO_2 slurry was partially filled with a silicious sol formed by hydrolysis and polycondensation of tetramethoxysilane and ethyltrimethoxysilane. After gelling and aging at room temperature, the column was dried with supercritical fluid (CO_2). Using a 21–29 cm $\times$ 75 μm I.D. monolithic column containing 9% sol-gel bonded 5-μm ODS particles, a mixture of PAHs was separated, and approximately 1.3×10^5 theoretical plates/meter were achieved.

In CE, the choice of mobile phase extends from a purely aqueous phase to a totally nonaqueous mobile phase. Lister et al. [158] investigated different solvents and water for their ability to support current flow without an added electrolyte. A sixteen-PAH standard was separated in acetonitrile–water (80 : 20) with efficiencies between 3.2×10^5 and 3.6×10^5 theoretical plates/meter.

In order to be able to determine PAHs present in very small amounts, LIF detection has been used instead of UV absorbance detection because it is much more sensitive [159–161]. Yan et al. [159] separated a mixture of PAHs using fused-silica capillary columns ranging in size from 50- to 150-μm I.D. packed (20-cm to 40-cm sections) with 3-μm particles. An intracavity-doubled argon ion laser operating at 257 nm was used to detect the PAHs. The limits of detection for individual PAHs ranged between 10^{-17} and 10^{-20} moles (10^{-9} to 10^{-11} M), with a linear response spanning four orders of magnitude in concentration. The same workers later reported the use of gradient elution to separate the PAHs [162]. Dynamic gradients with submicroliter per minute flow rates were generated by merging two electroosmotic flows that were regulated by computer-controlled voltages. Using capillary columns packed with 3-μm ODS particles, the sixteen PAHs were separated in less than 90 min. Xin and Lee [164] explored the use of voltage programming to accelerate the elution of late-eluting components. Voltage programming was demonstrated to be an effective alternative to composition gradient programming, and a mixture of 14 PAHs was separated in under 17 min.

2.2.5 Coupled HPLC–GC–MS: on-line HPLC clean-up

On-line coupling of HPLC to GC has been shown to be a powerful tool for on-line pre-separation and analysis for a variety of applications. The application of on-line HPLC–GC for the analysis of PAHs in a variety of samples has been described [176–180]. Specifically, analysis of PAHs in vegetable oils [176], soot [179] and heavy oils [180] has been investigated.

Ostman et al. [181,182] developed a fully automated system, comprising an HPLC system coupled on-line to GC by means of a loop interface for the isolation and analysis of PAH in lubricating oil and air sample. An autosampler was utilized for sample injection into the HPLC. By using a back-flush technique in conjunction with an aminopropylsilica-packed column, the PAHs were isolated by HPLC. A concurrent solvent evaporation injection technique was then used for on-line transfer of the PAH fraction to the GC, where the PAH analysis was completed. Compared with ordinary off-line HPLC clean-up followed by GC analysis, the sensitivity was increased by a factor of 50–100, yielding a detection limit of a few nanograms for individual PAHs when using flame ionization detection. Further, irreproducible losses of low-molecular-weight PAH as a result of solvent evaporation steps in off-line clean-up procedures were eliminated. Reproducibility of retention times and relative peak areas was high, facilitating component identification and quantification.

2.2.6 On-line LC–LC coupled methods

Van Stijn et al. [43] set up an on-line method for PAH analysis involving LC–LC coupling. After clean-up of the sample on a pre-concentration donor–acceptor complex chromatography (DACC) HPLC column, the PAHs were transferred to and separated on a conventional analytical HPLC column. With a DACC column, PAHs could be extracted from different matrices. PAHs are electron donors (π-electrons) and the strong interaction of the PAHs with an electron acceptor stationary phase results in retention of the PAHs and elution of (the bulk of) the other components in the sample. The quantification limits in the described work were 0.1 μg/kg for individual PAHs.

2.3 SAMPLE PREPARATION AND EXTRACTION

In general, for environmental samples, satisfactory analysis cannot be achieved by direct sample introduction due to the low analyte concentration pertaining to such samples, and also to the complexity of the matrix. It is therefore necessary that the sample be pretreated carefully to extract and enrich the target organic analytes from their host matrix. Sample clean-up or sample preparation remains the most time-consuming and labor-intensive work in the whole analytical scheme. The potential for analyte loss is also highest during this step. Thus, the care and rigor with which sample preparation is performed determines the soundness and quality of the final analytical results. In recognition of the need for effective, robust and reliable sample preparation, many procedures have been developed, normally based on the aim of carrying out this step to achieve fast, simple and if possible, especially in recent years, solvent-free or solvent-minimized operations.

The major problems associated with the analysis of PAHs are as follows [39]: (a) most PAHs are present in trace amounts (parts-per-billion or parts-per-trillion levels) in the environment (air, water and soil) or foods; they may also undergo losses during extraction or sample preparation; (b) many organic compounds can be coextracted with PAHs and interfere with subsequent separation, identification and quantification; (3) most PAHs are structurally similar; this can make separation and unequivocal identification difficult.

Various sample preparation approaches have been developed to address some or all of the above problems. Classical techniques for PAHs include liquid–liquid extraction (LLE) and Soxhlet extraction, depending on the matrix. Often, LLE is carried out as part of a clean-up process to isolate the PAH fraction. Although still useful, the classical methods which are time-consuming, tedious and normally require comparatively large amounts of solvents are being superseded by more modern techniques. The most widely used of the latter and some non-conventional ones are described below.

2.3.1 Solid-phase extraction

Solid-phase extraction (SPE), first introduced in the 1970s, remains an effective and useful sample preparation procedure for PAHs present in a variety of matrices. Previous approaches involving usually in-house packed columns (small-scale classical column chromatography), have in the past two decades or so given way to disposal, commercially available polypropylene or glass cartridges packed with a multitude of stationary phases. A more recent development is disk-based SPE in which the stationary phase is incorporated within a PTFE or cellulose-based membrane. This format is especially suitable for the extraction of PAHs from large amounts of aqueous samples. For both types of SPE, many companies have designed vacuum manifolds to process multiple samples simultaneously, thus increasing throughput and precision in PAH analysis.

Some representative SPE applications for the analysis of PAHs include the extraction of these compounds, using cartridges or disks, from water samples [183,184], soils [185–187], particulates [61,185,188] and sediments [61], although the technique is used for general sample clean-up in many of the applications already mentioned in this article. Although silica and C-18 are commonly used as the extraction materials, others such as $-NH_2$, Florisil, Alumina and XAD-2 have also been evaluated [61,186,187,190,191].

The following is a typical set of conditions for the extraction of PAHs from water, using a cartridge packed with C-18 (500 mg). The cartridge is first conditioned with 6-ml 2-propanol–deionized water (15 : 85). It is then rinsed with 6 ml of water; the packing should be allowed to become dry before the next step. 100 ml of the aqueous sample is applied. The cartridge is then washed with 2-propanol–deionized water (15 : 85). Dichloromethane (1 ml) is then used to elute the extract.

For SPE with a membrane disk (47-mm PTFE impregnated with C-18), the following set of conditions is typical. The membrane is conditioned with 10-ml methanol, 10-ml methyl-*t*-butyl ether (MTBE), 15-ml methanol and 15-ml deionized water in that order. The sample is passed through at a flow rate of 100 ml/min. Elution is carried out with 10 ml methanol and 10 ml of MTBE. The eluate is concentrated to 1 ml, then reconstituted to the appropriate volume with acetonitrile–water (50 : 50) (for HPLC analysis).

2.3.2 Supercritical fluid extraction

Supercritical fluid extraction (SFE) is now widely used in the extraction of low polarity pollutants from environmental samples. Compared with conventional methods, SFE saves time and does not consume huger amounts of solvents. With SFE, it is also not normally necessary to have, after the extraction itself, further extensive concentration and clean-up to prepare the extract for analysis.

A supercritical fluid affords the combination of gas-like characteristics (that permits highly efficient mass transfer) and liquid-like solvating ability, important considerations for an extraction solvent. Conventional solvents cannot measure up to supercritical fluids in terms of these advantages.

The majority of SFE work has focused on the use of supercritical CO_2 because of its reasonably moderate critical properties (critical temperature and pressure of 31.3°C and 7386.5 kPa (72.9 atm), respectively), low toxicity, ease of handling, affordability, chemical inertness, and the fact that it decompresses to a gas, leaving behind no residue. Although it is an excellent solvent for non-polar organics, CO_2 does have a limitation in that its polarity is often too low to obtain efficient extractions, either because the analytes lack sufficient solubility or the extractant has a poor ability to displace the analytes from active matrix sites in the sample. This can be obviated by having a mixed-fluid system, e.g. a mixture of methanol–CO_2 (see below).

One obvious approach to improving SFE efficiencies is to choose supercritical fluids that are more polar and selective than CO_2. For example, the use of supercritical N_2O

has been shown to increase the extraction efficiency of high-molecular-weight PAHs from fly ash and sediment [191]. Other polar fluids such as $CHClF_2$ (Freon-22) have been shown to be highly efficient in the extraction of nitrated and non-nitrated PAHs from Diesel-exhaust particulate matter [192]. A comparison of the use of Freon-22, N_2O and CO_2 for the SFE of native pollutants including PAHs from petroleum waste sludge, and those from railroad-bed soil showed that $CHClF_2$ yielded the highest extraction efficiencies, most likely because of its high dipole moment [193]. Supercritical ammonia would also be an attractive solvent, but it is chemically reactive and likely to be too dangerous for routine use.

Due to the practical difficulties involved in using pure polar fluids for SFE, the extraction of highly polar analytes has usually been carried out using CO_2 containing a few percent of added organic modifier to increase the polarity. Most SFE applications use methanol as modifier, but in some cases other solvents such as hexane, aniline, toluene or diethylamine have been shown to be more efficient [194]. The modifier can either be added to the sample in the extraction cell prior to SFE or be premixed with the CO_2.

2.3.2.1 Extraction Modes

SFE can either be dynamic or static. In the dynamic mode of operation, the extraction vessel is continuously flushed with fresh solvent. In static mode, the sample is inserted in the extraction vessel that is usually sealed after introduction of the supercritical fluid to perform the extraction. After extraction, the extract is dissolved in a conventional solvent suitable for the subsequent analysis. In practice, static extraction is less common than the dynamic mode.

SFE can be performed off-line and on-line.

2.3.2.1.1 Off-line SFE. Off-line SFE includes the extraction and solute collection steps. For PAH extraction, temperature is the most important parameter. In general, higher temperatures and higher pressures give better recoveries of PAHs since the higher temperature increases the vapor pressure of PAHs and accelerate the mass transfer between phases.

In order to achieve quantitative SFE, it is critical to ensure not only complete extraction but also that solute collection is quantitative. Three types of off-line trapping systems can be used with high recoveries when pure CO_2 is employed.

(1) The depressurized supercritical fluid passes through a cryogenically cooled inert material (usually stainless steel or glass beads). The solutes precipitate on the cooled surfaces before being dissolved by using an appropriate solvent.

(2) The trapping material used is solid adsorbent (generally C-18 silica) instead of inert material to achieve much more efficient trapping.

(3) Trapping the extracted solute directly in a liquid solvent. The outlet of the tube leading from the extraction chamber goes directly into the latter such that the PAHs are directly dissolved.

2.3.2.1.2 On-line SFE. With a suitable interface, SFE can be coupled to GC–flame ionization detection [195], GC [196], HPLC and SFC systems directly to achieve on-line sample preparation and separation. The advantage of an on-line SFE system is

TABLE 2.4

COMPARISON OF CONVENTIONAL EXTRACTION AND OFF-LINE AND ON-LINE SFE IN THE ANALYSIS OF PAHs IN URBAN DUST [197]

Parameter	Conventional extraction	Off-line SFE	On-line SFE
Sample size	1000 mg	20 mg	2 mg
Extraction time	48 h	1 h	15 min
Shortest possible total analysis time (one sample)	3 days	2 h	1 h

the speed with which extraction/separation can be carried out. It can also deal with limited amounts of sample (Table 2.4 [197]). Excellent extractions can be obtained by optimizing the flow by adjusting the temperature and pressure.

2.3.3 Subcritical water extraction

Water has a critical temperature and pressure of 374°C and 22088.4 kPa (218 atm). Thus, considerably more drastic conditions are needed to create supercritical water. Such a high critical temperature and pressure make supercritical water very corrosive [198–201]. This necessitates the use of special alloys such as Hasteroi for the plumbing in systems utilizing this fluid. Compared to supercritical water, subcritical water is much less reactive and thus can be used in analytical chemistry. Another advantage of subcritical water is its widely tunable dielectric constant, surface tension, and viscosity, since they decrease significantly by raising water temperature under moderate pressures to maintain water in the liquid state. At high temperature, water behaves like an organic solvent. Extraction efficiencies for polar and non-polar organics using water as solvent depend primarily on the temperature of the extraction as long as sufficient pressure is used to maintain the extractant in the liquid state.

Yang and Li [202] and Hageman et al. [203] have used subcritical water extraction to extract PAHs from environmental solids. The instrument and procedure for the extraction are very simple. All water extractions were performed using a 64-mm long, 7-mm I.D. (12-mm O.D.) stainless steel pipe with national pipe thread (npt) end caps which were rated by the supplier for a maximum pressure of 49,600 kPa. One end of the cell was closed using an end cap and a single layer of Teflon tape. The sample was then weighed into the cell, and the cell was filled with HPLC-grade water (ca. 3.5 ml) which had previously been purged with clean nitrogen for ca. 2 h to move dissolved oxygen. After being capped (again using a single layer of Teflon tape on the pipe threads), the cell was placed vertically in a gas chromatographic oven that had been preheated to the desired temperature. No attempt was made to mix the sample and extractant water during the heating step. After the heating was completed, the cell was immediately removed from the oven and cooled under tap water. The top cap was then removed, and 1.8 ml of the supernatant water was pipetted into a 2-ml autosampler vial containing a clean Teflon-coated stir bar. The vial was immediately sealed with a Teflon-lined cap to avoid the loss of more volatile components.

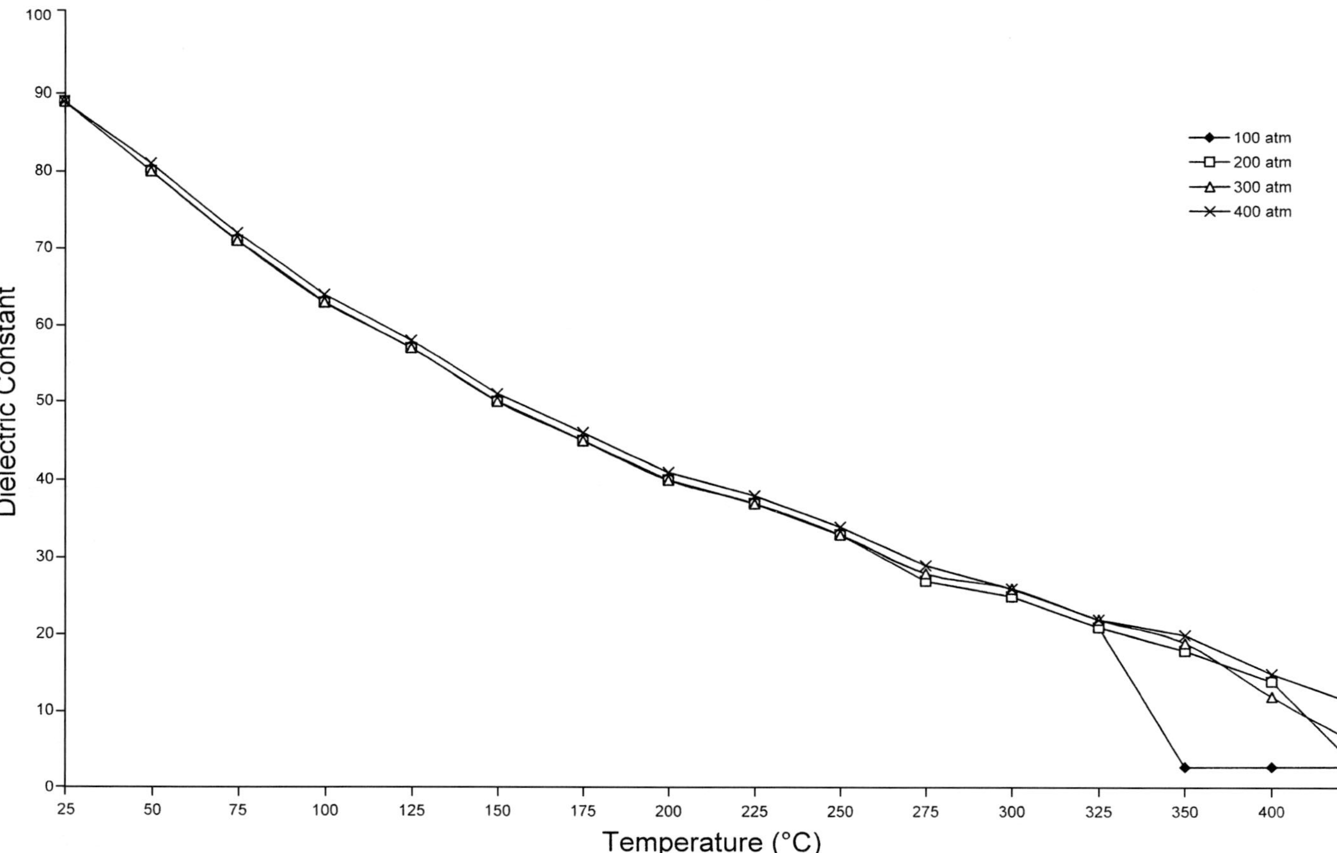

Fig. 2.7. Graph showing the variation of the dielectric constant of subcritical water with temperature at various pressures. (Modified from ref. [204].)

It is worth noting that the highest pressure expected during the heating step is the steam–water equilibrium pressure of 8600 kPa, which would occur at the highest temperature tested (300°C); this pressure is much lower than the vessel rating of 49,600 kPa. However, care should be taken to avoid the extraction of samples which may react with water to yield higher pressures.

Yang and Li [202] and Hageman et al. [203] found that temperature had a great effect on the subcritical water extraction of PAHs from soil, and extraction of all the PAHs was enhanced by a higher temperature. Since the dielectric constant, surface tension, and viscosity of water are decreased by raising temperature, the extraction efficiency for PAHs was also significantly enhanced by increasing the temperature. This was especially true for the higher-molecular-weight PAHs. Fig. 2.7 shows the variation of dielectric constant with temperature at various pressures [204]. A temperature of 250°C was necessary to achieve good recoveries of the higher-molecular-weight PAHs while lower temperatures were sufficient to extract lower-molecular-weight PAHs [205–207]. Water pressure was found not to affect PAH extraction efficiency [202].

2.3.4 Microwave extraction

Microwaves are high frequency (2.45 GHz) electromagnetic waves which can be strongly absorbed by a polar molecule and interact weakly with non-polar solvent, resulting in the accelerated extraction (through elevated temperatures) of polar compounds from various matrices into a non- or weak polar solvent. In a 1995 review, microwave heating was described as the first fundamentally new heating technique since the discovery of fire [208]. The efficiency of extraction with microwaves lies in the ability of the bulk material to transform electromagnetic radiation into heat without the disadvantages of convection and conduction, the conventional mode of heat distribution in a sample. The popular vision of a block of butter being heated in a domestic microwave oven collapsing unto itself explains vividly how microwave heating overcomes the limitations of conduction and convection heating. The heating begins from the interior of the bulk sample, thus going a long way towards preventing loss of analytes that occur commonly in conventional heating in which, to generate sufficiently high and optimum temperatures for extraction in the interior of the sample matrix, the surface of the bulk sample has to be subjected to even greater heating, over a prolonged period of time, resulting in enhanced losses of analytes.

Parameters that affect extraction include temperature, extraction time, solvent nature, sample size, moisture in the sample and microwave power. Generally, prolonged and high temperature extraction leads to poor recovery of PAHs, due to thermal degradation. The primary benefit of microwave heating is that the desired extraction temperature can be reached rapidly; thus exposure of the analytes to extended heating is minimized.

The solvent plays important role in the extraction of PAHs. As can be expected, solvents of different polarities lead to different extraction efficiencies. Polar solvent such as alcohols, ketones and esters absorb microwaves strongly, while non-polar solvents such as benzene, xylene and other aliphatic hydrocarbons do so weakly. Commonly used solvents for PAH extraction with microwaves are hexane, acetone, dichloromethane,

Diffused microwave system

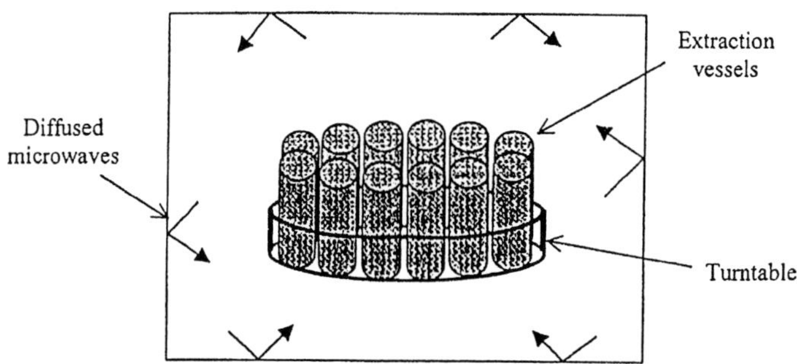

Focused microwave system

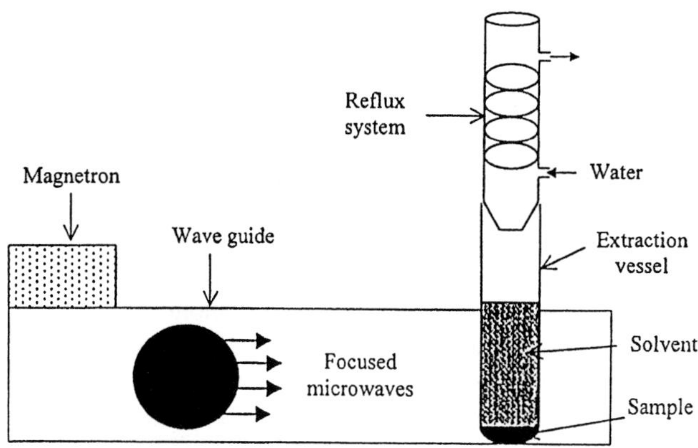

Fig. 2.8. Scheme showing alternative microwave systems using diffused, or focused microwaves. (Reproduced with permission from ref. [213].)

chloroform–methanol, acetone–petroleum ether and water. Since water is a good absorbent of microwaves, the presence of water or moisture can increase localized heating and give better extraction efficiencies.

There are various modes of microwave extraction techniques: closed vessel microwave technology [209], focused microwave or open vessel technology [210,211], and on-line microwave system [212]. The microwaves can also be diffused or focused, see Fig. 2.8 [213]. Due to possible losses of volatile analytes during microwave heating and the possibility of oven contamination in the open vessel extraction, closed vessels and on-line vessel system are recommended in extracting volatile compounds. Fig. 2.9 depicts a gas chromatogram of PAHs extracted by microwaves from airborne particulates collected during the severe 1994 haze episode in Southeast Asia [24]. Table 2.5 lists some additional applications of microwave extraction in the analysis of PAHs. A more

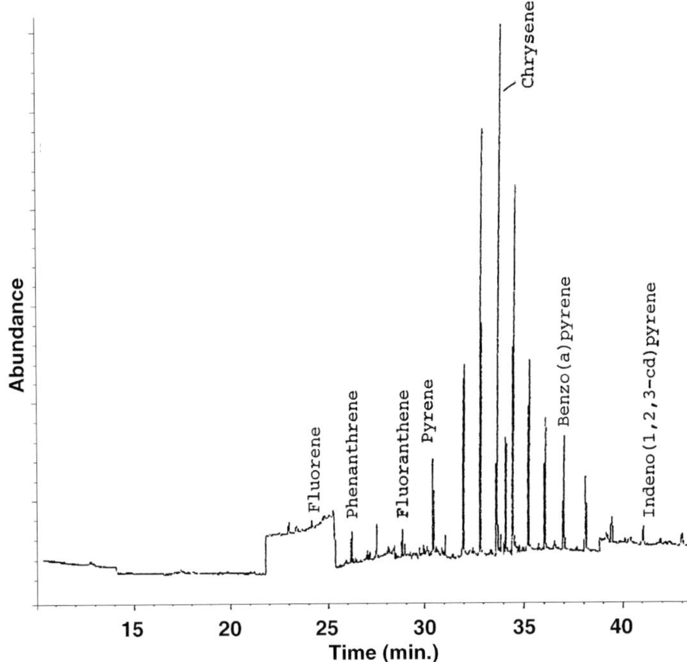

Fig. 2.9. Gas chromatogram of PAHs extracted by microwaves from airborne particulates collected during the 1994 Southeast Asian haze episode. (Reproduced with permission from ref. [24].)

TABLE 2.5

APPLICATION OF MICROWAVE EXTRACTION TO THE ANALYSIS OF PAHs

Sample	Procedure	Reference
Certified sediments	2 g sample, 10 min, 30 W, 30 ml dichloromethane, focused microwaves	[214]
	5 g sample, 5 min, 30 ml 1 : 1 acetone–hexane, closed vessel	[20]
Air particulate samples	1 : 1 acetone–hexane, closed vessel	[24]
Marine sediments	10 ml micellar medium (Polyoxyethylene 10 lauryl ether), 5 min, closed vessel	[215]
Wood samples	1–4 g samples, 20 min, 30 ml acetonitrile, closed vessel	[216]

comprehensive list encompassing extraction of these compounds from vegetation, soil and sediments, and water has recently been published [213].

2.3.5 Solid-phase microextraction

Solid-phase microextraction (SPME) [203,217–222] is a simple, fast and solvent-free extraction technique, suitable for a wide range of environmental pollutants, including

PAHs. It was developed in the late 1980s, and is now a commercially available product although it does seem to be rather expensive at this time. It represents an approach in the analytical sciences that focuses on miniaturization and environmental-friendliness (organic solvent-free and multiple-usage device).

Typically, a fused-silica fiber, coated with a thin layer of polymeric stationary phase, is used in SPME to extract analytes from water, soil and gaseous samples. The extracted analytes are then thermally desorbed in the injector of a gas chromatograph and swept into the chromatographic column for analysis.

The principle of SPME is based on the partitioning of analytes between a sample matrix and the stationary phase. The larger the partition coefficient of an analyte between the coating and the matrix, the greater the amount of analyte extracted.

Hageman et al. [203], Langenfeld et al. [220] and Chen and Pawliszyn [221] have used poly(dimethylsiloxane)-coated fibers to extract PAHs from water. Langenfeld et al. [220] successfully used SPME to determine low-part-per-billion concentrations of aromatic hydrocarbons in water samples containing very high concentrations (part-per-thousand) of matrix organics and suspended solids, although some of the spiked components were adsorbed by the suspended solids. The procedure offers sub- to low-part-per-billion sensitivity (even when not at equilibrium), has a wide linear range (three to six orders of magnitude), only requires small sample sizes (e.g. 2 ml), and is sensitive to trace determinations in the presence of high amounts of other organic interferences. Good agreement was obtained between a 45-min SPME, and a conventional dichloromethane extraction for the determination of PAH concentrations in creosote-contaminated water, demonstrating that SPME is a useful technique for the rapid determination of hydrocarbons in complex water matrices.

Since SPME involves analyte adsorption initially, and subsequently thermal desorption is the most convenient means of removing the analytes from the fiber, it is only natural that most SPME applications are combined with GC. There are now several vendors that offer automated SPME–GC systems. However, when used with a modified injection valve, SPME can also be combined with HPLC [222], in which the analytes are desorbed by the mobile phase before being directed into the separation column. Fig. 2.10 shows an example of an SPME–HPLC system [221].

Recently, there have also reports on the analysis of PAHs by SPME coupled with cyclodextrin-modified capillary electrophoresis [137]. Satisfactory reproducibility with respect to migration time and peak area was obtained using the same separation capillary; only the extraction fiber was discarded after each analysis.

2.3.6 Accelerated solvent extraction

Accelerated solvent extraction (ASE) or pressurized-fluid extraction, as it has also been termed, is another relatively new extraction technique. Its advantage over classical techniques is faster extraction (5–15 min), and relatively low consumption of organic solvents. It can also be easily automated to permit sequential extractions. As expected, ASE has been applied to PAH extractions, and indeed the USEPA has a method (Method 3545) for extracting these compounds from solid waste [223]. The proposed

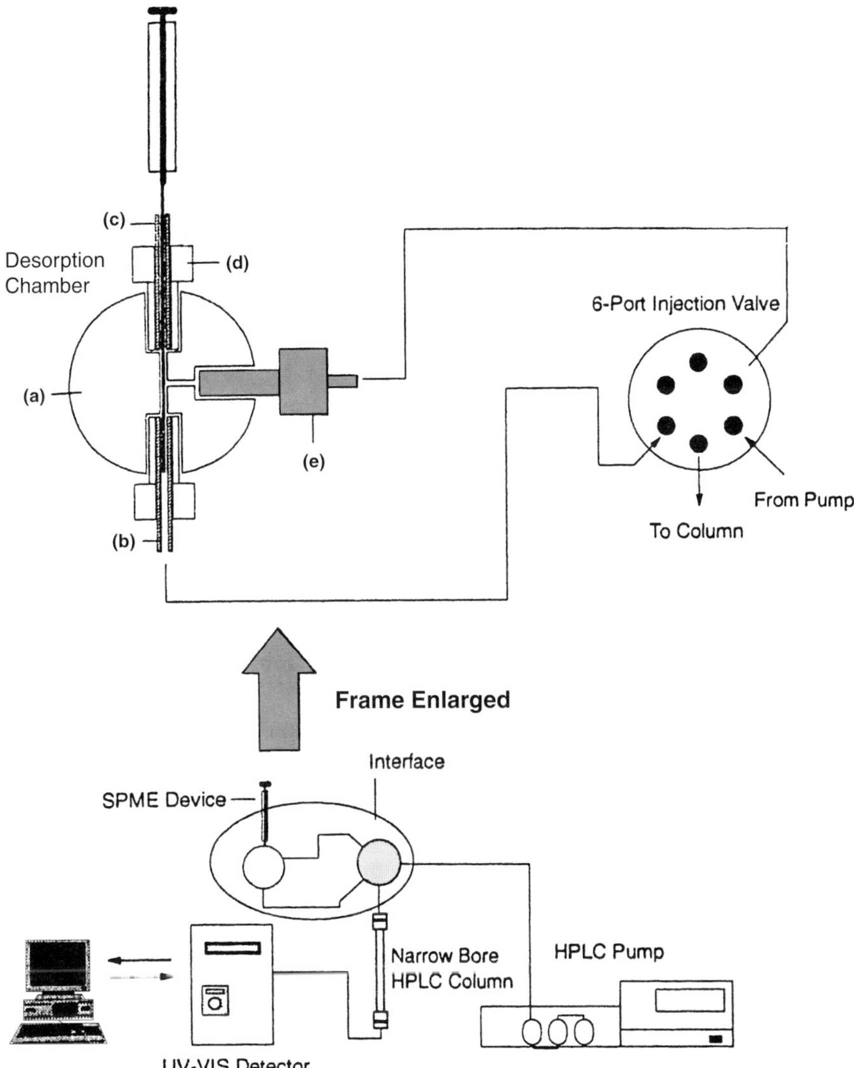

Fig. 2.10. An example of a solid-phase microextraction–high-performance liquid chromatography system. (a) Stainless steel tee joint; (b) stainless steel tubing; (c) PEEK tubing; (d) finger-tight PEEK union; (e) PEEK tubing with PEEK union. (From Chen and Pawliszyn [221]. Reproduced with permission of the American Chemical Society.)

conditions for extraction are: solvent, acetone–dichloromethane (1 : 1); pressure, 14 MPa; temperature, 100°C; extraction time, 5 min + 5 min equilibration time.

Heemken et al. [30] has also used ASE to extract marine particulate matter for its PAH content, and compared it to several techniques including SFE, Soxhlet extraction, sonication and extraction after methanolic saponification. Both ASE and SFE gave comparable recoveries and precision with these other methods. Saim et al. [224] developed an experimental design based on USEPA Method 3545 to extract PAHs from contaminated soils using ASE. Temperature was found to be the most important extraction parameter; the polarity of solvents was not found to be influential on the extraction recoveries.

ASE is an emerging technique; thus, the commercially available equipment is still very expensive. It will take some time before it becomes a routine tool in the general analytical laboratory.

2.3.7 Cloud-point extraction

The aqueous solutions of many nonionic surfactants undergo phase separation above a certain temperature that is also known as the cloud-point temperature [225,226]. The temperature at which the phase separation occurs depends on the surfactant concentration as well as on the pressure, amount, and type of organic additives. These dynamic micellar entities having a non-polar core possess the capacity to interact with non-polar species by hydrophobic interaction. During the cloud-point precipitation process, these micellar vesicles aggregate into a surfactant-rich phase, and any bound non-polar species concentrate in the surfactant-rich phase. The ability of this process to concentrate and separate non-polar target species from aqueous matrices has been demonstrated in analytical chemistry and separation science.

Sirimanne et al. [166,227] developed a cloud-point extraction method using the nonionic surfactants Triton X-100 [227] and Genapol X-080 [166], to extract PAHs from human serum. Triton X-100 (a *tert*-octylphenoxy poly(oxyethylene) ether (critical micelle concentration = 0.29 mmol/l; cloud-point temperature, 63.7°C), is widely used as a cloud-point surfactant. The surfactants belonging to the Triton X series show a strong dependence of the cloud-point temperature on the number of hydrophilic oxyethylene groups attached to the hydrophobic octylphenyl residue.

In general, the procedure for the cloud-point extraction of aqueous micellar solutions involves adding the surfactant to a concentration above its critical micelle concentration whilst maintaining the temperature below the cloud-point of the micellar solution. This procedure is then followed by induction of a phase separation by raising the temperature of the solution. The two phases, aqueous and surfactant-rich, are then usually separated by centrifugation.

To induce phase separation in aqueous micellar solution, additives are always used in the micellar solution. Some of the additives that have been used to effect a phase separation with Triton X-100 include urea, sodium chloride, sodium azide, and potassium chloride. Sirimanne et al. [227] used sodium chloride, to induce phase separation. Above a sodium chloride concentration of 2.5 M, the micellar serum samples undergo a phase separation, presumably by a salting-out mechanism. The incubation temperature, sodium chloride concentration and the concentration of Triton X-100 are the primary factors affecting the extraction efficiency.

2.3.8 Liquid–liquid extraction–gas chromatography

Baltussen et al. [228] has developed an extension to liquid–liquid extraction (LLE). Their approach involves the extraction and determination of aqueous PAHs by coupling LLE with GC. In this method, the water sample is passed through a sorption cartridge containing particles consisting of 100% polydimethylsiloxane (PDMS). The PDMS phase appears to be a solid but its sorptive characteristics are in fact similar to that of a liquid phase. Retention of the analytes is based on the adsorption of the solutes onto the surface of the PDMS material, or rather, the solutes dissolve (partition) into the bulk of this high viscosity liquid phase. Unlike the situation in SPE in which solutes are desorbed by a solvent, in this case, thermal desorption is used to transfer the analytes onto the GC column. In this way the consumption of organic solvents is minimized and maximum sensitivity is attained since all solutes trapped from the sample are actually introduced in the GC column. It is clear that the difference between this LLE method and SPME is in that it is not an equilibrium method but an exhaustive extraction procedure.

The most powerful aspect of the technique is the ability to transfer all the analytes from 100 ml of water to the GC column, resulting in excellent sensitivities. In this respect, quantification is somewhat more convenient to handle than in SPME which requires very careful attention to the establishment of valid calibration plots for each analyte of interest, under identical extraction conditions and parameters.

2.4 CONCLUDING REMARKS

PAHs have been the focus of attention in environmental analysis for several decades now. In all likelihood, as long as fossil fuels and other organic matter are burned in the course of anthropogenic or other activities, interest in these compounds, because of their biological activities, will be maintained. As scientists develop novel, newer or improved approaches in the analytical sciences, be they in sample preparation or extraction of a variety of matrices, separation and detection, it is without question that PAHs will continue to be used as test substances for the evaluation and validation of these techniques.

ACKNOWLEDGEMENTS

The author thanks C. Basheer, Y. He, M.L. Lesaicherre and L. Zhu for assistance in the preparation of the manuscript.

2.5 REFERENCES

1   Biologic Effects of Atmospheric Pollutants: Particulate Polycyclic Organic Matter, National Academy of Sciences, Washington, D.C., 1972.

2   (a) M.L. Lee, PhD thesis, Indiana University, Bloomington, IN, 1978; (b) M.L. Lee, K.D. Bartle and M. Novotny, Anal. Chem., 47 (1975) 540; (c) M.L. Lee, M. Novotny and K.D. Bartle, Anal. Chem., 48 (1976) 1566.

3   A. Bjorseth, J. Knutzen and J. Skei, Sci. Total Environ., 13 (1979) 71.

4   G. Schomburg, R. Dielmann, H. Borwitzky and H. Husmann, J. Chromatogr., 167 (1978) 337.

5   J.A. Schmit, R.A. Henry, R.C. Williams and J.F. Dieckman, J. Chromatogr. Sci., 9 (1971) 645.

6   K. Ogan and E. Katz, J. Chromatogr., 188 (1980) 115.

7   S.A. Wise, S.N. Chesler, H.S. Hertz, L.R. Hilpert and W.E. May, Anal. Chem., 49 (1977) 2306.

8   K. Ogan, E. Katz and W. Slavin, J. Food Sci., 35 (1977) 146.

9   F.L. Joe Jr., J. Salemme and T. Fazio, J. Assoc. Off. Anal. Chem., 67 (1984) 1076.

10   K. Takatsuki, S. Suzuki and S.N. Sato, J. Assoc. Off. Anal. Chem., 68 (1985) 945.

11   L.C. Sander and S.A. Wise, Anal. Chem., 61 (1989) 1749.

12   M.W. Dong, J.X. Duggan and S. Stefanou, LC–GC, 11 (1993) 802.

13   S.A. Wise, L.C. Sander and W.E. May, J. Chromatogr., 642 (1993) 329.

14   E.A. Gomaa, J.I. Gray, S. Rabie, C. Lopez-Bote and A.M. Booren, Food Addit. Contam., 10 (1993) 503.

15   H.Y. Yabiku, M.S. Martins and M.Y. Takahashi, Food Addit. Contam., 10 (1993) 399.

16   A. Liberti, G.P. Cartoni and V. Cantuti, J. Chromatogr., 15 (1964) 141.

17   J.C. Fetzer, in: T. Vo-Dinh (Ed.), Chemical Analysis of Polycyclic Aromatic Compounds, John Wiley, New York, 1988, pp. 59–101.

18   K.P. Naikwadi, P.P. Wadgaonkar, D. LeBlanc, R. Boyd and J. Curtis, in: Proceedings of the 3rd Biennial International Conference on Monitoring and Measuring of the Environment, Ottawa, ON, Canada, May 8–11, 2000, pp. 229–234.

19   A. Bjorseth and G. Eklund, in: W. Bertsch, W.G. Jennings and R.E. Kaiser (Eds.), Recent Advances in Capillary Gas Chromatography, Alfred Huthig Verlag, Heidelberg, 1981, pp. 477–490.

20   K.K. Chee, M.K. Wong and H.K. Lee, J. Chromatogr. A, 723 (1996) 259.

21   J.W.M. Wegener, W.P. Cofino, E.A. Maier and G.N. Kramer, Trends Anal. Chem., 18 (1999) 14.

22   K.K. Chee, M.K. Wong and H.K. Lee, Anal. Chim. Acta, 330 (1996) 217.

23   I.J. Barnabas, J.R. Dean, I.A. Fowlis and S.P. Owen, Analyst, 120 (1995) 1897.

24   K.K. Chee, M.K. Wong and H.K. Lee, Environ. Monitor. Assess., 44 (1997) 391.

25   C.M. Reddy and J.G. Qinn, Mar. Pollut. Bull., 38 (1999) 126.

26   J.J. Langenfeld, S.B. Hawthorne, D.J. Miller and J. Pawliszyn, Anal. Chem., 65 (1993) 338.

27   Z. Zhang and J. Pawliszyn, Anal. Chem., 66 (1996) 844A.

28   R.F. Shore, J. Wright, J.A. Horne and T.H. Sparks, Mar. Pollut. Bull., 38 (1999) 509.

29   B.W. Wright, C.W. Wright, R.W. Gale and R.D. Smith, Anal. Chem., 59 (1987) 38.

30   O.P. Heemken, N. Theobald and B.W. Wenclawiak, Anal. Chem., 69 (1997) 2171.

31   M. Notar and H. Leskovsek, Fresenius J. Anal. Chem., 358 (1997) 623.

32   X. You, X. Wang, R. Bartha and J.D. Rosen, Environ. Sci. Technol., 24 (1990) 1732.

33   S.B. Hawthorne and D.J. Miller, Anal. Chem., 59 (1987) 1705.

34   V. Lopez-Avila, R. Young and N. Teplitsky, J. AOAC Int., 79 (1996) 142.

35   R.M. Lancas, M.H.R. Matta, L.J. Hayasida and E. Carriho, J. High Resolut. Chromatogr., 14 (1991) 633.

36   K.K. Chee, M.K. Wong and H.K. Lee, Int. J. Environ. Stud., 56 (1996) 689.

37   M.M. Rhead and C.J. Trier, Trends Anal. Chem., 11 (1982) 255.

38   A. Bemgard, A. Colmsj and B.O. Lundmark, J. Chromatogr., 595 (1992) 247.

39   B.H. Chen, C.Y. Wang and C.P. Chiu, J. Agric. Food Chem., 44 (1996) 2244.

40   L. Zhu, Y. Takahashi, T. Amagai and H. Matsushita, Talanta, 45 (1997) 113.

41   R. Reupert and G. Brausen, Acta Hydrochim. Hydrobiol., 22 (1994) 202.

42   S. Hartik, J. Lehotay, M. Chakrt and R. Brandstetr, J. Liq. Chromatogr., 18 (1995) 4149.

43   F. Van Stijn, M.A.T. Kerkhoff and B.G.M. Vandeginste, J. Chromatogr. A, 750 (1996) 263.

44   H.P. Nirmaier, E. Fischer, A. Meyer and G. Henze, J. Chromatogr. A, 730 (1996) 169.

45   E. Veigl, W. Posch, W. Lindner and P. Tritthart, Chromatographia, 38 (1994) 199.

46   G. Ignesti, M. Lodovici, P. Dolara, P. Lucia and D. Grechi, Bull. Contam. Toxicol., 48 (1992) 809.

47   C. Venkataraman, J.M. Lyons and S.K. Friedlander, Environ. Sci. Technol., 28 (1994) 555.

48　C. Venkataraman and S.K. Friedlander, Environ. Sci. Technol., 28 (1994) 563.

49　E. Manoli and C. Samara, Chromatographia, 43 (1996) 135.

50　C.P. Chiu, Y.S. Lin and B.H. Chen, Chromatographia, 44 (1997) 497.

51　S. Moret, L. Conte and D. Dean, J. Agric. Food Chem., 47 (1999) 1367.

52　G.A. Perfetti, P.J. Nyman, S. Fisher, F.L. Joe Jr. and G.W. Diachenko, J. AOAC Int., 75 (1992) 872.

53　A. Kurganov, K.K. Unger and F. Eisenbeiß, Chromatographia, 39 (1994) 175.

54　Y. Saito, H. Ohta, H. Terasaki, Y. Katoh, H. Nagashima, K. Jinno and K. Itoh, J. High Resolut. Chromatogr., 18 (1995) 569.

55　M.N. Kayali, S. Rubio-Baroso and L.M. Polo Diez, J. Liq. Chromatogr. Rel. Technol., 19 (1996) 759.

56　L.C. Sander and S.A. Wise, J. Chromatogr., 316 (1987) 163.

57　L.C. Sander and S.A. Wise, Adv. Chromatogr., 25 (1986) 139.

58　L.C. Sander and S.A. Wise, Crit. Rev. Anal. Chem., 18 (1988) 299.

59　J.F. Lawrence and D.F. Weber, J. Agric. Food Chem., 32 (1984) 794.

60　W.G. Lan, K.K. Chee, M.K. Wong, H.K. Lee and Y.M. Sin, Analyst, 120 (1995) 281.

61　K. Peltonen and T. Kuljukka, J. Chromatogr. A, 710 (1995) 93.

62　H.K. Lee, J. Chromatogr. A, 710 (1995) 79–92.

63　G.W. Schieffer, J. Chromatogr., 319 (1985) 317.

64　B.A. Tomkins, R.A. Jenkins, W.H. Griest and R.R. Reagen, J. Assoc. Off. Anal. Chem., 68 (1985) 935.

65　P. Simko and B. Brunckova, Food Addit. Contam., 10 (1993) 257.

66　J.L. Beltran, J. Guiteras and R. Ferrer, Anal. Chem., 70 (1998) 1949.

67　M.N. Kayali, S. Rubio-Barros and L.M. Polo-Diez, J. Chromatogr. Sci., 33 (1995) 18.

68　S.O. Baek, M.E. Goldstone, P.W.W. Kirk, J.N. Lester and R. Perry, Environ. Sci. Technol., 12 (1991) 107.

69　M.W. Dong and A. Greenberg, J. Liq. Chromatogr., 11 (1988) 1887.

70　J.F. Coetzee, G.H. Katzi and J.C. Spurgeon, Anal. Chem., 48 (1976) 2170.

71　M.T. Galceran and E. Moyano, Talanta, 40 (1993) 615.

72　M.T. Galceran and E. Moyano, J. Chromatogr. A, 731 (1996) 75.

73　M.T. Galceran and E. Moyano, J. Chromatogr. A, 683 (1994) 9.

74　C.H. Marvin, R.W. Smith, D.W. Bryant and B.E. McCarry, J. Chromatogr. A, 863 (1999) 13.

75　C.H. Marvin, B.E. McCarry, J. Villella, D.W. Bryant and R.W. Smith, Polycyl. Aromat. Comp., 9 (1996) 193.

76　T. Letzel, U. Posch, E. Rosenberg, M. Grasserbauer and R. Niessner, Rapid Commun. Mass Spectrom., 13 (1999) 2456.

77　M. Castillo, A. Oubina and D. Barcelo, Environ. Sci. Technol., 32 (1998) 2180.

78　(a) H. Moriwaki, A. Imaeda and R. Arakawa, Anal. Commun., 36 (1999) 53; (b) H. Moriwaki, Analyst, 125 (2000) 417.

79　B.A. Mansoori, Rapid Commun. Mass Spectrom., 12 (1998) 712.

80　L. Bonfanti, M. Careri, A. Mangia, P. Manini and M. Maspero, J. Chromatogr. A, 728 (1996) 359.

81　C.M. Pace and L.D. Betowskild, J. Am. Soc. Mass Spectrom., 6 (1995) 597.

82　R.M. Smith (Ed.), Supercritical Fluid Chromatography, Royal Society of Chemistry, London, 1988.

83　M.L. Lee and K.E. Markides (Eds.), Analytical Supercritical Fluid Chromatography and Extraction, Chromatography Conferences Inc., Provo, UT, 1990.

84　K. Jinno, T. Hoshino, T. Hondo, M. Saito and M. Senda, Anal. Chem., 58 (1986) 2696.

85　P. Sim, C. Elson and M. Quillaim, J. Chromatogr., 445 (1988) 239.

86　D.W. Later, D.J. Bornhop, E.D. Lee, J.D. Henion and R.C. Wieboldt, LC–GC, 5 (1987) 804.

87　P.A. Peadon, J.C. Fjeldsted, M.L. Lee, S.R. Springston and M. Novotny, Anal. Chem., 54 (1982) 1090.

88　E. Moyano, E. McCullagh, M.T. Galceran and D.E. Games, J. Chromatogr. A, 777 (1997) 167.

89　D.R. Baker, Capillary Electrophoresis, John Wiley, New York, 1995.

90　S. Terabe, K. Otsuda and T. Ando, Anal. Chem., 834 (1985) 57.

91　S. Terabe, Y. Ishihama, H. Nishi, T. Fukuyama and K. Otsuka, J. Chromatogr., 545 (1991) 359.

92   K. Otsuka, M. Higashimori, R. Koike, K. Karuhaka, Y. Okada and S. Terabe, Electrophoresis, 15 (1994) 1280.

93   O. Brüggemann and R. Freitag, J. Chromatogr., 717 (1995) 309.

94   T. Imasaka, K. Nishitani and N. Ishibashi, Anal. Chim. Acta, 256 (1992) 3.

95   Y.F. Yik, C.P. Ong, S.B. Khoo, H.K. Lee and S.F.Y. Li, J. Chromatogr., 589 (1992) 333.

96   H. Nishi and M. Matsuo, J. Liq. Chromatogr., 14 (1991) 973.

97   S. Terabe, Y. Miyashita, Y. Ishihama and O. Shibata, J. Chromatogr., 636 (1993) 47.

98   K. Jinno and Y. Sawada, J. Liq. Chromatogr., 18 (1995) 3719.

99   B. Jiménez, D.G. Patterson, J. Grainger, Z. Liu, M.J. González and M.L. Marina, J. Chromatogr. A, 792 (1997) 411.

100   R.O. Cole, M.J. Sepaniak, W.L. Hinze, J. Gorse and K. Oldiges, J. Chromatogr., 557 (1991) 113.

101   T. Kaneta, T. Yamashita and T. Imasaka, J. Chromatogr., 299 (1995) 371.

102   J. Cai and Z. El Rassi, J. Chromatogr., 608 (1992) 31.

103   J.T. Smith and Z. El Rassi, J. Chromatogr. A, 685 (1994) 131.

104   J.T. Smith, W. Nashabeh and Z. El Rassi, Anal. Chem., 66 (1994) 1119.

105   C.P. Palmer, M.Y. Khaled and H.M. McNair, J. High Resolut. Chromatogr., 15 (1992) 756.

106   C.P. Palmer and S. Terabe, J. Microcolumn Sep., 8 (1996) 115.

107   C.P. Palmer and S. Terabe, Anal. Chem., 69 (1997) 1852.

108   S.A. Shamsi, C. Akbay and I.M. Warner, Anal. Chem., 70 (1998) 3078.

109   C. Akbay, I.M. Warner and S.A. Shamsi, Electrophoresis, 20 (1999) 145.

110   N. Tanaka, T. Fukutome, T. Tanigawa, K. Hosoya, K. Kimata, T. Araki and K.K. Unger, J. Chromatogr. A, 699 (1995) 331.

111   N. Tanaka, H. Iwasaki, T. Fukutome, K. Hosoya and T. Araki, J. High Resolut. Chromatogr., 20 (1997) 529.

112   N. Tanaka, T. Fukutome, K. Hosoya, K. Kimata and T. Araki, J. Chromatogr. A, 716 (1995) 57.

113   N. Tanaka, K. Nakagawa, H. Iwasaki, K. Hosoya, K. Kimata, T. Araki and D.G. Patterson, J. Chromatogr. A, 781 (1997) 139.

114   C. Akbay, S.A. Shamsi and I.M. Warner, Electrophoresis, 18 (1997) 253.

115   S. Yang, J.G. Bumgarner and M.G. Khaledi, J. High Resolut. Chromatogr., 18 (1995) 443.

116   J.P. Quirino and S. Terabe, Anal. Chem., 71 (1999) 1638.

117   B. Göttlicher and K. Bächmann, J. Chromatogr. A, 768 (1997) 320.

118   K. Bächmann and B. Göttlicher, Chromatographia, 45 (1997) 249.

119   S. Terabe, Y. Miyashita, O. Shibata, E.R. Barnhart, L.R. Alexander, D.G. Patterson, B.L. Karger, K. Hosoya and N. Tanaka, J. Chromatogr., 516 (1990) 23.

120   C.L. Copper and M.J. Sepaniak, Anal. Chem., 66 (1994) 147.

121   W.C. Brumley and W.J. Jones, J. Chromatogr. A, 680 (1994) 163.

122   P.G. Muijselaar, H.B. Verhelst, H.A. Claessens and C.A. Cramers, J. Chromatogr. A, 764 (1997) 323.

123   W. Ding and J.S. Fritz, Anal. Chem., 69 (1997) 1593.

124   Y. Walbroehl and J.W. Jorgenson, Anal. Chem., 58 (1986) 479.

125   S. Nie, R. Dadoo and R.N. Zare, Anal. Chem., 65 (1993) 3571.

126   Y. Shi and J.S. Fritz, J. High Resolut. Chromatogr., 17 (1994) 713.

127   Y. Shi and J.S. Fritz, Anal. Chem., 67 (1995) 3023.

128   R. Jankowiak, D. Zamzow, W. Ding and G.J. Small, Anal. Chem., 68 (1996) 2549.

129   J.H.T. Luong and Y. Guo, Electrophoresis, 19 (1998) 723.

130   J.H.T. Luong, Electrophoresis, 19 (1998) 1461.

131   W. Ding and J.S. Fritz, Anal. Chem., 70 (1998) 1859.

132   J. Li and J.S. Fritz, Electrophoresis, 20 (1999) 84.

133   O.H.J. Szolar, R.S. Brown and J.H.T. Luong, Anal. Chem., 67 (1995) 3004.

134   M.J. Sepaniak, C.L. Copper, K.W. Whitaker and V.C. Anigbogu, Anal. Chem., 67 (1995) 2037.

135   K.W. Whitaker, C.L. Copper and M.J. Sepaniak, J. Microcolumn Sep., 8 (1996) 461.

136   R.S. Brown, J.H.T. Luong, O.H.J. Szolar, A. Halasz and J. Hawari, Anal. Chem., 68 (1996) 287.

137   A.-L. Nguyen and J.H.T. Luong, Anal. Chem., 69 (1997) 1726.

138   K. Bächmann, A. Bazzanella, I. Haag and K.-Y. Han, Fresenius J. Anal. Chem., 357 (1997) 32.

139   I.S. Lurie, J. Chromatogr. A, 792 (1997) 297.

140   J.L. Miller, M.G. Khaledi and D. Shea, Anal. Chem., 69 (1997) 1223.
141   J.L. Miller, M.G. Khaledi and D. Shea, J. Microcolumn Sep., 10 (1998) 681.
142   X. Xu and R.J. Hurtubise, J. Chromatogr. A, 829 (1999) 289.
143   K. Bächmann, A. Bazzanella, I. Haag, K.-Y. Han, R. Arnecke, V. Böhmer and W. Vogt, Anal. Chem., 67 (1995) 1722.
144   V. Pretorius, B.J. Hopkins and J.D. Schieke, J. Chromatogr., 99 (1974) 23.
145   J.W. Jorgenson and K.D. Lukacs, J. Chromatogr., 218 (1981) 209.
146   J.H. Knox, Chromatographia, 26 (1988) 329.
147   R.M. Seifar, W.Th. Kok, J.C. Kraak and H. Poppe, Chromatographia, 46 (1997) 131.
148   A. Maruska and U. Pyell, Chromatographia, 45 (1997) 229.
149   M.M. Robson, S. Roulin, S.M. Shariff, M.W. Raynor, K.D. Bartle, A.A. Clifford, P. Myers, M.R. Euerby and C.M. Johnson, Chromatographia, 43 (1996) 313.
150   S.E. van den Bosch, S. Heemstra, J.C. Kraak and H. Poppe, J. Chromatogr. A, 755 (1996) 165.
151   C. Fujimoto, Y. Fujise and E. Matsuzawa, Anal. Chem., 68 (1996) 2753.
152   J.-L. Liao, N. Cheng, C. Ericson and S. Hjertén, Anal. Chem., 68 (1996) 3468.
153   Y. Guo and L.A. Colón, Anal. Chem., 67 (1995) 2511.
154   R. Asiae, X. Huang, D. Farnan and C. Horváth, J. Chromatogr. A, 806 (1998) 251.
155   Q. Tang, B. Xin and M.L. Lee, J. Chromatogr. A, 837 (1999) 35.
156   G. Chirica and V.T. Remcho, Electrophoresis, 20 (1999) 50.
157   H. Sawada and K. Jinno, Electrophoresis, 20 (1999) 24.
158   A.S. Lister, J.G. Dorsey and D.E. Burton, J. High Resolut. Chromatogr., 20 (1997) 523.
159   C. Yan, R. Dadoo, H. Zao, R.N. Zare and D.J. Rakestraw, Anal. Chem., 67 (1995) 2026.
160   H. Rebscher and U. Pyell, J. Chromatogr. A, 737 (1996) 171.
161   R. Dadoo, R.N. Zare, C. Yan and D.S. Anex, Anal. Chem., 70 (1998) 4787.
162   C. Yan, R. Dadoo, R.N. Zare, D.J. Rakestraw and D.S. Anex, Anal. Chem., 68 (1996) 2726.
163   A.S. Lister, C.A. Rimmer and J.G. Dorsey, J. Chromatogr., 828 (1998) 105.
164   B. Xin and M.L. Lee, J. Microcolumn Sep., 11 (1999) 271.
165   B. Xin and M.L. Lee, Electrophoresis, 20 (1999) 67.
166   S.R. Sirimanne, J.R. Barr and D.G. Patterson, J. Microcolumn Sep., 11 (1999) 109.
167   H. Knox and I.H. Grant, Chromatographia, 32 (1991) 317.
168   K.W. Whitaker and M.J. Sepaniak, Electrophoresis, 15 (1994) 1341.
169   H. Rebscher and U. Pyell, Chromatographia, 42 (1996) 171.
170   M.M. Dittmann and G.P. Rozing, J. Chromatogr. A, 744 (1996) 63.
171   C. Ericson, J.-L. Liao, K. Nakazato and S. Hjertén, J. Chromatogr. A, 767 (1997) 33.
172   M.M. Robson, M.G. Cikalo, P. Myers, M.R. Euerby and K.D. Bartle, J. Microcolumn Sep., 9 (1997) 357.
173   P.B. Wright, A.S. Lister and J.G. Dorsey, Anal. Chem., 69 (1997) 3251.
174   J.J. Pesek and M.T. Matyska, Electrophoresis, 18 (1997) 2228.
175   R.J. Dadoo, C. Yan, R.N. Zare, D.S. Anex, D.J. Rakestraw and G.A. Hux, LC–GC, 15 (1997) 630.
176   J.J. Vreuls, G.J. De Jong and U.A.Th. Brinkman, Chromatographia, 31 (1991) 113.
177   I.L. Davies, M.W. Raynor, P.T. Williams, G.F. Andrews and K.D. Bartle, Anal. Chem., 59 (1987) 2579.
178   D. Duquet, C. Dewaele and M. Verzele, HRC CC, 11 (1988) 252.
179   G.S. Heo and J.K. Suh, HRC CC, 13 (1990) 748.
180   S. Matsuzawa, P. Garrigues, O. Setokuchi, M. Sato, T. Yamamoto, Y. Shimizu and M. Tamura, J. Chromatogr., 498 (1990) 25.
181   C. Ostman, A. Bemgard and A. Colmsjo, J. High Resolut. Chromatogr., 15 (1992) 438.
182   C. Ostman and U. Nilsson, J. High Resolut. Chromatogr., 15 (1992) 745.
183   D. Eastwood, M.E. Domingues, R.L. Lidberg and E.J. Poziomek, Analusis, 22 (1994) 305.
184   M.C. Hennion, Trends Anal. Chem., 10 (1991) 317.
185   V. Librando, G. D'Arrigo and D. Spampinato, Analusis, 22 (1994) 340.
186   F.J. Gonzalez-Vila, J.L. Lopez, F. Martin and J.C. Del Rio, Fresenius J. Anal. Chem., 339 (1991) 750.
187   T. Spitzer, J. Chromatogr., 643 (1993) 43.

188   C.A. Menzie, B.B. Potocki and J. Santodonato, Environ. Sci. Technol., 26 (1992) 1278.

189   S.L. Simonich and R.A. Hites, Environ. Sci. Technol., 28 (1994) 939.

190   J. Dumont, F. Larocque-Lazure and C. Iorio, J. Chromatogr. Sci., 31 (1993) 371.

191   G. Castello and T.C. Gerbine, J. Chromatogr., 612 (1993) 351.

192   T. Paschke, S.B. Hawthorne, D.J. Miller and B. Wenclawiak, J. Chromatogr., 609 (1992) 333.

193   A.L. Howard and L.T. Taylor, Abstract, Pittsburgh Conference, New Orleans, LA, 1993, p. 1181.

194   Y. Yang, A. Gharaibeh, S.B. Hawthorne and D.J. Miller, Anal. Chem., 67 (1995) 641.

195   J. Vejtoster, P. Karasek and J. Planeta, Anal. Chem., 71 (1991) 905.

196   J.A. Field, D.J. Miller, T.M. Field, S.B. Hawthorne and W. Griger, Anal. Chem., 64 (1992) 3161.

197   S.B. Hawthorne, Anal. Chem., 62 (1990) 633A.

198   R.W. Shaw, T.B. Brill, A.A. Clifford, C.A. Eckert and E.U. Franck, Chem. Eng. News, Dec, 23 (1991) 26.

199   T.B. Thomason and M. Modell, Hazard. Waste, 1 (1984) 453.

200   T. Horth and E.U. Franck, Ber. Bunsenges. Phys. Chem., 97 (1993) 1091.

201   L.X. Li, E.F. Gloyna and J.E. Sawicki, Water Environ. Res., 65 (1993) 250.

202   Y. Yang and B. Li, Anal. Chem., 71 (1999) 1491.

203   K.J. Hageman, L. Mazeas, C.B. Grabanski, D.J. Miller and S.B. Hawthorne, Anal. Chem., 68 (1996) 3892.

204   L. Haar, J.S. Gallagher and G.S. Kell, National Bureau of Standards/National Research Council Steam Tables: Thermodynamic and Transport Properties and Computer Program for Vapor and Liquid States of Water in SI Units, Hemisphere Publishing Corp., Washington, D.C., 1984.

205   S.B. Hawthorne, Y. Yang and D.J. Miller, Anal. Chem., 66 (1994) 2912.

206   Y. Yang, S. Bowadt, S.B. Hawthorne and D.J. Miller, Anal. Chem., 67 (1995) 4571.

207   Y. Yang, S.B. Hawthorne and D.J. Miller, Environ. Sci. Technol., 31 (1997) 430.

208   A. Zlotorzynski, Crit. Rev. Anal. Chem., 25 (1995) 43.

209   V. Lopez-Avila, R. Young and R. Kim, J. Chromatogr. Sci., 33 (1995) 481.

210   M. Letellier, H. Budzinski, L. Charrier, S. Capes and A.M. Porthe, Fresenius J. Anal. Chem., 364 (1999) 228.

211   L.E. Garcia, M. Ayuso, A. Sanchez, F. de Alba and M.D. Luque de Castro, Anal. Chem., 70 (1998) 2426.

212   V. Karanassios, F.H. Li, B. Lu and E.D. Salin, J. Anal. At. Spectrom., 6 (1991) 457.

213   V. Camel, Trends Anal. Chem., 19 (2000) 229.

214   H. Budzinski, A. Papineau, P. Baumard and P. Garrigues, Anal. Chem., 65 (1995) 321.

215   L. Pensado and R. Cela, J. Chromatogr. A, 869 (2000) 505.

216   V. Pina and V. Gonzalez-Diaz, J. Chromatogr. A, 869 (2000) 515.

217   Z. Zhang, M. Yang and J. Pawliszyn, Anal. Chem., 66 (1994) 844A.

218   J. Pawliszyn, Trends Anal. Chem., 14 (1995) 113.

219   D. Louck, S. Motlagh and J. Pawliszyn, Anal. Chem., 64 (1992) 1187.

220   J.J. Langenfeld, S.B. Hawthorne and D.J. Miller, Anal. Chem., 68 (1996) 144.

221   J. Chen and J. Pawliszyn, Anal. Chem., 67 (1995) 2530.

222   (a) R. Eisert and K. Levsen, J. Am. Soc. Mass Spectrom., 6 (1995) 1119; (b) R. Eisert and K. Levsen, Fresenius J. Anal. Chem., 351 (1995) 555.

223   Test Methods in Evaluating Solid Waste, Method 3545, USEPA SW-846, 3rd ed., Update III, USGPO, Washington, D.C., 1995.

224   N. Saim, J.R. Dean, M.P. Abdullah and Z. Zakaria, Anal. Chem., 70 (1998) 420.

225   H. Watanabe, in: K.L. Nfittal and E.J. Fender (Eds.), Solution Behaviour of Surfactants, Plenum Press, New York, 1982, pp. 1305–1316.

226   W.L. Hinze and E. Pramauro, Crit. Rev. Anal. Chem., 24 (1993) 133.

227   S.R. Sirimanne, J.R. Barr, D.G. Patterson Jr. and L. Ma, Anal. Chem., 68 (1996) 1556.

228   E. Baltussen, H.G. Janssen, P. Sandra and C. Cramers, J. High Resolut. Chromatogr., 20 (1997) 395.

W. Kleibömer (Ed.), *Environmental Analysis*
Handbook of Analytical Separations, Vol. 3

CHAPTER 3

Separation methods in the analysis of polycyclic aromatic sulfur heterocycles

Jan T. Andersson

*Department of Analytical Chemistry, University of Münster, Wilhelm-Klemm-Strasse 8,
D-48149 Münster, Germany*

3.1 INTRODUCTION

3.1.1 Polycyclic aromatic sulfur heterocycles

In chemical jargon, the expression 'polycyclic aromatic hydrocarbons (PAH)' has been used for so long that it was on the verge of being accepted for all polycyclic aromatic compounds (PAC). This would have been improper since compounds containing heteroatoms are not hydrocarbons; thus polyaromatic heterocycles and also substituted PAHs (such as nitro- and hydroxy-derivatives, often found in environmental samples) are not really covered by the term PAH. In this chapter, the expression PAH will be used for hydrocarbons only. Compounds containing a sulfur atom in an aromatic ring are derivatives of thiophene and are called Polycyclic Aromatic Sulfur Heterocycles (PASH). Occasionally the expression 'thiaarene' is used synonymously with PASH, but strictly speaking includes the one-ring aromatic compound thiophene also, which, however, is not covered by the term PASH.

PASHs are based on a thiophene ring which is annealed with one or more other aromatic rings (Fig. 3.1). Conceptually they can be derived from the PAHs by replacing one or more of the benzene rings with a thiophene ring which is another way of saying that a CH=CH function has been replaced by a sulfur atom which is part of the aromatic system. The simplest member of this class of compounds is benzothiophene which corresponds to naphthalene. (A second isomer, benzo[*c*]thiophene, typifies the quinoid thiophenes that will not be included in this treatment since they are practically unknown in real-world samples.) Two other PASHs also correspond to naphthalene, namely thieno[2,3-*b*]thiophene and thieno[3,2-*b*]thiophene, but compounds with two sulfur atoms are not frequently found in natural samples. However, some matrices like high-molecular weight fossil materials may contain considerable amounts of such compounds as well as compounds with other heteroatoms in addition to the sulfur

References pp. 96–98

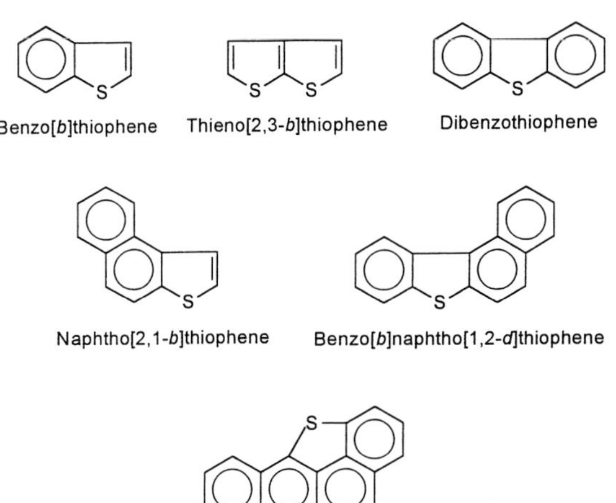

Fig. 3.1. Examples of non-substituted polycyclic aromatic sulfur heterocycles (PASH).

atom. For instance, in a Qatar crude oil PASHs with two sulfur atoms were detected by GC–MS [1].

As more rings are added, the number of possible isomers rises quickly, even if only PASHs are counted which contain no more than one sulfur atom. The equivalents of the three-ring PAHs phenanthrene and anthracene are dibenzothiophene and the three isomeric naphthothiophenes, whereas there are five cata-condensed four-ring PAHs and 13 PASHs [2]. Furthermore, the introduction of a sulfur atom leads to a loss of symmetry for the parent compound which means that there are more non-equivalent carbon atoms and consequently more isomeric substitution products possible for the PASHs than for the PAHs. Thus there are only two methylnaphthalenes but six methylbenzothiophenes, and eight monomethylated derivatives of the three-ring PAHs but 28 of the three-ring PASHs [2].

The larger number of isomers is one reason why the analysis of PASHs can pose considerably more difficult separation problems than that of the PAHs. Other reasons include the lower concentration, compared to the PAHs, that are found in many (but certainly not all) samples. In fact, it is not uncommon among crude oils, and therefore in environmental samples derived from these, that the dibenzothiophenes dominate over the phenanthrenes. An example for such a case is a crude oil from the Kirkuk field in Iraq whose carbon- and sulfur-selective gas chromatograms are shown in Fig. 3.2 [3]. The gas chromatographic patterns in the two selective traces among the three-ring aromatics, starting with dibenzothiophene (DBT), show great similarities. In a study of the polycyclic aromatic compounds in eleven teas, the concentration of the PASHs ranged from 8 to 108% of the concentration of the PAHs [4].

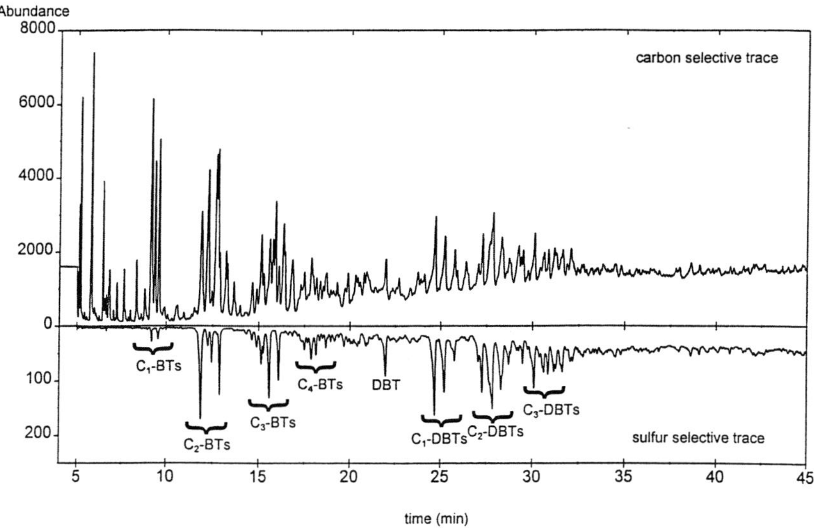

Fig. 3.2. The aromatic fraction from a Kirkuk crude oil analyzed on a 25-m 5% phenyl–95% methyl siloxane GC phase. Atomic emission detection at 181 nm (sulfur) and 193 nm (carbon). BT, benzothiophene; DBT, dibenzothiophene. (From Bobinger and Andersson [3]. Reproduced with permission of Elsevier Science.)

3.1.2 Why PASHs?

The sulfur-containing aromatic compounds occur practically everywhere where PAHs are found so that the same two main sources are obvious, namely fossil fuels (petrogenic) and combustion (pyrogenic sources). In traditional PAH analytical work, PASHs are therefore often found although rarely included in routine determinations since these tend to ignore other PACs than the 16 PAHs listed as priority pollutants by the US Environmental Protection Agency.

While petrogenic samples are well investigated, those resulting from pyrogenic sources have not been treated in depth to the same extent. Thermal treatment of organic material under oxygen-poor conditions, such as pyrolysis, may well lead to the formation of PASHs, e.g. in cigarette smoke [5]. PASHs were also identified in private residences after fire accidents [6]. Even such mild a treatment as boiling an aqueous solution of L-cysteine and glucose leads to detectable amounts of benzothiophene and thieno[3,2-*b*]thiophene [7].

The impetus for a directed effort toward PASH determination comes from several directions. The lowering of the legal sulfur limit in gasoline and heating oil has spurred the oil industry to intense efforts to develop processes to remove even traces of sulfur [8]. Although many sulfur forms are fairly easily reduced, e.g. through a hydrotreatment, the aromatic sulfur can be particularly recalcitrant. Powerful analytical separation and identification methods are necessary to determine which isomers are unaffected and at what concentration levels they are present.

Much less is known about the carcinogenic and mutagenic effects of PASHs than of PAHs although some representatives are known to be biologically very active [9]. Thus benzo[2,3]phenanthro[4,5-*bcd*]thiophenc (Fig. 3.1) was found to be more carcinogenic than benzo[*a*]pyrene. This area remains largely unexplored, though.

In organic geochemistry, PASHs and their alkyl derivatives yield information on important parameters such as the maturity of a crude oil. The methyldibenzothiophene ratio was introduced as an easily measured number and shown to correlate well with the maturity of crude oils [10,11]. The thermodynamically more stable methyldibenzothiophenes are enriched over the less stable ones on maturation of an oil. This can be used as an indicator for the depth of burial of a petroleum since the dibenzothiophene maturity index seems to vary systematically with this depth [12].

Environmental applications are manifold and demonstrate how PASH analysis can be useful in quite different areas. Identification of the source of an oil spill can be aided through a correlation of the PASH pattern of the spill with that of suspected sources [13]. Although it is known that dibenzothiophenes are degraded microbially [14] and photochemically [15] in the environment, these processes seem to be slower than for the corresponding PAHs and therefore the sulfur aromatics may be a more useful compound class to study, also in cases where the biomarker patterns are too similar to differentiate between various crudes [13].

A last illustration of the possible practical uses for PASH analysis is their use as pollution source tracers in urban air [16]. The PASH profile in air particulate varied significantly depending on the wind direction, since different sources of air pollution were affected: in one direction coke oven emissions predominated (with a ratio of alkyldibenzothiophenes to dibenzothiophene less than unity) and when the winds blew from a different direction, diesel exhaust emissions prevailed (with a ratio greater than unity). Such source apportionment studies are of large importance in environmental studies.

The above examples are included to show some of the many facets of PASH studies and some of the possible practical uses that can be made of PASH data. With more awareness of this potential and of more knowledge of the analytical procedures for PASH analysis, many new applications should be expected in the future.

3.1.3 Literature

A book summarizing the knowledge up to 1986 on the synthesis, spectral properties, biological activity and analytical aspects has been published [9]. A recent review gives more details on the sources and occurrence of PASHs [17]. A survey of all sulfur compounds described in the literature as occurring in fossil fuels was published in 1983 and comprises many hundred PASHs [18]. The occurrence, toxicity, and biodegradation of PASHs found in petroleum was the subject of a recent review [19].

In this contribution, the emphasis will be on various separation problems with respect to PASHs and only some general information on this class of substances will be given by way of introduction. When speaking about the separation of PASHs, two things can be implied, namely either the class separation of the PASHs from the PAHs or the separation of the PASH compounds from each other. Those two tasks obviously involve different techniques and will both be covered in this review. Works where PASHs have been analyzed incidentally together with PAHs but where no special attention is paid to them will not be discussed.

This chapter is organized along the lines normally used in an analysis, viz. sample workup, group separations and finally the high-resolution chromatographic step, including detectors, that normally also involves a quantification of the aromatic analytes. Finally sources of standard reference materials and of individual PASHs will be discussed.

3.2 SAMPLE WORKUP FOR PASH ANALYSIS

The physico-chemical properties of PAHs and PASHs are so similar that the two classes of compounds will be found together after the usual workup schemes employed for the analysis of aromatic compounds. Specifically, there is no group separation when a mixture is taken through a normal-phase chromatographic column such as silica, alumina or a bonded material like aminopropyl silica. In such experiments, the members of the two classes of compounds are separated mainly according to the number of unsaturated carbon atoms. A sulfur atom contributes as much to the retention as one to two aromatic carbon atoms, somewhat dependent on the position of the sulfur atom in the molecule (see below).

Polar bonded phases, and among them especially aminopropylsilane phases, are extensively used in normal-phase liquid chromatography of PACs [20]. There is no class separation of PAHs and PASHs but some characteristics are noticeable. Sulfur heterocycles with a terminal thiophene ring elute somewhat later than the corresponding PAHs. Thus benzothiophene elutes after naphthalene, and the three naphthothiophenes, possessing a terminal heterocyclic ring, appear after the corresponding PAH, phenanthrene. However, dibenzothiophene with an internal thiophenic ring elutes ahead of phenanthrene and the four-ring benzonaphthothiophenes display somewhat shorter retention times than the four-ring PAH chrysene [21]. This can be used to separate for instance the dibenzothiophenes from the three-ring PAHs; however, the other three-ring PASHs (naphthothiophenes) will elute together with the PAHs.

Alkylation has only a small effect on the retention on aminopropylsilane phases, smaller than on silica and alumina [20]. With hexane as mobile phase, benzothiophene had a retention index of 2.29 (naphthalene = 2.00, phenanthrene = 3.00) and 15 alkylated derivatives appeared between 1.86 and 2.17 [21]. Under the same conditions, the retention index for dibenzothiophene was 2.84 and for six alkylated derivatives it varied between 2.63 and 2.77. In another study with pentane as mobile phase, 11 mono- to tetramethyldibenzothiophenes showed retention factors in the range of 0.86–1.19, compared to 0.92 for dibenzothiophene and 1.06 for phenanthrene [22].

An example of a fractionation on aminopropylsilane is shown in Fig. 3.3. The use of the refractive index detector reveals that the aliphatic components are retained to a negligible extent only. The aromatic components, detected by a UV detector, are separated according to the number of aromatic rings. Often a semi-preparative column is used to separate the aromatic compounds into such fractions which are collected for further investigations or, as in [23] and similar works [24], introduced on-line into a gas-chromatographic injector.

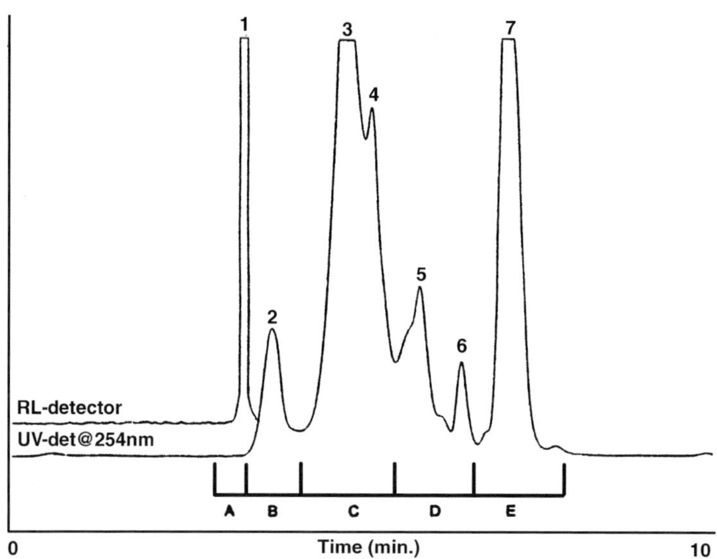

Fig. 3.3. HPLC chromatogram of a synthetic Nigerian cut (160–350°C) on an aminosilane bonded silica column with heptane as mobile phase. Upper trace: refractive index detection, lower trace: UV detection at 254 nm. 1, Saturates; 2, monoaromatics; 3, naphthalenes; 4, biphenyls and benzothiophenes; 5, fluorenes; 6, dibenzothiophenes; 7, three-ring aromatics. (From Beens and Tijssen [23]. Reproduced with permission.)

Stationary phases based on charge–transfer interactions have been used for the clean-up of samples although generally this remains a rather unexplored approach. The electron-rich PACs are retained on phases containing electron-deficient aromatic compounds, usually of the type (poly)nitro- or polychlorobenzenes or -fluorenones, through a kind of electron donor–acceptor (also called charge–transfer) interactions. A nitrophenylpropyl silica phase was found to be useful in the determination of PASHs containing at least four aromatic rings in workplace air from an aluminum melting factory [25]. On 2,4-dinitroanilinopropylsilica sulfur aromatic compounds were not separated by number of aromatic rings in the same way as PAHs [26]. Dibenzothiophenes (but not benzothiophenes) are so strongly complexed by tetranitrofluorenone that they are precipitated in the form of a charge–transfer complex from a gas oil [27]. Coprecipitation of PAHs such as fluorenes and phenanthrenes was also observed, although there is a selectivity in favor of the three-ring sulfur compounds.

3.3 CLASS SEPARATION OF PAHs AND PASHs

Except for the rarest of samples, such as a north-German crude oil [2] where nearly every aromatic compound contains sulfur, the sulfur aromatics are present together with a complex mixture of PAHs and this causes considerable analytical problems. In such a situation, two principle solutions can be used for the study of the PASHs, namely a physical separation of the two classes of compounds or the use of a selective detector in chromatographic separations. Since both paths are used in practice, they will be discussed here together with their advantages and disadvantages.

Two main routes have been employed for the group separation of PAHs and PASHs, namely: (a) the oxidation of the sulfur aromatics to their sulfones followed by a chromatographic separation in the normal-phase mode; and (b) the complexation with palladium chloride deposited on silica gel. Both methods show drawbacks, the main one being that PASHs with a terminal thiophene ring behave differently from those with an internal heterocyclic ring in that they resemble the PAHs more than those PASHs with an internal ring. They can therefore easily be overlooked or even lost.

3.3.1 Oxidation to sulfones

Normal phase chromatography can be used for group separations if the PASHs are first derivatized to the sulfones (PASHO$_2$) through oxidation. These products are sufficiently more polar than the PAHs so that the desired group separation on silica or alumina is easily achieved. This procedure has been used in many instances although there are certain limitations to its use. Even with the most suitable oxidant described so far, a certain selectivity during the oxidation is recorded. Furthermore, PAHs are destroyed to varying degrees. Several oxidants have been tested out of which *meta*-chloroperbenzoic acid has shown the greatest utility. As will be demonstrated below, hydrogen peroxide should be avoided.

In Fig. 3.4, a flow chart is shown illustrating both the principle of oxidation of the PASHs (here: for the two-ring compounds) and separation of PAHs and PASHs on palladium chloride/silica (here: for the three-ring compounds) that will be discussed in detail in Section 3.3.2.3.

3.3.1.1 Oxidation with hydrogen peroxide

The oxidation with hydrogen peroxide in an organic solvent was suggested for analytical work in 1967 [29] and re-endorsed in 1981 [30]. It was up picked by many authors for various kinds of samples, like coal gasification tar, emissions from coal-fired furnaces, synthetic and fossil fuels, crude oil, etc. However, a later critical study [2] of the procedure concluded that it is totally useless due to a severe distortion of the pattern of PASHs through destruction of a large part of the aromatic compounds. Although an earlier paper had noticed some of the disadvantages, stressing side reactions during the reduction with lithium aluminum hydride back to the PASHs (see below) [31], later authors have continued to apply the oxidation with hydrogen peroxide so that it would appear to be necessary to look at it again.

The procedure calls for boiling the sample in an acetic acid/benzene mixture for 16 h, followed by a column chromatographic separation on silica. Non-oxidized material is eluted with benzene and in a second fraction the sulfones come with benzene : methanol (1 : 1) [30]. The oxidized fraction is then subjected to a reduction with lithium aluminum hydride in ether which is supposed to lead cleanly to the original PASHs. Recoveries of between 0 and 41% [32] for 3- to 5-ring PASHs were later reported, low enough that a reinvestigation should be warranted. A careful analysis [2] showed that the

References pp. 96–98

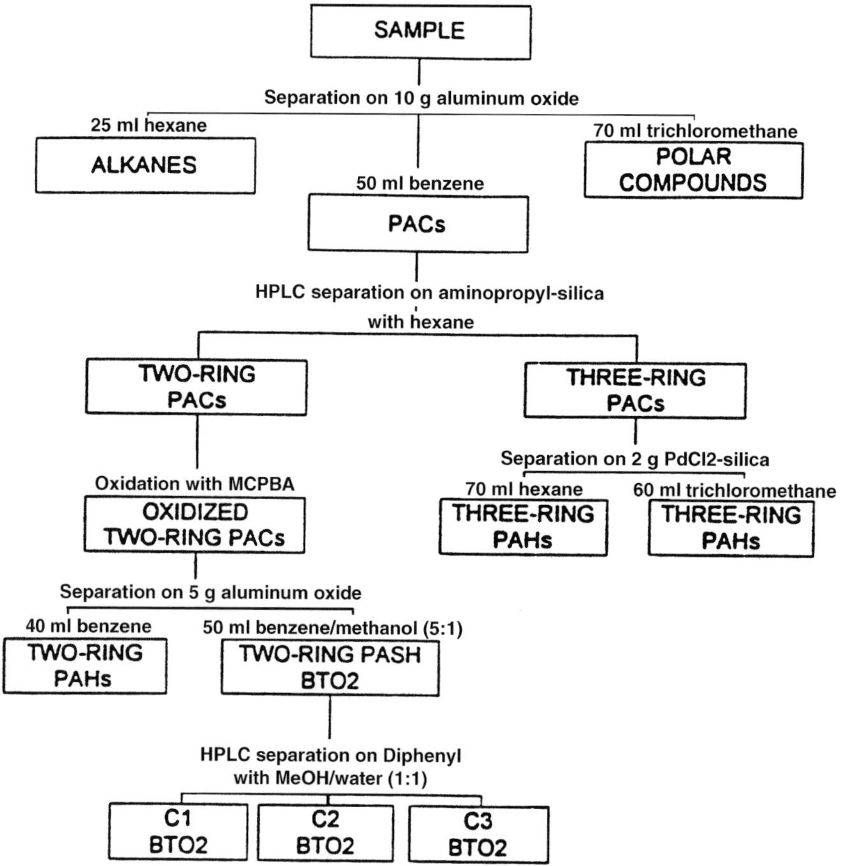

Fig. 3.4. Separation scheme for PASHs involving oxidation of the sulfur aromatics to their sulfones for the two-ring fraction and separation on palladium chloride for the three-ring fraction. (From Andersson and Schmid [28]. Reproduced with permission of Elsevier Science.)

oxidation of many PASHs leads to other products than the sulfones since the non-sulfur parts of the molecules can be oxidized to hydroquinones, anhydrides etc. Complicating the situation is the possibility that terminal sulfones can undergo a Diels–Alder type dimerization which leads to larger ring systems after spontaneous elimination of sulfur dioxide [2]. Finally, the reduction with lithium aluminum hydride of compounds with a terminal sulfone ring often produces dihydrothiophenes through hydrogenation of the carbon–carbon double bond in the thiophenic part of the molecule before the sulfone group is reduced [2,30,33]. All these reactions severely change the PASH pattern to the point of complete removal of many of them from the sample. The degree of distortion is evident from the chromatograms in Fig. 3.5 that shows a standard mixture of nine PASHs that have been taken through the oxidation/reduction sequence. Only some of these compounds are recovered and in most instances in low yield. Especially alkylated dibenzothiophenes may be less of a problem since usually their recoveries are better

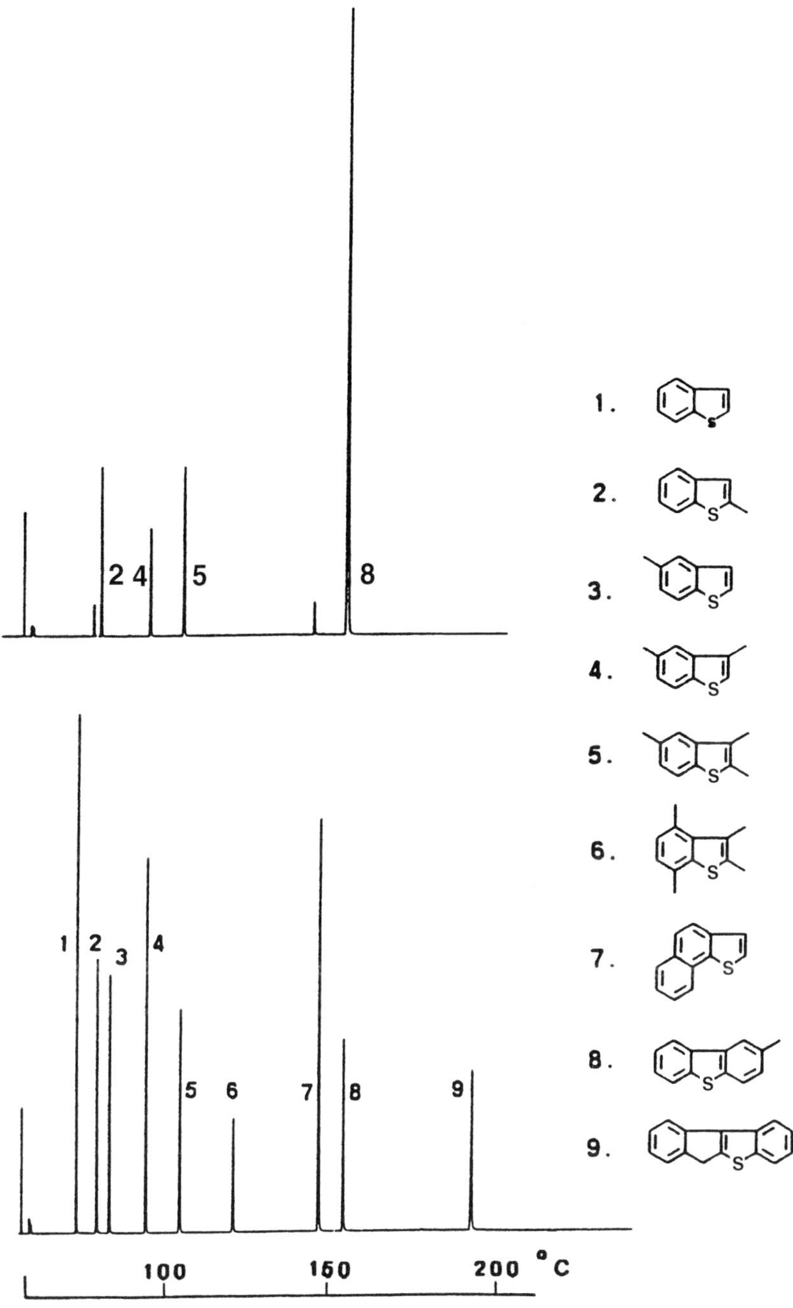

Fig. 3.5. Standard mixture of PASHs before (bottom) and after (top) oxidation with hydrogen peroxide in acetic acid followed by reduction with lithium aluminum hydride. (From Andersson [2]. Reproduced with permission of Gordon and Breach Science Publishers.)

than those of other PASHs. This was also noted in a modified hydrogen peroxide procedure in dichloromethane for diesel fuels [34].

It should be stressed here that the use of this oxidation in order to remove interferences due to PASHs in studies of the PAHs [35,36] is equally disastrous since many PAHs will be transformed into polar products by the oxidant and therefore be lost before the quantification step [2].

3.3.1.2 Oxidation with m-chloroperbenzoic acid

Obviously, the oxidation must be carried out with another oxidant than hydrogen peroxide in order to be useful. Since no reducing agent is known that cleanly cleaves the sulfur–oxygen double bonds in a sulfone functional group [2], the reduction should be avoided. Conceivably a better alternative should be the sulfoxides since they are more polar than the sulfones and also much more easily reduced to the PASHs, but no oxidant has been found that cleanly effects this oxidation; over-oxidation to the sulfone is frequently a serious side reaction. Also, in contrast to the sulfones, the sulfoxides are not thermally stable enough to be easily used in gas chromatography [37,38].

With this background *m*-chloroperbenzoic acid (MCPBA) was employed with better success: side reactions occur to a much lesser degree and the reaction is carried out under milder conditions. Furthermore, after a column chromatographic separation of the sulfones from the rest of the material, they can be further analyzed as sulfones without a reduction back to the thiophenes. This has the added advantage that the sulfones, which are perfectly stable under GC conditions [37], in many instances are better resolved on gas chromatographic columns than the thiophenes themselves [39].

The oxidation with MCPBA is most commonly carried out in a chlorinated solvent like dichloromethane. After a reaction time of ca. 30 min at room temperature, the mixture can be worked up after destruction of excess peracid. Column chromatography as described above is used to separate non-oxidized material from the sulfones (see Fig. 3.4).

MCPBA shows some selectivity towards certain PASHs since preferentially sulfur in internal thiophene rings, e.g. in dibenzothiophene, is transformed. Although benzothiophene is oxidized in the same fashion, naphtho[1,2-*b*]- and naphtho[2,1-*b*]thiophene are not [40]. Naphtho[2,3-*b*]thiophene is not recovered but it is doubtful if it is oxidized to its sulfone [40]; more probably a quinone is formed [2]. Similar behavior was found for four-ring PASHs containing terminal or internal thiophene rings [40].

3.3.1.3 Oxidation with other oxidants

2-Benzylsulfonyl-3-(*p*-nitrophenyl)-oxaziridine has also been investigated as an oxidant and is supposed to lead to the sulfoxide, but was found to be less selective than MCPBA and to produce sulfones rather than sulfoxides [40]. The milder oxidants *tert*-butyl hydroperoxide [41], singlet oxygen [42] and tetrabutylammonium periodate [43] do not oxidize aromatic sulfur. Only sulfides whose sulfur atom is not part of an aromatic ring form sulfones and sulfoxides.

3.3.2 Separation through complexation with metal ions

Since organic sulfides have been known for a long time to form complexes with heavy metals, it was only natural to extend such studies to PASHs with a view to separate them from the non-complexing PAHs. Several metals that are known to have an affinity for sulfur have been investigated, usually in the form of a metal salt deposited on a chromatographic support material.

3.3.2.1 Silver

Unlike olefins and nitrogen heterocycles, dibenzothiophene did not elute faster from a C-18 column when silver ions were included in the mobile phase which means that a complexation interaction between the PASH and the silver ions was not observed [44]. However, if the sulfur atoms are not part of an aromatic system, e.g. in thianthrene, a considerable complexation effect was found and the compound showed a much reduced retention time.

Silver nitrate can be used to impregnate a silica phase. One- and two-ring aromatics from a shale oil fraction were separated into fractions enriched in the sulfur heterocycles. An enrichment factor of about 16 was claimed for benzothiophene compared to naphthalene [45]. However, a reinvestigation showed that while there might be some preference for two-ring sulfur heterocycles over the hydrocarbons, this effect diminishes for the three-ring systems [46]. Furthermore, alkylation also increased the retention of PAHs, making methylnaphthalenes elute together with the methylbenzothiophenes.

A more promising system seems to be the dual column procedure of a base-washed silica gel followed by a Ag-loaded propylsulfonic acid bonded onto silica [47]. The polyaromatic concentrate of a 200-–425°C neutral oil was fractionated on this system into seven fractions at 0°C and a generally satisfying separation of PAHs and PASHs was reported. This procedure does not seem to have been explored in more detail later.

3.3.2.2 Other metals

In a study using Hg, Zn, Cd, Cu and Ag salts on silica, dibenzothiophene behaved essentially as anthracene [48,49]. When a carboxylic cation exchange resin was loaded with copper ions, sulfides (but not benzothiophene) were strongly retained [50].

3.3.2.3 Palladium

The one metal that has been shown to possess useful properties for the ligand exchange chromatography of PASHs is palladium. Originally it was used in the form of its chloride deposited on silica [51]. This approach was quickly picked up and developed into a practical method for separating PAHs and PASHs containing from two to six aromatic rings in a coal liquid [52]. It was noticed that compounds containing a terminal thiophene ring were largely lost and this was ascribed to the compounds reacting with the metal salt. It was also noticed that the sulfur aromatics eluted as complexes with palladium. When the eluate was injected into a gas chromatographic injector, the compounds were desulfurized with the carrier gas hydrogen acting as reagent and palladium as catalyst. This complication was circumvented by destroying the PASH/PdCl$_2$ complexes with diethylamine prior to GC injection. Even so, the recoveries for five PASH standards varied between 35 and 90%.

Another study improved considerably on this [53]. Instead of the diethylamine addition to destroy the complexes in the isolated fraction, a small amount of aminopropyl-bonded silica was incorporated into the separation column so that the decomplexation occurred in situ at the outlet of the column. The fast elution of heterocycles with a terminal thiophene ring was also investigated; benzothiophene and naphtho[1,2-*b*]thiophene showed the same retention as phenanthrene and fluorene and would therefore elute together with the PAH fraction. However, alkylated derivatives of the same compounds were considerably more strongly retained and were collected in the PASH fraction. PAHs showed a weak retention on the material. Chrysene was reported to have a retention factor of 4.6, similar to that of several monomethylbenzothiophenes. In Fig. 3.6, the separation of the preisolated aromatic fraction of a shale oil is shown with a flame photometric detector for sulfur selective detection. Fig. 3.6a shows the fraction obtained from the palladium chloride column with hexane as eluent. Supposedly only the PAHs should elute but it is immediately obvious from the figure that there are many sulfur aromatics present, too. They are mainly alkylated benzothiophenes, but the chromatogram depicted in Fig. 3.6b reveals that there is a large number of three-ring PASHs also, mainly naphthothiophenes that do not have sufficiently high retention on the PdCl$_2$ column. When the eluent was changed to 20% chloroform in hexane, the more strongly complexed PASHs eluted from the PdCl$_2$ column as shown in Fig. 3.6c. The peaks for the dibenzothiophenes are prominent but there are also many benzothiophenes alkylated with two or more carbon atoms as visualized in Fig. 3.6d. The strength of the interaction between palladium ions and terminal thiophene groups is obviously not strong enough to afford a clean separation between PAHs and such PASHs and this probably explains why the original investigation [52] failed to obtain high recoveries for all the test compounds.

On the whole, the strength of the sulfur–palladium interaction is correlated with the Hückel π-electron density [54]. Benzothiophene has the density 1.663 and dibenzothiophene 1.722 [54]; their retention factors were 2.1 and 37.3, respectively [53]. Benzo[*b*]naphtho[2,1-*d*]thiophene is intermediate with an electron density of 1.697 which agrees with its less strong retention by PdCl$_2$ compared to dibenzothiophene. Sulfides in which the sulfur is not part of the aromatic ring system show even higher Lewis basicity and therefore are more strongly retarded [54].

Other systems have also been described which rely on the use of complexation with palladium. 2-Amino-2-cyclopentene-1-dithiocarboxylic acid was bonded onto silica gel and complexed with silver or palladium [55]. Silver was found to be unsuitable due to a similar retention of PAHs and PASHs. Palladium displayed strong selectivity for PASHs which could be eluted either in the back flush mode or through the addition of isopropanol to the mobile phase. However, no applications have been published on this method.

Palladium can also be bonded in the form of a phenyl sulfonate attached covalently to a silica support [56]. At room temperature the PASHs were strongly retained and back flush at elevated temperatures with methyl *t*-butyl ether had to be employed to wash the sulfur heterocycles from the column. Nitrogen heterocycles were irreversibly retained but oxygen heterocycles eluted with the sulfur analogues.

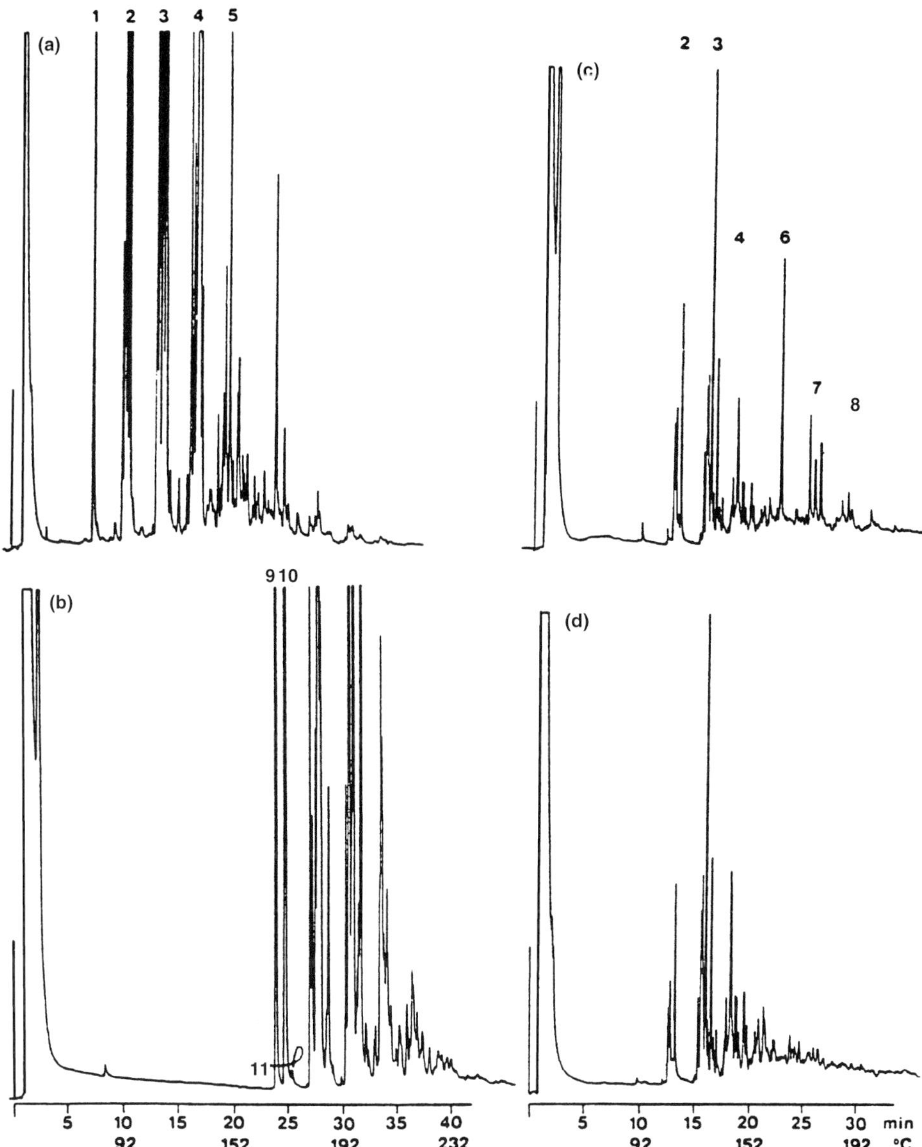

Fig. 3.6. Gas chromatogram on a 30-m DB-5 capillary with flame photometric detection of the aromatic fraction of a shale oil separated on a $PdCl_2$ liquid chromatographic column. (a) Eluate with hexane. (b) The three-ring compounds of the fraction shown in a as obtained on an aminopropyl silica column. (c) Eluate with 20% chloroform in hexane. (d) The two-ring compounds of the fraction shown in c as obtained on an aminopropyl silica column. 1, Benzothiophene (BT); 2, C_1-BT; 3, C_2-BT; 4, C_3-BT; 5, C_4-BT; 6, dibenzothiophene (DBT); 7, C_1-DBT; 8, C_2-DBT; 9, naphtho[1,2-*b*]thiophene; 10, naphtho[2,1-*b*]thiophene; 11, naphtho[2,3-*b*]thiophene. (From Andersson [53]. Reproduced with permission of the American Chemical Society.)

References pp. 96–98

The palladium chloride on silica separation method is the best studied method to date for the separation of PAHs and PASHs but is obviously not an ideal system. Especially the weak complexation of the terminal thiophenes should be improved. Possibly this could be achieved through a lowering of the temperature since it has been shown that interactions between a metal ion and a compound capable of acting as a ligand are more sensitive to temperature changes than the interactions of PAHs [47]. A procedure for the recycling of the $PdCl_2$/silica material would also be desirable.

3.4 LIQUID CHROMATOGRAPHIC PROPERTIES OF PASHs

In PASH analytical work, liquid chromatographic steps in the normal-phase mode are mainly used for group separations as discussed above. The lack of a sulfur-selective detector in liquid chromatography means that PAHs and PASHs need to be separated before the latter are investigated further by liquid chromatographic methods. Although the mass-selective detector is now gaining ground in HPLC work, it is not truly a sulfur-selective but a mass-selective detector which has to be tuned to expected masses. In Section 3.5.3.3 this detection principle in gas chromatography and some of the problems involved are discussed.

It has been demonstrated that reversed-phase LC has a high separation power for the unsubstituted PASH parent systems [57], although it is doubtful if this is sufficient to make it competitive with capillary GC. It has in fact not found any major applications for the separation of real sample components. An example for a determination of sulfur heterocycles in a solvent refined coal liquid by reversed-phase HPLC on the isolated PASH fraction has been published together with retention data for 38 four- and five-ring PASHs in both normal and reversed phase modes [58].

3.5 GAS CHROMATOGRAPHIC PROPERTIES OF PASHs

An alternative to the class separation of PASH discussed above, which can be coupled with chromatography with a universal detector for identification and quantification of individual compounds, is the gas chromatographic separation of the whole aromatic fraction coupled with a selective detector. This saves time and reduces the risk of contamination and loss of PASH components compared to the isolation methods described above, but, on the other hand, occasionally introduces other problems.

As in all analytical work, no analyte can be quantified without first being un-equivocally identified. Identification of unknown compounds is best carried out by comparison with an authentic standard. Nowadays a fairly good selection is available commercially (see Section 3.9). Lacking standards, a comparison of GC retention data with published data can be of great aid but can only be considered to be a helpful tool and not conclusive proof. The literature abounds with papers in which dozens of PASHs in different kinds of samples have been found by GC–mass spectrometry and whose masses are listed along with a general suggestion of a structure such as 'C_2-dibenzothiophene' or '$C_{16}H_{10}S$-thiarene, methylated'. Since only m/z ratios are involved in this conclusion, a C_2-naphthothiophene could just as well have given rise to the peak labeled 'C_2-dibenzothiophene'. In fact, there are cases where seven GC peaks are labeled 'methyldibenzothiophene' although only four isomers are possible. With the variety of standards now available and with more retention data being published, this undesirable lack of identification of sample components is increasingly avoidable. Similarly, for accurate quantification the compound in question should be available in pure form in order to determine its response factor. Neglect of these factors has been shown to lead to quantification errors for PASHs in real samples of up to 30% [59]. Finally, the frequent observation of coelution also on high-resolution columns must be kept in mind. Before a component is quantified, it must be ascertained that there is no interference from coeluting compounds (see Section 3.5.2).

In the following, we will first look at the gas chromatographic separation of PASHs and then investigate the properties and uses of the various selective detectors that are available for sulfur-selective detection.

3.5.1 Gas chromatographic retention indices of PASHs

A very comfortable way to record and compare retention data is to calculate the retention indices. For aromatic compounds, the most useful form is the following:

$$RI = 100(t_{Rx} - t_{Rz})(t_{Rz+1} - t_{Rz})^{-1} + 100z$$

where t_R is the retention time, x the compound of interest, and z and $z + 1$ the number of aromatic rings in the marker compound eluting immediately prior to and immediately after the compound, respectively. For PAC work, the markers are commonly naphthalene, phenanthrene, chrysene and picene ($z = 2, 3, 4,$ and 5, respectively). This equation can be applied when a universal detector is employed but not for sulfur-selective detectors since they do not visualize the marker compounds mentioned. For such purposes, a similar retention scale was introduced based on benzothiophene, dibenzothiophene, benzonaphtho[2,1-*d*]thiophene and benzophenanthro[2,1-*d*]thiophene as markers for retention indices, RI_S (with S for sulfur), 200, 300, 400, and 500, respectively [39]. It should be noted that in all work with these RI scales, all compounds must elute during the temperature program and not in an isothermal portion of the chromatogram. If one knows the retention times for the markers belonging to the PAH scale, the indices can be recalculated to the PASH scale and vice versa [39].

In the literature, some fairly extensive compilations of retention indices on different stationary phases can be found. In Table 3.1, some of the more important tabulations of information on PASH retention indices are listed.

3.5.2 Stationary phases

The choice of stationary phase is of the greatest importance for a successful chromatographic analysis. What is often overlooked is the severity of coelution of components in complex mixtures. Theoretical calculations show that a chromatogram must be 95%

TABLE 3.1

SOURCES FOR GC RETENTION INDICES OF PASHs

Compounds	Phases	Reference
26 Three- to five-ring PASH	SE-52	[60]
C_1- and C_2-substituted benzothiophenes	DB-5, Carbowax 20M, CP Sil 88	[39]
C_1- and C_2-substituted dibenzothiophenes	Smectic	[61]
69 Two- to four-ring PASHS	DB-5, SB-30-Biphenyl, SP-2331	[28]
80 Three- and four-ring PASHs	DB-5ms, DB-17, SB-Smectic	[62]
68 Benzo- and dibenzothiophenes	HP-PONA	[63]

vacant to ensure with a 90% probability that a peak does not coelute with other components [64]. Such a situation is very rarely attained in reality and coelution should always be presumed to occur in real-world samples. How severe it can be in complex samples was shown for petroleum by using a two-dimensional gas chromatographic separation. Peaks that in the first dimension appeared pure could be separated into ten discrete peaks in the second dimension [65]. No single stationary phase will be able to provide a complete separation of all components in such complex mixtures as are often analyzed for PASHs. This limitation in the separation capabilities of even high-resolution chromatographic systems should not be overlooked, although this question is much too rarely addressed.

Commonly non-polar stationary phases are employed for the analysis of PACs, e.g. such that contain 5% phenyl–95% methylsiloxanes, and consequently these are the phases routinely used also for PASHs. Such phases are thermally sufficiently stable to allow the analysis of large ring systems also. A drawback is that several important compounds are not resolved on them. The six methylbenzothiophenes elute in four peaks, the four methyldibenzothiophenes in three (with 2- and 3-methyldibenzothiophene coeluting). Dibenzothiophene coelutes with the isomeric naphtho[1,2-*b*]thiophene. In a sample from the workplace air of an aluminum smelter, about 80% of the PASH components were said to be resolved (at a resolution factor of 0.8) on a 50-m column with a 5% phenyl–95% methylsiloxane phase with 16% of the components showing marked coelution [66]. However, it was not stated what concentration levels were considered for this calculation. Because of this coelution problem, several other phases have been investigated for PASH separations and it seems that phases containing 50% phenyl–50% methylsiloxane show a better separation but retain many of the attributes of the somewhat less polar phases with only 5% phenyl groups, such as high temperature stability and good resistance to oxidation. Such phases will separate dibenzothiophene from the three isomeric naphthothiophenes [62] and the four methyldibenzothiophenes are resolved from each other. Likewise it shows a better separation of dimethyldibenzothiophenes and 13 of 15 tested isomers (the sixteenth isomer, 1,9-dimethyldibenzothiophene, was not available) were resolved. On a 5% phenylsiloxane, only seven isomers were usefully resolved. This separation is illustrated in Fig. 3.7. It can therefore be expected that in the future such phases will receive much more attention for this kind of analysis.

More polar phases such as those based on cyanopropyl siloxanes also show a better resolution of many PASHs. The four isomeric three-ring PASHs as well as

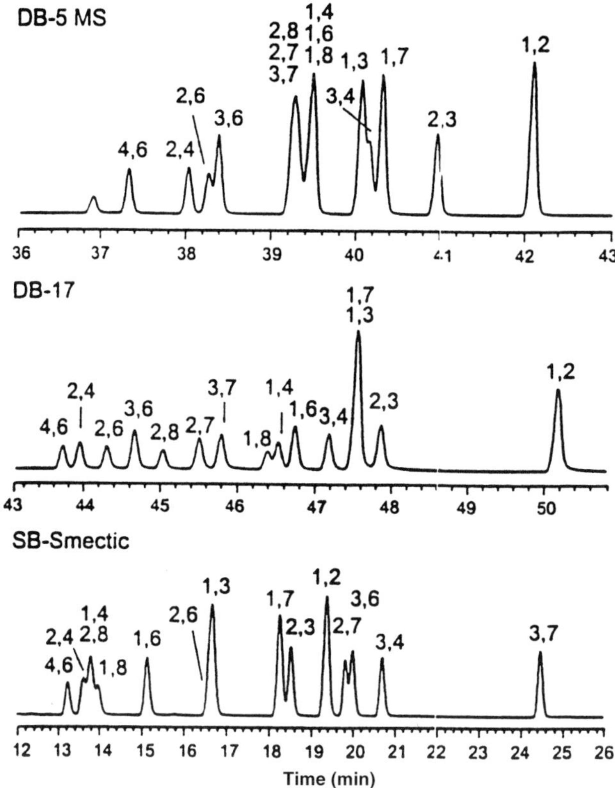

Fig. 3.7. Gas chromatographic separation of 15 (of the 16 possible) isomers of dimethyldibenzothiophene on the three stationary phases DB-5MS, DB-17 and SB-Smectic. Numbers identify the position of the methyl groups in the molecule. (From Mössner et al. [62]. Reproduced with permission of Elsevier Science.)

the methyldibenzothiophenes are well resolved on them [28]. A terminal (but not an internal) thiophenic ring considerably increases the retention index on cyanopropyl and thus improves the selectivity. However, due to the lower thermal stability of these phases, compounds that need higher elution temperatures will be hard to analyze on them.

Unique selectivities are often seen on smectic liquid crystalline phases. A good correlation is seen between the easily calculated length to breadth ratio (L/B) and the retention indices of many classes of compounds; in general, the larger this ratio is, the longer the retention time. For PASHs, a stationary phase containing a solution of liquid crystals in SE-52 was shown to give a considerably better separation of the four-ring (but not the five-ring) PASHs than pure SE-52 [60]. More detailed investigations of alkylated dibenzothiophenes showed that 12 out of 17 C_2-dibenzothiophenes were separated on a commercial column (illustrated in Fig. 3.7) and this result was applied to a crude oil sample [61]. Later, more extensive data showed the smectic phase to be superior to the phenyl–methylsiloxane phases in overall separation of the 30 methyl-derivatives of the four-ring PASHs benzonaphtho[1,2-*d*]-, benzonaphtho[2,1-*d*]-, and benzonaphtho[2,3-*d*]thiophene [62]. A commercial MPMS phase has the capability to resolve several PASH pairs in SRM 1648 air particulate that are not resolved on the commonly used isotropic stationary phases [67].

Despite its powerful separation characteristics, smectic phases have not found much routine use due to such drawbacks as a lower temperature limit than phenyl–methyl-siloxanes and a — often frustrating — change in column selectivity with use [68].

3.5.3 Selective detection in gas chromatography

Selective detectors of importance for PASH work include those that respond to the element sulfur and mass-selective detectors. In the former group, we find several commercial detectors, such as the flame photometric detector (FPD), the electrolytic detector (ELD), the sulfur chemoluminescence detector (SCD) and the atomic emission detector (AED). Although the detection characteristics of the other detectors may be competitive [69], it seems that only the FPD and the AED have gained much practical importance of the detectors listed and only these two will therefore be treated here. For an evaluation of the performance of six detectors for sulfur determination, see [70]. It is noteworthy that ASTM method 5623-94 (sulfur compounds in light petroleum liquids by gas chromatography and sulfur selective detection) recommends the use of either an AED or an SCD, but not the FPD [71].

3.5.3.1 The flame photometric detector (FPD)

The flame photometric detector has been in use for over 30 years and is still frequently relied on in sulfur-selective detection [72]. It is based on the combustion of the analytes from the capillary column in a hydrogen-rich flame, whereby a part of the sulfur is reduced to the S_2 molecule in an excited state. The excess energy can be emitted as light in a broad band. Undesirable wavelengths are removed through a filter and the remaining light falls on a photomultiplier and a signal is registered. The signal is not a linear function of the sulfur concentration in the sample but obeys an exponential law whose exponent can depend on many factors but ideally should equal 2. Added to this is the possibility of quenching of the emitted light due to large amounts of coeluting (but sulfur-free) compounds. Therefore this detector should be used for quantification only when very careful attention is paid to all the different influences on the detection [73,74], although this is frequently not the case. These drawbacks have led to the development and commercialization of a novel FPD design, called the pulsed FPD [75]. Its response is still quadratic, but the minimum detectable quantity (below 1 pg S/s) is lower than that of the traditional FPD and quenching phenomena are said not to be observed. Also beneficial is the equimolar response of different sulfur compounds [75].

3.5.3.2 Atomic emission detector (AED)

Sulfur is an element with an excellent minimum detectable quantity in atomic emission detection, namely in the low picogram range. Furthermore, the selectivity versus carbon is very high, ca $3.5 x 10^4$ [71] and the linear range, independent of sulfur species, is on the order of 10^4 [71]. Since several elements can be monitored in one chromatographic run (see Fig. 3.2 with simultaneous carbon and sulfur selective detection), the AED is a multi-element selective detector. In the AED, the analytes are fed from the separation column into a microwave induced helium plasma and atomized. Some of the atoms gain enough energy to become excited and when giving off this excess energy, light at certain wavelengths that are characteristic for the element is emitted.

Despite being considerably more expensive than the other detectors and costing more to run as well as putting higher demands on the user, it has found such wide acceptance for sulfur determinations that applications for sulfur are among the most numerous for any element with atomic emission detection.

3.5.3.3 Mass-selective detection

Mass-selective detectors (MSD) can be said to be universal as well as selective at the same time. Unless the PASHs have been separated from other components of the sample, an MSD will typically be run in the selected ion mode (SIM) for such compounds. This means that those m/z numbers that are typical for the PASH being looked for will be monitored to the exclusion of all others. Thus a MSD does not specifically respond to the element sulfur in the way the sulfur-selective detectors described above do but rather to a predetermined molecular mass.

One can therefore not speak of a selectivity of an MSD for PASHs but only of a selectivity for a certain ion. The degree of selectivity for PASHs therefore depends on what other sample components give rise to the signal monitored. For instance, dibenzothiophene would be found at 184 but this is also the m/z for tetramethylnaphthalenes and this can cause interference problems since the GC elution range is similar for these compounds. An example that well illustrates this non-selectivity for PASHs in a real sample when the MSD is used is shown in Fig. 3.8 [13] where only the last peak in the cluster is the sulfur-containing aromatic compound dibenzothiophene. It is true throughout all numbers of aromatic rings that a PASH has the same molecular mass as the PAH with one ring less but with four side-chain carbons more.

Electron impact (EI) ionization is nearly always employed but there are indications that other ionization techniques might give better results. Chemical ionization (CI) with isobutane was found to lead to a better signal to noise (S/N) ratio for alkylated dibenzothiophenes than for other petroleum components because of their higher proton affinity versus isobutane [76]. For dibenzothiophene itself, the S/N ratio declined to half but it improved threefold for 1,3-dimethyldibenzothiophene. Coupling CI with the MS/MS technique led to still much better S/N values for all dibenzothiophenes to as high as 35 times better for 1,3-dimethyldibenzothiophene than using EI alone. Negative CI with ammonia has been reported to strongly enhance the signals for sulfur compounds, including thiophenes, relative to hydrocarbons in gasoline [77].

3.6 HETEROATOM-SUBSTITUTED PASHs

In some samples, PASHs substituted with hydroxyl or amino groups can be found. Aminodibenzothiophenes were found in a coal liquid [78]. The compounds were isolated from the solvent-refined coal heavy distillate through adsorption chromatog-

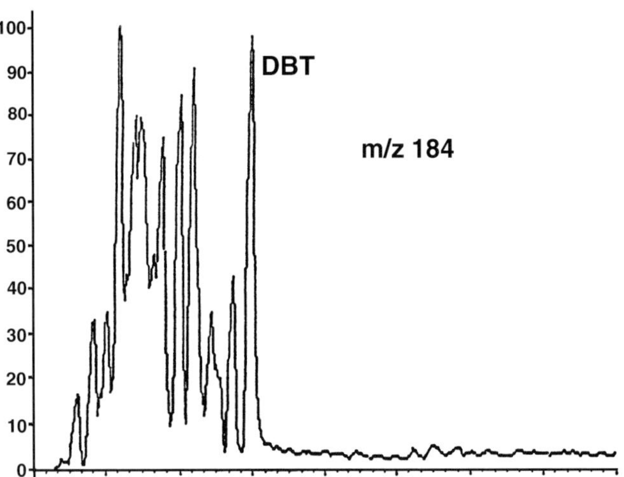

Fig. 3.8. Gas chromatographic determination of the aromatic fraction of a North Sea crude oil. Mass selective detection at m/z 184. The last major peak is dibenzothiophene, the cluster immediately prior to it are signals from C_4-substituted naphthalenes. From Tibbetts and Large [13]. Reproduced with permission of John Wiley and Sons.)

raphy on alumina and silica and extraction with sulfuric acid. A 25% biphenyl methylpolysiloxane stationary phase provided baseline resolution of the four aminodibenzothiophenes and several alkylsubstituted aminodibenzothiophenes. The analysis of these compounds is relevant because of the significantly higher mutagenicity of 2- and 3-aminodibenzothiophene than that of benzo[*a*]pyrene [78].

3.7 COMPOUNDS WITH ONE OTHER HETEROCYCLIC RING

Especially in fossil fuel-related materials, polycyclic aromatic compounds have been detected that, apart from benzo rings, possess both a thiophene ring and an aromatic ring containing another hetero element than sulfur, usually nitrogen [18]. When the aminodibenzothiophenes in a solvent-refined coal heavy destillate were analyzed [78,79], several isomers of azadibenzothiophenes were also isolated and their general structure confirmed through MS data. In a commercial anthracene oil, azadibenzothiophenes, azanaphthothiophenes as well as their C_1- and C_2-alkylated derivatives were found in addition to C_0- and C_1-azaphenanthro[4,5-*bcd*]thiophenes and an azanaphthobenzothiophene. The concentrations ranged up to 250 ppm for individual PASNHs [80]. In all cases, the mixed sulfur and nitrogen heterocycles were separated from the nitrogen-free components through extraction with sulfuric acid.

Compounds with both sulfur and oxygen in different rings seem to occur more rarely, but have been identified in crude oils and syncrudes [18].

3.8 PASHs IN REFERENCE MATERIALS

No reference material has been brought on the market explicitly for the analysis of PASHs but for some such compounds certified values have been issued together with those for PAHs. The Joint Research Centre of the European Commission offers two BCR reference materials with a certified amount of benzo[*b*]naphtho[2,1-*d*]thiophene together with several PAHs. They are CRM 088 Sewage sludge (0.42 μg/kg) and CRM 524 Contaminated industrial soil (3.8 μg/kg). The two harbor sediments HS-3B and HS-4B from the National Research Council Canada have certified concentrations of 1.19 and 0.11 mg/kg, respectively, for dibenzothiophene.

The materials characterized in greatest detail are those from the National Institute of Standards and Technology (NIST) in the USA that offers three standard reference materials (SRM) of interest, namely SRM 1597 'Complex Mixture of Polycyclic Aromatic Hydrocarbons from Coal Tar', SRM 1582 'Petroleum Crude Oil' and SRM 1580 'Shale Oil'.

SRM 1597 is a combustion-related aromatic mixture from a crude coke oven tar. The concentrations for 12 PAHs have been certified (results from two independent analytical methods agree) and non-certified concentration values (based on results from one analytical method) are given for two PASHs, benzothiophene and dibenzothiophene. A value of 27.3 μg/g was assigned to benzothiophene and 23.0 μg/g to dibenzothiophene after determination by GC/FID [81]. These values were later questioned [59] and it was shown that the use of wrong relative response factors (RRF) for both compounds led to false quantifications and that the concentration of benzothiophene was more likely 35.8 μg/g. In the case of dibenzothiophene the coelution with naphtho[1,2-*b*]thiophene on the non-polar stationary phase used (DB-5) complicated the determination further. More polar stationary phases, like cyanopropyl siloxanes (see Section 3.5.2), led to a complete resolution of those two three-ring PASHs and with the corrections for this and the RRF a concentration of 18.2 μg/g was found [59]. Ten more PASHs, including methylbenzothiophenes and -dibenzothiophenes, were quantified using the atomic emission detector [59]. Later work by NIST has confirmed the corrections mentioned and through the use of several stationary phases (see below for SRM 1582) the quantification of the four three-ring PASHs, the four methyldibenzothiophenes and the three benzonaphthothiophenes succeeded [82].

A very complex PASH mixture is found in SRM 1580 'Shale Oil' with a predominance of substituted two-ring PACs. NIST has not provided concentration values for any PASHs but a carefully designed workup including separation of two- and three-ring fractions and an oxidation of the two-ring compounds to their sulfones, analogous to that in Fig. 3.4, made the quantification of eleven PASHs by GC/AED possible [59]. The total concentration of the three naphthothiophenes is high enough to equal that of dibenzothiophene.

The crude oil SRM 1582 contains a certified amount of 33 μg/g of dibenzothiophene [82]. The methyldibenzothiophenes could also be determined through chromatography on a biphenyl stationary phase [59]. A very detailed analysis of the PASHs of this sample has been carried out and 57 PASHs were quantified. This was possible only through an initial HPLC separation on an aminopropyl phase of the aromatics into four

References pp. 96–98

fractions depending on the number of aromatic carbon atoms followed by GC analysis on three stationary phases, a non-polar, a semi-polar and a smectic phase [82].

Other SRMs have been investigated for PASH although the data are not certified. PASHs from SRM 1648 air particulate were separated on a liquid crystalline column with a flame photometric detector and shown to consist mainly of three- and four-ring heterocycles [67].

3.9 COMMERCIAL SOURCES OF PASHs

Access to reference compounds is a necessity for the analyst since they can be used to check an analytical procedure for extraction recoveries and the response factors, retention times, spectral features etc. of the individual analytes. From the large chemical supply houses only a few unsubstituted and methylated PASHs are available. Two sources provide a good selection of reference solutions of PASHs for analytical work. A German group [83] offers toluene solutions of 53 compounds, out of which 22 are benzothiophenes, 19 dibenzothiophenes and other three-ring systems and 12 other compounds, mainly larger ring systems. A Norwegian company [84] advertizes 32 PASHs in isooctane solution, viz. 15 benzothiophenes, 12 dibenzothiophenes and five larger systems.

3.10 REFERENCES

1   G. Grimmer, J. Jacob and K.-W. Naujack, Fresenius Z. Anal. Chem., 314 (1983) 29–36.
2   J.T. Andersson, Int. J. Environ. Anal. Chem., 48 (1992) 1–15.
3   S. Bobinger and J.T. Andersson, Chemosphere, 36 (1998) 2569–2579.
4   M. Imanaka, M. Kadota, N. Ogawa, K. Kumashiro and T. Mori, J. Jpn. Soc. Nutr. Food Sci. 45 (1992) 61-70; Chem. Abstr. 117 (1992) 24858.
5   E.R. Schmid, G. Bachlechner, K. Varmuza and H. Klus, Fresenius Z. Anal. Chem., 322 (1985) 213–219.
6   M. Wobst, H. Wichmann and M. Bahadir, Chemosphere, 38 (1999) 1685–1691.
7   R.A. Scanlan, S.G. Kayser, L.M. Libbey and M.E. Morgan, J. Agric. Food Chem., 21 (1973) 673–675.
8   K.S. Betts, Environ. Sci. Technol., 34 (2000) 161A.
9   J. Jacob, Sulfur analogues of polycyclic aromatic hydrocarbons (thiaarenes), Cambridge University Press, Cambridge, 1990.
10   M. Radke and H. Willsch, Geochim. Cosmochim. Acta, 58 (1994) 5223–5244.
11   S. Chakhmakhchev, M. Suzuki and K. Takayama, Org. Geochim., 26 (1997) 483–490.
12   J.D. Payzant, T.W. Mojelsky and O.P. Strausz, Energy Fuels, 3 (1989) 449–454.
13   P.J.C. Tibbetts and R. Large, in: G.B. Crump (Ed.), Petroanalysis '87, John Wiley and Sons, New York, 1988, pp. 45–57.
14   K.G. Kropp, J.T. Andersson and P.M. Fedorak, Environ. Sci. Technol., 31 (1997) 1547–1554.
15   F. Traulsen, J.T. Andersson and M.G. Ehrhardt, Anal. Chim. Acta, 392 (1999) 19–28.
16   B.E. McCarry, L.M. Allan, A.E. Legzdins, J.A. Lundrigan, C.H. Marvin and D.W. Bryant, Polycyclic Arom. Comp. , 11 (1996) 75–82.
17   A.A. Herod, in: A.H. Neilson (Ed.), PAHs and Related Compounds, The Handbook of Environmental Chemistry, 3-1, Springer, Berlin, 1998, pp. 271–323.
18   C.-D. Czogalla and F. Boberg, Sulfur Rep., 3 (1983) 121–167.
19   K.G. Kropp and P.M. Fedorak, Can. J. Microbiol., 44 (1998) 605–622.

20　S.A. Wise, in: A. Björseth (Ed.), Handbook of Polycyclic Aromatic Hydrocarbons, Dekker, New York, 1983, pp. 183–256.

21　J.T. Andersson, Fresenius Z. Anal. Chem., 326 (1987) 425–433.

22　F. Barthou and Y. Dreano, J. High Resolut. Chromatogr. Chromatogr. Commun., 11 (1988) 706–712.

23　J. Beens and R. Tijssen, J. Microcolumn Separations, 7 (1995) 345–354.

24　I.L. Davis, K.D. Bartle, P.T. Williams and G.E. Andrews, Anal. Chem., 60 (1988) 204–209.

25　G. Becker, U. Nilsson, A. Colmsjö and C. Östman, J. Chromatogr., A 826 (1998) 57–66.

26　J.S. Thomson, P.L. Grizzle and J.W. ReynoldsReport NIPER-324 (1988), Chem. Abstr., 110 (1989) 117884.

27　A. Milenkovic, E. Schulz, V. Meille, D. Loffreda, M. Forissier, M. Vrinat, P. Sautet and M. Lemaire, Energy Fuels, 13 (1999) 881–887.

28　J.T. Andersson and B. Schmid, J. Chromatogr. A, 693 (1995) 325–338.

29　H.V. Drushel and A.L. Sommers, Anal. Chem., 39 (1967) 1819–1829.

30　C. Willey, M. Iwao, R.N. Castle and M.L. Lee, Anal. Chem., 53 (1981) 400–407.

31　P.J. Arpino, I. Ignatiadis and G. de Rycke, J. Chromatogr., 390 (1987) 329–348.

32　R.C. Kong, M.L. Lee, M. Iwao, Y. Tominaga, R. Pratap, R.D. Thompson and R.N. Castle, Fuel, 63 (1984) 702–708.

33　D.S. Rao, Abstract of Papers, National Meeting 137, American Chemical Society, Washington, DC, 1960, 26 O.

34　J. Bundt, W. Herbel and H. Steinhart, J. High Resolut. Chromatogr., 15 (1992) 682–685.

35　M. Radke, D.H. Welte and H. Willsch, Org. Geochem., 10 (1986) 51–63.

36　P. Jadaud, M. Caude, R. Rosset, X. Duteurtre and J. Henoux, J. Chromatogr., 464 (1989) 333–342.

37　J.T. Andersson, J. High Resolut. Chromatogr. Chromatogr. Commun., 7 (1984) 334–335.

38　P.M. Fedorak and J.T. Andersson, J. Chromatogr., 591 (1992) 362–366.

39　J.T. Andersson, J. Chromatogr., 354 (1986) 83–98.

40　L.M. Allan, B.E. McCarry and C. Li, Polycyclic Aromatic Compounds, submitted for publication.

41　J.- M Ruiz, B.M. Carden, L.J. Lena, E.J. Vincent and J.-C. Escalier, Anal. Chem., 54 (1982) 688–691.

42　J.D. Payzant, D.S. Montgomery and O.P. Strausz, Org. Geochem., 9 (1986) 357–369.

43　J.D. Payzant, T.W. Mojelsky and O.P. Strausz, Energy Fuels, 3 (1989) 449–454.

44　B. Vonach and G. Schomburg, J. Chromatogr., 149 (1978) 417–430.

45　W.F. Joyce and P.C. Uden, Anal. Chem., 55 (1983) 540–543.

46　J.T. Andersson, Fresenius Z. Anal. Chem., 327 (1987) 327.

47　J.W. Vogh, J.W. Reynolds Report, NIPER-41, Natl. Inst. Pet. Energy Res., 1985; Chem. Abstr. 103 (1985) 162920.

48　T. Kamimai and A. Matsunaga, Anal. Chem., 50 (1978) 268–270.

49　W.L. Orr, Anal. Chem, 38 (1966) 1558–1562.

50　J.W. Vogh and J.E. Dooley, Anal. Chem., 47 (1975) 816–821.

51　K.D. Gundermann, H.P. Ansteeg and A. Glitsch, in: Proceedings of the International Conference on Coal Science, Pittsburgh, PA, 1983, p. 63.

52　M. Nishioka, R.M. Campbell, M.L. Lee and R.N. Castle, Fuel, 65 (1986) 270–273.

53　J.T. Andersson, Anal. Chem., 59 (1987) 2207–2209.

54　M. Nishioka, Energy Fuels, 2 (1988) 214–219.

55　U. Pyell, S. Schober and G. Stork, Fresenius J. Anal. Chem., 359 (1997) 538–541.

56　G. Félix, M. Liu, P. Ithurralde, J. Goupy, J.M. Colin and M. Bouquet, Analysis, 21 (1993) 153–156.

57　S.A. Wise, in: A. Björseth and T. Ramdahl (Eds.), Handbook of Polycyclic Aromatic Hydrocarbons, Vol. 2, Dekker, New York, 1985, pp. 113–191.

58　S.A. Wise, R.M. Campbell, W.E. May, M.L. Lee and R.N. Castle, in: M. Cooke and A.J. Dennis (Eds.), Polynuclear Aromatic Hydrocarbons: Formation, Metabolism and Measurement, Battelle Press, Columbus, OH, 1983, pp. 1247–1266.

59　B. Schmid and J.T. Andersson, Anal. Chem., 69 (1997) 3476–3481.

60　R.C. Kong, M.L. Lee, Y. Tominaga, R. Pratap, M. Iwao, R.N. Castle and S.A. Wise, J. Chromatogr. Sci., 20 (1982) 502–510.

61　H. Budzinski, P. Garrigues and J. Bellocq, J. Chromatogr., 590 (1992) 297–303.

62   S.G. Mössner, M.J. Lopez de Alda, L.C. Sander, M.L. Lee and S.A. Wise, J. Chromatogr. A, 841 (1999) 207–228.

63   G.A. Depauw and G.F. Froment, J. Chromatogr. A, 761 (1997) 231–247.

64   J.M. Davis and J.C. Giddings, Anal. Chem., 55 (1983) 418–424.

65   R.B. Gaines, G.S. Fryisinger, M.S. Hendrick-Smith and J.D. Stuart, Environ. Sci. Technol., 33 (1999) 2106–2112.

66   G. Becker, U. Nilsson, A. Colmsjö and C. Östman, J. Chromatogr. A, 826 (1998) 57–66.

67   P. Fernández, C. Porte, D. Barceló, J.M. Bayona and J. Albaigés, J. Chromatogr., 456 (1988) 155–164.

68   L.C. Sander, M. Schneider, S.A. Wise and C. Woolley, J. Microcol. Sep., 6 (1994) 115–125.

69   S.E. Eckert-Tilotta, S.B. Hawthorne and D.J. Miller, J. Chromatogr., 591 (1992) 313–323.

70   H.P. Tuan, H.-G.M. Janssen, C.A. Cramers, E.M. Kuiper-van Loo and H. Vlap, J. High Resol. Chromatogr., 18 (1995) 333.

71   B.D. Quimby, D.A. Grudoski and V. Giarrocco, J. Chromatogr. Sci., 36 (1998) 435–443.

72   G. Guiochon and C.L. Guillemin, Quantitative Gas Chromatography for Laboratory Analyses and On-line Process Control, Elsevier, Amsterdam, 1988, p. 463–466.

73   R.E. Rebbert, S.N. Chesler, F.R. Guenther and R.M. Parris, J. Chromatogr., 284 (1984) 211–217.

74   P. Burchill, A.A. Herod and E. Pritchard, J. Chromatogr., 242 (1982) 51–64.

75   A. Amirav and H. Jing, Anal. Chem., 67 (1995) 3305–3318.

76   J.B. Edwards and S.T. Fannin, The Analysis of Trace Level Components in Petrochemical Matrices by GC–MS/MS, Finnigan, Austin, TX, 1997.

77   J. Guieze, G. Devant and D. Loyaux, Int. J. Mass Spectrom., 46 (1983) 313–316.

78   M. Nishioka, R.M. Campbell, W.R. West, P.A. Smith, G.M. Booth, M.L. Lee, H. Kudo and R.N. Castle, Anal. Chem., 57 (1985) 1868–1871.

79   M. Nishioka, P.A. Smith, G.M. Booth, M.L. Lee, H. Kudo, D.R. Muchiri, R.N. Castle and L.H. Klemm, Fuel, 65 (1986) 711–714.

80   P. Burchill, A.A. Herod and E. Pritchard, J. Chromatogr., 242 (1982) 65–76.

81   S.A. Wise, L.R. Hilpert, R.E. Rebbert, L.C. Sander, M.M. Schantz, S.N. Chesler and W.E. May, Fresenius Z. Anal. Chem., 332 (1988) 573–582.

82   S.G. Mössner and S.A. Wise, Anal. Chem., 71 (1999) 58–69.

83   www.uni-muenster.de/Chemie/AC/anders/Pash/welcome.html

84   www.chiron.no

CHAPTER 4

Polycyclic aromatic hydrocarbon (PAH) metabolites

A. Höner

Technische Universität Berlin, Strasse des 17. Juni 135, D-10623 Berlin, Germany

4.1 INTRODUCTION

Polycyclic aromatic hydrocarbons (PAHs) represent an important class of environmental pollutants and have gained special attention because some of them are strong mutagens and carcinogens [4–6]. They are present in fossil fuels and are formed upon incomplete combustion and pyrolysis of organic matter. PAHs are generated for example in forest fires and volcanic eruptions and are natural constituents of crude oil and other petrochemical products. In addition they are formed in industrial facilities such as coke plants and gasworks, in domestic heating, and in traffic. Furthermore they are one of several classes of carcinogenic chemicals in tobacco smoke and are also present in food. Their widespread generation respectively release, both from natural and anthropogenic sources, together with distribution and transport processes, is responsible for their ubiquitous occurrence. Humans are exposed to polycyclic aromatic hydrocarbons from environmental, occupational, medicinal (e.g. coal tar treatment) and from dietary sources (e.g. broiled and smoked foods) [5–14].

As already mentioned, some polycyclic aromatic hydrocarbons are potent carcinogens for humans and experimental animals [15–17]. However, carcinogenic properties vary widely for individual PAHs. In addition synergistic effects play an important role and may enhance the adverse effect on living organisms [18]. Therefore analytical determinations of polycyclic aromatic hydrocarbons usually include a (representative) number of analytes rather than focusing on one or two single compounds. The US Environmental Protection Agency for example classifies 16 PAHs as environmentally relevant [19]. More than half of them are carcinogenic and/or mutagenic. A large number of publications dealing with PAHs is focused on these so-called EPA-PAHs. Among these benzo[*a*]pyrene is, due to its high carcinogenic potential, often used as a marker compound. Concerning the analytical determination of parent PAHs please refer to Chapter 2, written by H.K. Lee.

4.2 METABOLISM

The carcinogenic potential of polycyclic aromatic hydrocarbons, which are mostly non-toxic as such, is caused by enzymatic activation. As parent PAHs are lipophilic substances, they need biotransformation to water-soluble derivatives before they can be efficiently excreted from the mammalian organism. This detoxification process converts them to a variety of reactive metabolites amongst which epoxides have a special importance.

The enzymes involved are classified in two categories: phase 1 enzymes, which catalyze oxidative reactions, and phase 2 enzymes, which catalyze conjugative reactions of oxidized PAHs with endogenous compounds such as sulfuric acid, glucuronic acid and glutathione.

Phase 1 metabolism involves the following steps. At first the polycyclic aromatic hydrocarbons are transformed by cytochrome P450-dependent monooxygenases to various arene oxides which then may rearrange to form phenols. Alternatively, epoxide hydrases catalyze the addition of water what may result in *trans*-dihydrodiols, the proximate carcinogens of polycyclic aromatic hydrocarbons. Secondly, primary metabolites may be further oxidized to e.g. *trans*-dihydrodiol epoxides, the ultimate carcinogens. The latter may covalently bind to nucleophilic sites of DNA.

In phase 2 metabolism phenols, phenol dihydrodiols, quinones, and dihydrodiols are conjugated to form sulfates or glucuronides. Spontaneous formation of conjugates is possible as well. The resulting sulfates, glucuronides or protein adducts are detoxifying products of hazardous PAH metabolites. DNA adducts, however, can lead to mutation and initiation of cancer cells if they are not repaired [1,11,20–25].

The various PAH metabolites may be excreted from the organism either as free or as conjugated compounds.

Structure–activity studies have revealed that the tumor initiating activities of dihydrodiol epoxides derived from hydrocarbons with bay regions, such as benzo[*a*]pyrene, with hindered bay regions, such as 5,6-dimethylchrysene, or with fjord regions, such as benzo[*g*]chrysene, are high [26,27]. Fig. 4.1 shows the structure of the mentioned examples.

In the following section the metabolic activation of benzo[*a*]pyrene will be discussed in more detail. Fig. 4.2 shows some benzo[*a*]pyrene metabolites.

Metabolic activation of benzo[*a*]pyrene is shown in Fig. 4.3. In the initial step an epoxide is formed, which is successively hydrolyzed to form a diol. In the following reaction step dihydroxy-epoxides can be formed, which in turn may decompose to

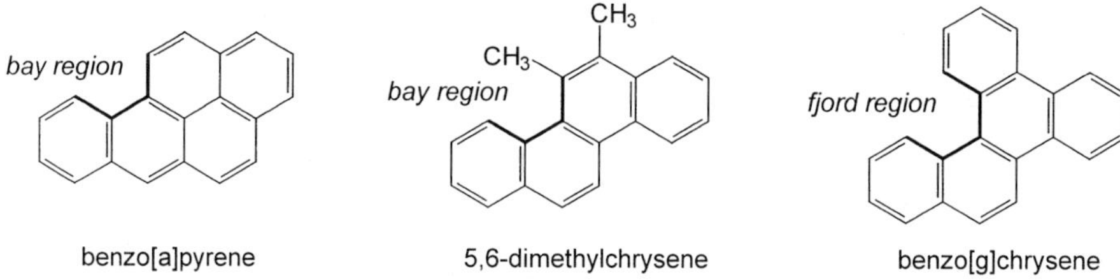

Fig. 4.1. Structure of benzo[*a*]pyrene, 5,6-dimethylchrysene, and benzo[*g*]chrysene.

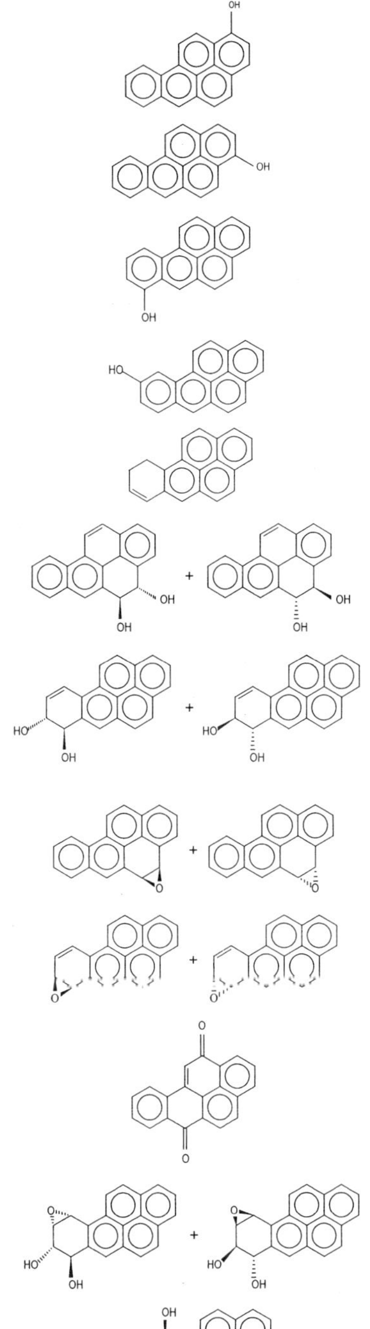

1-hydroxybenzo[a]pyrene

3-Hydroxybenzo[a]pyrene

7-Hydroxybenzo[a]pyrene

9-Hydroxybenzo[a]pyrene

9,10-Dihydrobenzo[a]pyrene

(±) Benzo[a]pyrene-*trans*-4,5-dihydrodiol

(±) Benzo[a]pyrene-*trans*-7,8-dihydrodiol

(±) Benzo[a]pyrene-4,5-dihydroepoxide

(±) Benzo[a]pyrene-7,8-dihydroepoxide

Benzo[a]pyrene-6,12-dion

(±) Benzo[a]pyrene-7,8-dihydrodiol-9,10-dihydroepoxide (anti)

Benzo[a]pyrene-r-7, *trans*-8,9,-cis-10-tetrahydrotetrol

Fig. 4.2. Structure of some benzo[*a*]pyrene metabolites.

References pp. 117–121

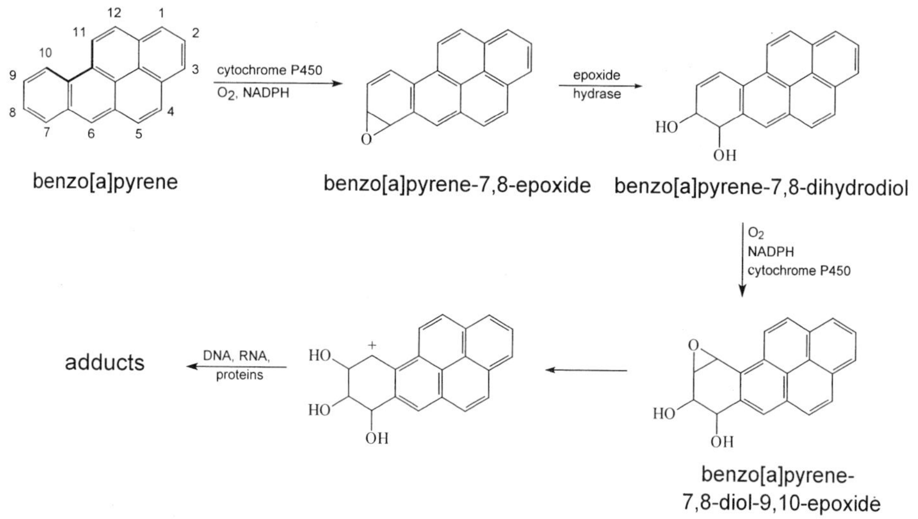

benzo[a]pyrene

benzo[a]pyrene-7,8-epoxide

benzo[a]pyrene-7,8-dihydrodiol

benzo[a]pyrene-
7,8-diol-9,10-epoxide

adducts

Fig. 4.3. Metabolic activation of benzo[*a*]pyrene [28].

strong electrophilic compounds capable to react with e.g. DNA. These DNA adducts can then, in case no reparation mechanisms start, cause tumorigenesis.

The figure shown above gives no information on the configuration of the molecules. However, each reaction step can lead to products with a different stereochemistry, which in turn complicates the analysis of the products. Fig. 4.4 shows the stereochemistry of the enzymatic formation of benzo[*a*]pyrene dihydrodiol epoxides more closely.

The shown benzo[*a*]pyrene-*anti*-7,8-dihydrodiol-9,10-epoxide is the ultimate carcinogen of benzo[*a*]pyrene. However, it is only one compound among several possible isomers which are formed upon metabolic activation of benzo[*a*]pyrene. In general every single PAH is transformed to a number of metabolites. These metabolites show structural similarities and sometimes differ from one another only in their stereochemistry.

Furthermore polycyclic aromatic hydrocarbons do not occur isolated. Consequently humans are usually exposed to quite complex mixtures of pollutants. This again leads to the formation of a huge number of different metabolites. A detailed analysis of all metabolites formed by living organisms is therefore almost impossible. For this reason it is necessary to choose, besides a number of relevant parent PAHs, suitable metabolites.

4.3 BIOLOGICAL MONITORING

As already mentioned before, human beings are usually exposed to a complex mixture of different polycyclic aromatic hydrocarbons. In view of serious health concerns, especially due to the carcinogenic properties of PAHs, it is therefore necessary to monitor the occurrence of polycyclic aromatic hydrocarbons in the environment. However, measurement of external (environmental and occupational) PAH concentrations does not necessarily reflect the internal exposure and thus the individual risk. In addition personal

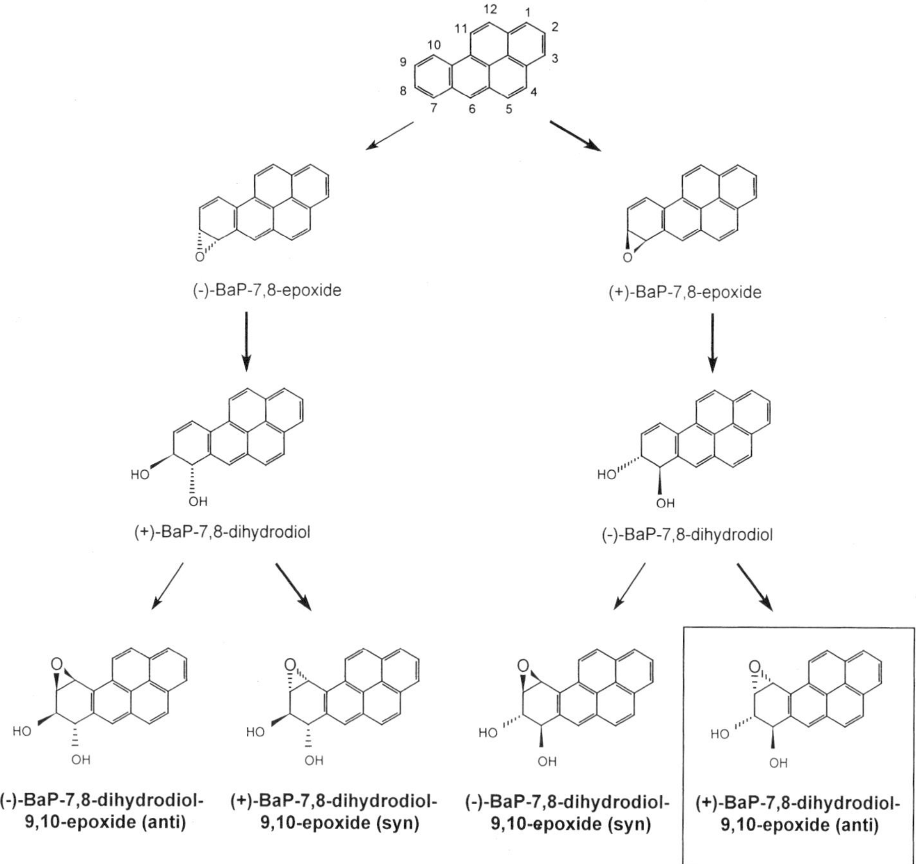

Fig. 4.4. Enzymatic formation of 4 benzo[*a*]pyrene-7,8-dihydrodiol-9,10-epoxides [6].

habits such as diet and smoking strongly influence the uptake of PAHs. It is therefore advisable to monitor certain biomarkers, the analytes as such or metabolites of them present in e.g. urine, blood or exhaled air instead of (or in addition to) parent PAHs in the surroundings [29,30]. This so-called biological monitoring and biochemical effect monitoring was recently reviewed by Angerer et al. in view of exposure to polycyclic aromatic hydrocarbons [11]. In the first part of their survey they assessed different analytical methods for measurement of DNA and protein adducts. Here they came to the conclusion that techniques such as ^{32}P postlabeling, immunoassays, and synchronous fluorescence spectroscopy lack specificity and accuracy. For this reason results from different methods are often not comparable which in turn makes data interpretation difficult. The second part of the review dealt with chromatographic separations of PAH metabolites and lists pyrene and phenanthrene metabolite concentrations determined in several occupational studies. Another recent article of Jongeneelen reviewed different analytical techniques for determination of biomarkers for PAH exposure with regard to

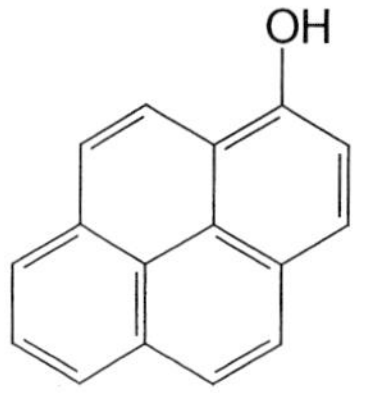

OH

1-hydroxypyrene

Fig. 4.5. Structure of 1-hydroxypyrene.

their applicability as routine methods [31]. He concluded that urinary 1-hydroxypyrene is currently the best suited biomarker because sensitive and reliable analysis methods are available at relatively low costs. Method details for its HPLC determination are given in Section 4.3.2.1 below.

Biological monitoring usually focuses on analysis of polycyclic aromatic hydrocarbon metabolites or their respective conjugates in urine samples. Parent PAHs, however, differing widely in their physical, chemical and toxicological properties, are transformed to an even more complex mixture of metabolites. As the larger PAH metabolites (more than three rings) are mainly excreted via the faeces, urinary concentrations are in general quite low. In addition only a limited number of metabolites are commercially available. These are the reasons why biomonitoring is often restricted to one or only few metabolites which occur in a comparatively high concentration and can be identified unambiguously. The most popular among them is 1-hydroxypyrene, the main metabolite of pyrene [31–34]. Fig. 4.5 shows the structure of 1-hydroxypyrene.

4.3.1 Sample preparation

To enable excretion of the water-insoluble polycyclic aromatic hydrocarbons from living cells, they are metabolized to oxygenated products. These metabolites are then excreted either as free or as conjugated compounds in the form of glucuronides and sulfates. Two examples are shown in Fig. 4.6. Sometimes the latter are analyzed as such, but in most cases they are hydrolyzed prior to analysis. This is mostly done enzymatically with β-glucuronidase/aryl sulfatase. Common reaction conditions are 16 h at 37°C in acidic medium. Another possibility to split the conjugates is acidic hydrolysis using

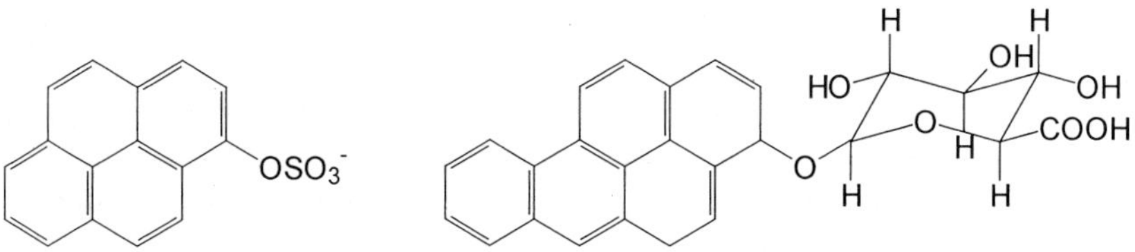

Fig. 4.6. Structure of 1-OH-pyrene sulfate and 3-OH-benzo[*a*]pyrene glucuronide.

hydrochloric acid at elevated temperatures such as 90°C. The reaction time is with approximately 60 min significantly lower than with enzymatic hydrolysis. The first alternative, though, is commonly preferred.

In case the urine samples are not analyzed directly, they are mostly stored frozen at approx. −20°C, sometimes at −70°C or around 0°C. Results are usually given in relation to urinary creatinine (μmol metabolite/mol creatinine) in order to account for varying urine volume.

Chromatographic analysis of 1-hydroxypyrene and other metabolites present in urine samples both from environmental and from occupational exposure is performed using high-performance liquid chromatography as well as gas chromatography [35–39]. As the clean-up procedures depend upon the kind of analysis that follows, they are dealt with in the respective section.

4.3.2 High-performance liquid chromatography (HPLC)

4.3.2.1 Analysis of 1-hydroxypyrene

Several researchers used high-performance liquid chromatography for separation and quantification of PAH metabolites. As pyrene is usually one of the principal components in PAH mixtures, 1-hydroxypyrene, the main urinary pyrene metabolite, has become a preferential biomarker in the past years. Several studies have shown that 1-hydroxypyrene is a valid and sound biomarker for occupational exposure to polycyclic aromatic hydrocarbons. An HPLC method for 1-hydroxypyrene in urine samples, including suitable enrichment and clean-up procedures, has been developed by Jongeneelen et al. [40–42].

A chromatogram obtained by Jongeneelen et al. is given in Fig. 4.7. They were able to identify 1-hydroxypyrene in the urine of PAH exposed workers using the following procedure.

(1) *Enzymatic hydrolysis.* An aliquot of urine (10–25 ml) is adjusted to pH 5.0 with 1.0 M hydrochloric acid and 0.1 M acetate buffer (pH 5.0) is added to a total volume of 30 ml. This mixture is incubated overnight (16 h) with 12.5 μl of β-glucuronidase/arylsulfatase at 37°C.

(2) *Extraction procedure.* A cartridge packed with C18 reversed-phase material (Sep-Pak C18 cartridge, Waters) is primed with 5 ml of methanol, followed by 10 ml of distilled water, and the hydrolyzed sample is passed through the cartridge at ca. 10 ml/min. Subsequently the cartridge is washed with 8 ml of distilled water. The retained solutes are eluted with 10 ml of methanol. The solvent is evaporated at 60°C under a gentle flow of nitrogen, and the residue is dissolved in 2.0 ml of methanol.

(3) *HPLC analysis.* A reversed phase C18 column (150 × 4 mm I.D., LiChrosorb RP-18, 5 μm, Merck) and a methanol–water eluent at 40°C is used. Starting conditions are methanol–water (46 : 54) for 5 min, a linear gradient in 35 min to methanol–water (94 : 6) follows, and final hold time is 10 min. 1-Hydroxypyrene is quantified using fluorescence detection at 242 nm (excitation wavelength)/388 nm (emission wavelength).

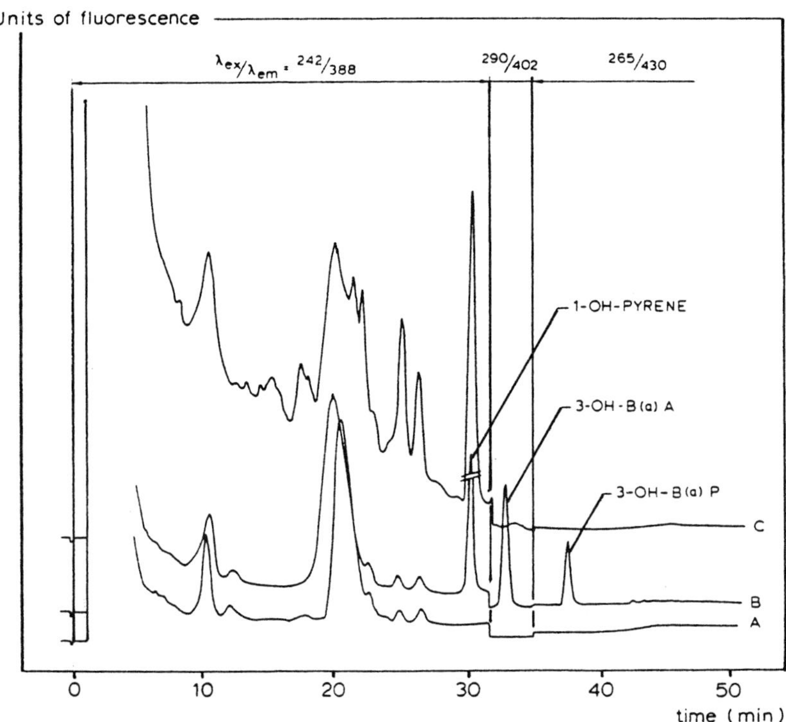

Fig. 4.7. HPLC profiles of extracts from (A) a blank urine, (B) a blank urine spiked to contain ca. 400 nmol/l of each of the three hydroxylated PAHs and (C) a urine sample from a worker exposed to creosote, a PAH-containing wood preservative. Peaks: 1-OH-pyrene = 1-hydroxypyrene; 3-OH-B(a)A = 3-hydroxybenz[*a*]anthracene; 3-OH-B(a)P = 3-hydroxybenzo[*a*]pyrene. Reprinted from [40], with permission of Elsevier Science.

The same or similar methods have been used by a multitude of other authors in order to assess PAH exposure [9,10,13,29,30,32–36,43–49]. These methods basically comprise the following steps: (1) enzymatic hydrolysis of glucuronides and sulfates; (2) clean-up and enrichment using C18 SPE-columns; (3) HPLC with fluorescence detection using methanol/water or acetonitrile/water eluents (isocratic or gradient elution).

Sample intake ranges between 5 and 25 ml and for 1-hydroxypyrene detection limits such as 0.1 nmol/l urine can be reached. In comparison with the sensitivity of GC methods (see Section 4.3.3 below) these detection limits are rather high. However, sample preparation and the chromatographic separation are relatively easy to perform, which has rendered the method quite popular for assessment of exposure to polycyclic aromatic hydrocarbons.

It has to be mentioned, though, that analysis of just one metabolite is not sufficient for a comprehensive exposure assessment because of varying PAH profiles in different surroundings.

4.3.2.2 Further metabolites

In order to allow for a more detailed estimation of individual risks associated with PAH exposure, the analysis is often extended to further metabolites.

To prevent decomposition of the sensitive metabolite standards, they are usually stored under nitrogen in the cold and in the dark. 1-hydroxy-benz[a]anthracene [50] and 3-hydroxybenzo[a]pyrene [51] have been identified in urine samples and the simultaneous determination of e.g. 1-hydroxypyrene, 3-hydroxybenzo[a]pyrene and 3-hydroxyphenanthrene or 3-hydroxybenzo[a]pyrene [37,40] present in urine samples is possible as well. Fig. 4.8 shows HPLC chromatograms of both spiked and native urine extracts.

Sometimes modifiers such as ascorbic acid are added to the mobile phase in order to overcome reproducibility problems [51]. Because of the high relevance of benzo[a]pyrene as an environmental carcinogen, a lot of work has been done in order to identify benzo[a]pyrene metabolites in various matrices and to develop suitable HPLC methods [52–54]. Separation of various benzo[a]pyrene phenols, dihydrodiols and quinones has been carried out using high-performance liquid chromatography.

In most cases fluorescence detection and sometimes UV detection [55] is employed, but there are also examples of e.g. mass spectrometric detection [56,70] or amperometric detection of PAH metabolites [57].

Fig. 4.9 shows the structure of 1-hydroxyphenanthrene, 1-hydroxypyrene, 1-hydroxybenzo[a]pyrene, and 3-hydroxybenzo[a]pyrene.

4.3.2.3 Column-switching techniques

HPLC column-switching techniques for clean-up and selective enrichment of PAH metabolites have been used as well. A copper phthalocyanine modified stationary phase has shown its suitability to retain the analytes in question. The method, originally developed by Lintelmann et al. [58,59], can be fully automated and has been used to analyze several pyrene, phenanthrene, benz[a]anthracene and benzo[a]pyrene hydroxides [60–63,82] in urine. A recent publication [64] showed that 3-hydroxybenzo[a]pyrene present in urine samples can be successfully analyzed after purification and concentration using an automated column-switching HPLC method employing conventional packing materials.

4.3.2.4 HPLC separations with chiral stationary phases or chiral mobile phase modifiers

One possibility to add further selectivity to a chromatographic method is the use of chiral stationary phases. Enantioselective separations of *trans*-dihydrodiol metabolites of several fjord-region PAHs such as benzo[c]phenanthrene, dibenzo[a,l]pyrene, benzo[c]chrysene, benzo[g]chrysene, naphtho[1,2-a]pyrene and naphtho[1,2-e]pyrene were performed using commercially available cellulose-based chiral stationary phases [65]. The corresponding chromatograms are given in Fig. 4.10.

Another approach to enhance the selectivity of a method is the addition of chiral modifiers to the mobile phase. In this case e.g. cyclodextrins (CDs) may be used. Cyclodextrins are water-soluble cyclic oligo-saccharides which consist of α-1,4-linked

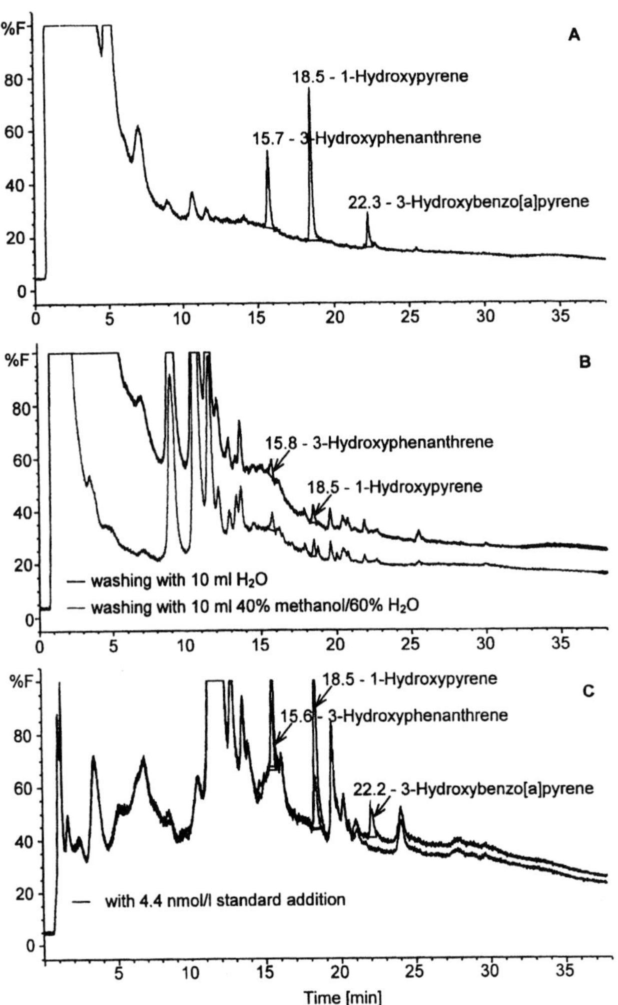

Fig. 4.8. Chromatograms of extracts of (A) pooled urine spiked with 3-hydroxyphenanthrene (3.3 nmol/l), 1-hydroxypyrene (2.9 nmol/l) and 3-hydroxybenzo[*a*]pyrene (2.4 nmol/l) (signal amplification of the detector: pmt gain 16), (B) native urine sample, with different washing of the SPE column during sample preparation (signal amplification of the detector: pmt gain 16), (C) native urine sample of a child with and without 4.4 nmol/l standard addition (signal amplification of the detector: pmt gain 17). Reprinted from [37], with permission of Elsevier Science.

glucopyranose subunits. The most common forms are α-CD, β-CD, and γ-CE, with six, seven, and eight subunits, respectively. They possess a toroidal structure with a non-polar interior cavity and can form host–guest inclusion complexes with many hydrophobic compounds. In reversed-phase HPLC β-cyclodextrin was used in combination with a C18 stationary phase for analysis of various benzo[*a*]pyrene metabolites [66].

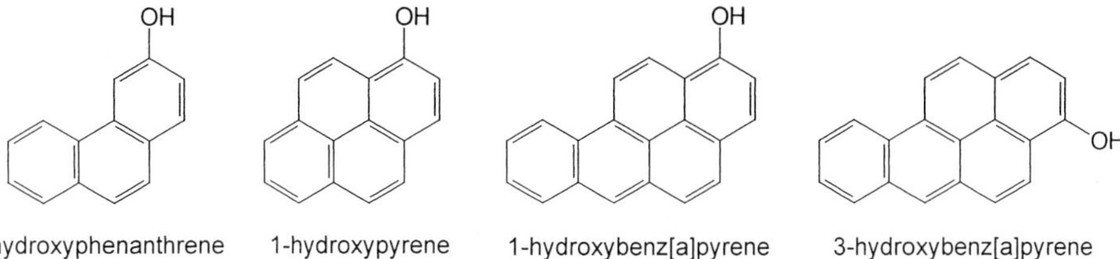

Fig. 4.9. Structure of 1-hydroxyphenanthrene, 1-hydroxypyrene, 1-hydroxybenzo[*a*]pyrene, and 3-hydroxy-benzo[*a*]pyrene.

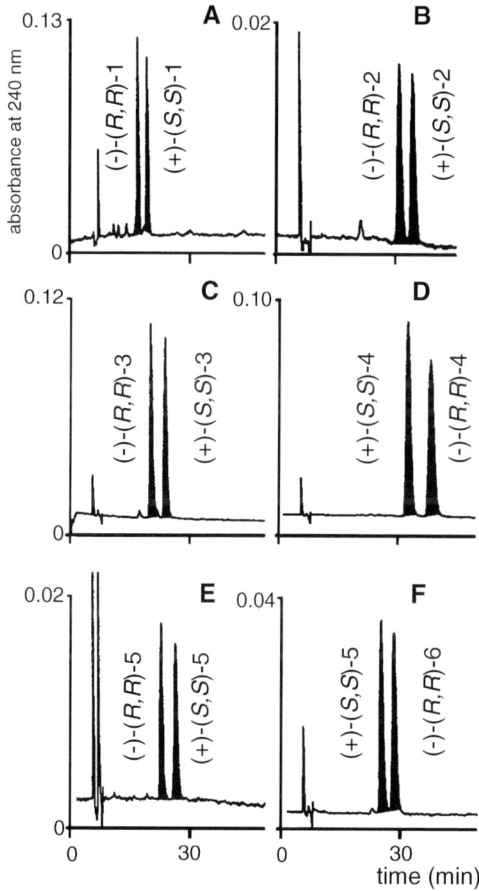

Fig. 4.10. Enantiomeric separation of *trans*-dihydrodiols of polycyclic aromatic hydrocarbons by CSP–HPLC on a Chiracel OD-H phase using a mixture of *n*-heptane–ethanol (9 : 1, v/v) as eluent at a flow-rate of 0.5 ml/min; benzo[*c*]phenanthrene-3,4-dihydrodiol (A) dibenzo[*a*,*l*]pyrene-11,12-dihydrodiol (B) benzo[*c*]chrysene-9,10-dihydrodiol (C) benzo[*g*]chrysene-11,12-dihydrodiol (D) naphtho[1,2-*a*]pyrene-9,10-dihydrodiol (E) and naphtho[1,2-*e*]pyrene-11,12-dihydrodiol (F). Reprinted from [65], with permission of Elsevier Science.

References pp. 117–121

4.3.2.5 Separation of conjugates

Commonly the excreted conjugates are hydrolyzed prior to analysis. However, it is possible to analyze the conjugates as such together with the free metabolites present in urine samples after enrichment and purification without hydrolysis. It was demonstrated that it is possible to analyze 1-hydroxypyrene glucuronide in urine samples after non-occupational exposure [67]. Simultaneous detection of 1-hydroxypyrene glucuronide, 1-hydroxypyrene sulfate and 1-hydroxypyrene in a single chromatographic analysis is possible as well [68]. Here it could be shown that 1-hydroxypyrene glucuronide is the major pyrene metabolite in the urine of occupationally exposed persons. Furthermore the fluorescence intensities of both the glucuronide and the sulfate are higher than that of 1-hydroxypyrene. This high sensitivity together with the missing enzymatic hydrolysis makes them especially attractive as potential biomarkers. Moreover a separation of 5 benzo[a]pyrene sulfate isomers present in standard solutions (1-, 3-, 6-, 7-, and 9-OH-benzo[a]pyrene sulfate) was demonstrated [69]. The respective chromatograms are shown in Fig. 4.11.

In urine of workers occupationally exposed to naphthalene the conjugates α-naphthylglucuronide and β-naphthylsulfate could be determined simultaneously [70]. The free metabolite α-naphthol was not detectable, though. A recent publication showed the separation of several pyrene and benzo[a]pyrene sulfates and glucuronides present in smokers' urine [71].

All separations mentioned above were performed using HPLC with fluorescence or mass spectrometric detection.

4.3.3 Gas chromatography (GC)

4.3.3.1 Gas chromatography with multistage clean-up procedures

Gas chromatographic separations from several PAH metabolites were described by Grimmer, Jacob and co-workers [39,72,73]. The conjugates present in urine samples undergo either acid or enzymatic hydrolysis to form the free metabolites. Liquid–liquid extraction with toluene (or alternatively benzene) is used to extract the metabolites from the matrix. Successively the organic phase is washed with water and reduced to a small volume. Phenols are derivatized with diazomethane to the corresponding methylethers. During the following clean-up procedure first silica gel and then Sephadex LH20™ is employed. Dihydrodiols present in the first toluene extract are converted into phenols and afterwards treated in the same way as already described. GC–MS (or GC–FID) is used for separation and quantification. The method is capable of analyzing several metabolites of phenanthrene, fluoranthene, pyrene, chrysene and benzo[a]pyrene and thereby allows the determination of individual metabolite profiles. The following GC chromatogram shows the separation of several isomeric benz[a]anthracene metabolites (Fig. 4.12).

The described GC method has the advantage of being highly sensitive and selective. In addition it covers a higher number of different analytes than liquid chromatographic methods. However, the clean-up procedures necessary to obtain clean extracts suitable

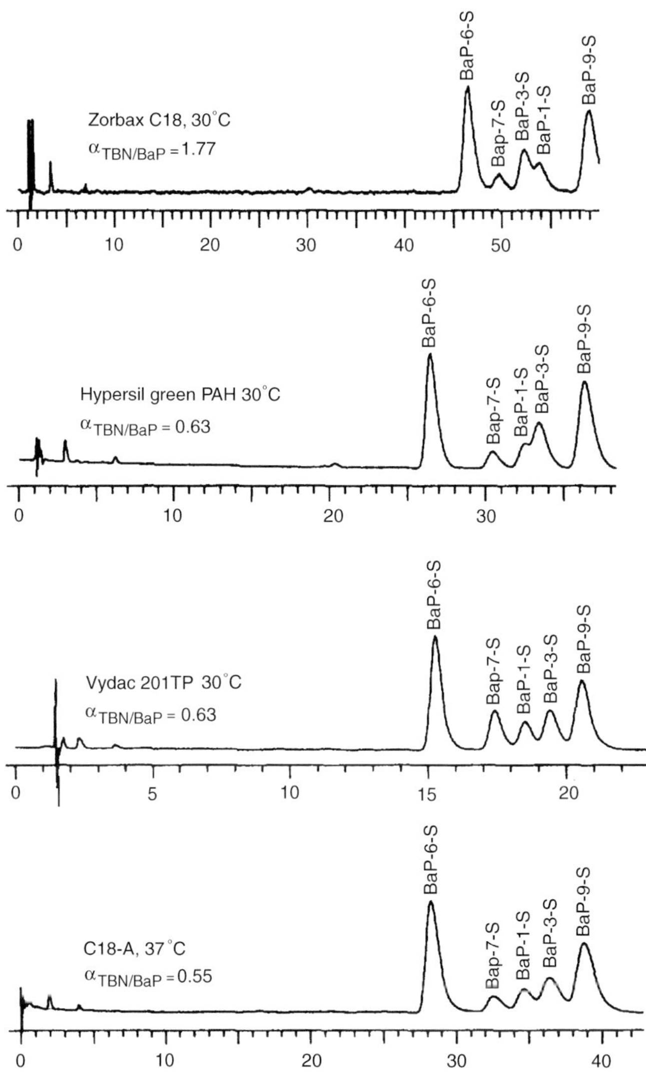

Fig. 4.11. Separation of benzo[*a*]pyrene sulfate standards on commercial and custom C18 columns. Reprinted from [69], with permission of Elsevier Science.

for GC-measurements are more complicated than those for HPLC-measurements and rather time-consuming.

Analysis of polycyclic aromatic hydrocarbon metabolites in other matrices has been performed as well. DNA adducts of benzo[*a*]pyrene metabolites for instance were analyzed after hydrolysis and derivatization to methoxy derivatives [74,75]. The derivatization procedure that was used in this case was different from that described above and employed methyl iodide as methylation reagent. After derivatization an HPLC clean-up was performed and the compounds were analyzed using GC–MS. The detection limit

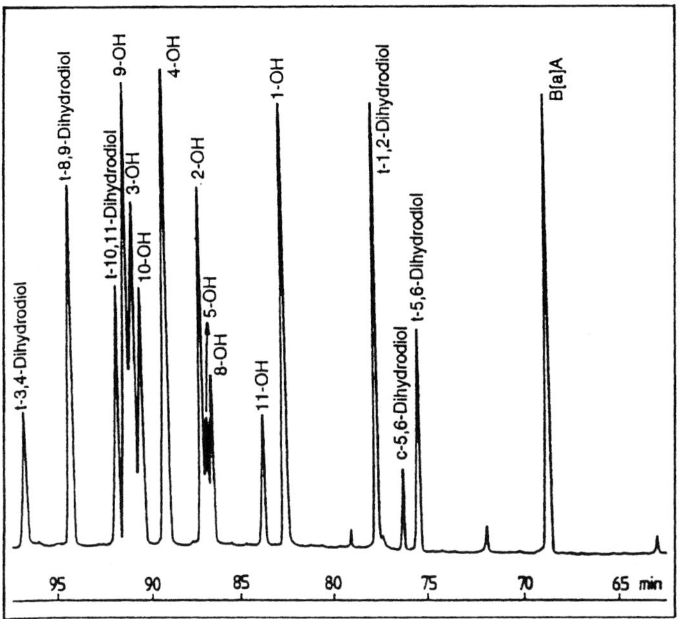

Fig. 4.12. Capillary GC separation of isomeric phenol and dihydrodiol benz[*a*]anthracene metabolites as OTMS-ether. Reprinted from [72].

for tetramethoxy derivatives of benzo[*a*]pyrene-tetraols released from DNA was one fmol/injection.

4.3.3.2 SPME

Another approach to perform sample preparation for gas chromatographic measurements is solid-phase microextraction. Gmeiner et al. [76] used a polyacrylate SPME fibre for analysis of hydroxynaphthalenes, hydroxyphenanthrenes and hydroxypyrenes in enzymatically hydrolyzed urine samples. After extraction (direct immersion, 45 min, 35°C) the fiber was transferred to the headspace of the derivatization agent (BSTFA: *N,O*-bis-(trimethylsilyl)-trifluoro-acetamide) for 45 min at 60°C. Subsequently the fiber was placed in the GC injector and the silylated PAHs were desorbed and analyzed. The method was able to detect hydroxylated metabolites of naphthalene, phenanthrene and pyrene in 5 ml urine samples of smokers. Chromatograms are shown in Fig. 4.13.

4.3.4 Capillary electrophoresis (CE)

During the past years capillary electrophoresis with its high separation power has found its place among the conventional chromatographic techniques such as HPLC and GC. While capillary zone electrophoresis (CZE) has emerged as one of the most efficient methods available for separation of complex mixtures, it is usually limited to the analysis of water-soluble charged species. Therefore its applicability to the analysis of PAHs

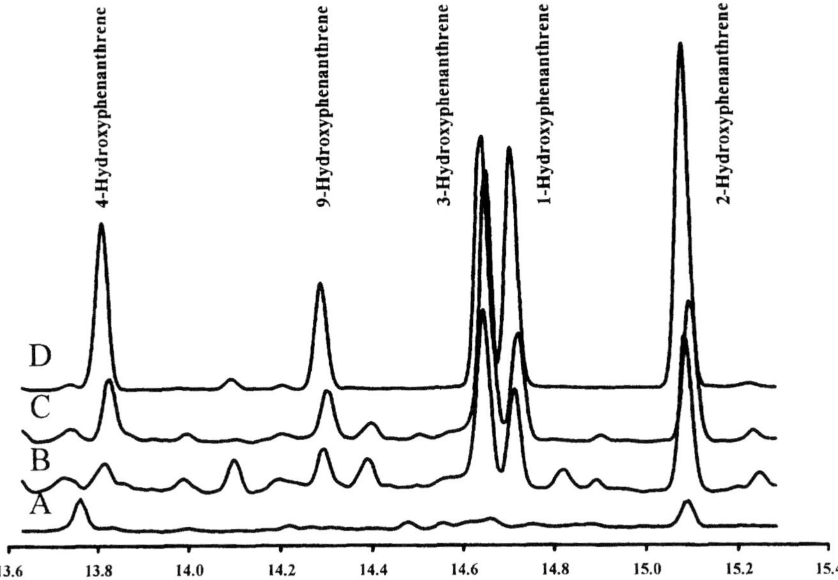

Fig. 4.13. Selected ion chromatogram, m/z 261 (trimethylsilyl-derivatives of hydroxyphenanthrenes) after SPME sampling and derivatization with BSTFA. (A) Blank water, (B) 5 ml enzymatically hydrolyzed urine, (C) 5 ml enzymatically hydrolyzed urine, spiked with 50 l stock solution and (D) 5 ml water spiked with 50 l stock solution. Reprinted from [76], with permission of Elsevier Science.

and their metabolites is limited. However, separation of PAH metabolites into compound classes has been proven to be possible using certain buffer systems with and without organic modifiers [77]. Fig. 4.14 shows an electropherogram of several PAH metabolites.

Further CE methods suitable for separating uncharged molecules have been developed, for example micellar electrokinetic capillary chromatography (MECC). MECC separation is based on analyte partition between an aqueous buffer and a charged, so-called pseudo-stationary micellar phase. This micellar phase enables the separation of non-polar substances as for instance parent polycyclic aromatic hydrocarbons. However, MECC separation of hydroxylated PAH metabolites is often not satisfactory. Because of their structural similarity the different isomers show nearly identical elution times. To overcome these problems, cyclodextrin modified MECC (CD-MECC) was employed for analysis of polycyclic aromatic hydrocarbon metabolites. γ-Cyclodextrin has been successfully used in order to introduce further selectivity depending on analyte size, shape, and chirality [78–81]. Various hydroxylated PAH metabolites could be separated within 20 min. Detection of the analytes can be carried out using either UV- or fluorescence detectors, with a superior sensitivity of the latter. Example electrochromatograms obtained with different γ-cyclodextrin concentrations present in the eluent are shown in Fig. 4.15.

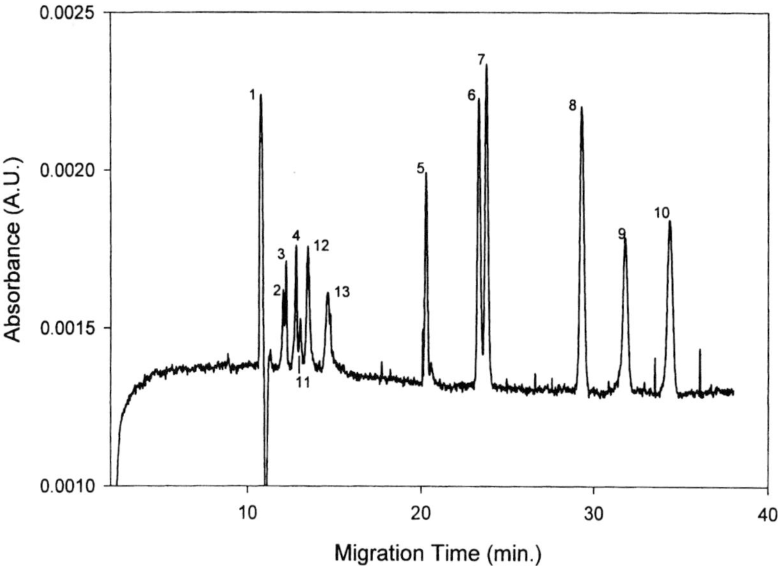

Fig. 4.14. Electropherogram of a mixture of three parent compounds and nine PAH metabolites with 40% MeOH in 100 mM CAPS at an apparent pH of 10.4. Voltage 25 kV, temperature 35°C. (1) EOF marker (MeOH); (2) benzo[*a*]pyrene-*r*-7-*trans*-8,9-*cis*-10-tetrahydrotetrol; (3) benzo[*a*]pyrene-*trans*-9,10-dihydrodiol; (4) benzo[*a*]pyrene-*trans*-7,8-dihydrodiol; (5) 3-hydroxybenz[*a*]anthracene; (6) 7-hydroxy-benzo[*a*]pyrene; (7) 9-hydroxybenzo[*a*]pyrene; (8) 12-hydroxybenzo[*a*]pyrene; (9) 3-hydroxybenzo[*a*]pyrene; (10) 1-hydroxypyrene; (11) pyrene; (12) benz[*a*]anthracene; (13) benzo[*a*]pyrene. Reprinted from [77], with permission of Elsevier Science.

4.3.5 Immunochemical methods

The use of immunochemical methods for analysis of PAH metabolites is relatively scarce in the literature. Recently a competitive ELISA on the basis of a polyclonal antiserum that had been raised against pyrenebutyric acid coupled to thyroglobulin was developed [82]. Urine samples of persons occupationally exposed to PAHs and of controls were simply diluted prior to analysis. 1-Hydroxypyrene, which showed the highest affinity towards the antiserum, was used to calibrate the ELISA. The binding of 1-hydroxypyrene glucuronide and the phenanthrols was acceptable as well, whereas it was low for 1-hydroxypyrene sulfate, 1-naphthol and 3-hydroxy-benzo[*a*]pyrene. In addition two different HPLC methods were used in order to measure 1-hydroxypyrene and the sum of 1-, 2-, 3-, 4-, and 9-hydroxyphenanthrenes after enzymatic hydrolysis. Although

Fig. 4.15 (right). Electrochromatograms showing separations of 12 hydroxy-PAHs. Separation conditions are: 37 cm × 50 μm I.D. capillary, 12 kV, 30 mM sodium borate, 60 mM SDS. (a) 5 mM γ-CD, (b) 20 mM γ-CD, (c) 40 mM γ-CD. The peaks are identified as follows: (1) 3-OHBaP, (2) 7-OHBaP, (3) β-naphthol, (4) α-naphthol, (5) 9-OHBaP, (6) 1-OHpyrene, (7) 3-OHchrysene, (8) 3-OHbenz[*a*]anthracene, (9) 1-OHBaP, (10) 1-OHbenz[*a*]anthracene, (11) 1-OHbenzo[*b*]fluoranthene, (12) 2-OHindeno[1,2,3-*cd*]pyrene. Reprinted from [79], with permission of Elsevier Science.

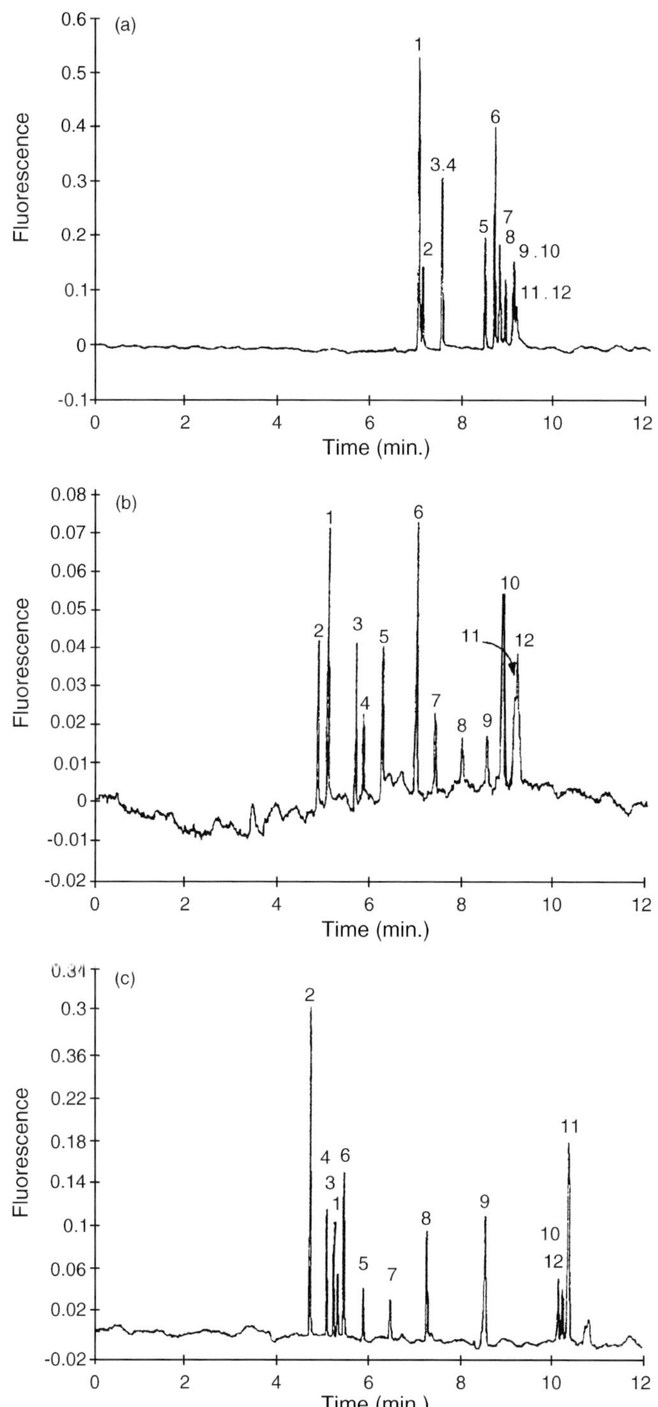

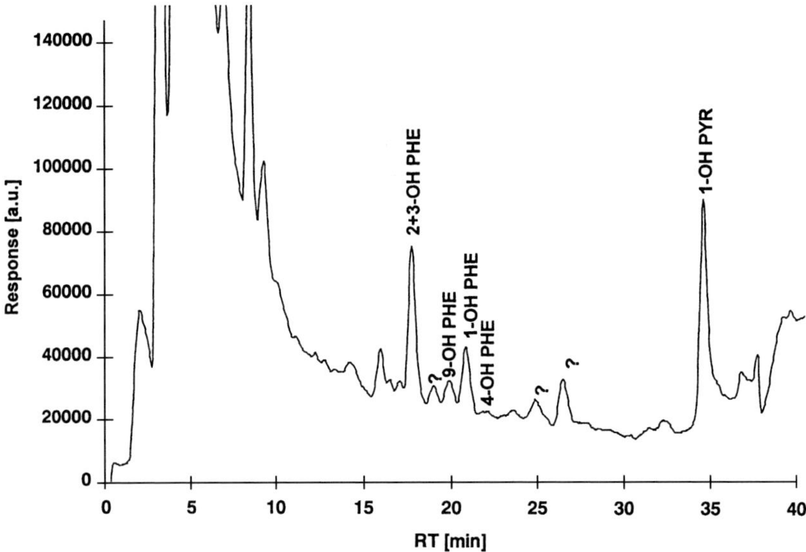

Fig. 4.16. Typical chromatogram with frequently observed unknown peaks (labelled with '?'). Reprinted from [82], with permission of Elsevier Science.

the ELISA results were usually higher than the corresponding 1-hydroxypyrene concentrations and lower than the sum of 1-hydroxypyrene and phenanthrols as determined by HPLC, no significant differences between ELISA and HPLC results could be observed. Therefore it was concluded that employment of an ELISA is useful for biomonitoring studies of PAH metabolites in diluted urine without hydrolysis, clean-up and/or derivatization steps. A typical chromatogram obtained from smokers urine is given in Fig. 4.16.

Another competitive ELISA, using an existing antibody that had been developed against 6-aminobenzo[*a*]pyrene coupled to bovine serum albumine, was employed for analysis of urine samples from persons exposed to diesel exhaust [83]. However, although the ELISA exhibited a certain cross-reactivity towards e.g. 1-hydroxypyrene, benzo[*a*]pyrene and pyrene, the excretion of metabolites as determined using the immunoassay was not related to 1-hydroxypyrene excretion as determined by HPLC.

In addition the use of immunoaffinity columns for purification of 1-hydroxypyrene glucuronide in urine samples has already been demonstrated [84,85,67,86].

Monoclonal antibodies with binding capacity to benzo[*a*]pyrene-tetrols have been developed as well [87]. Immunoaffinity columns were used for purification of hydrolyzed urine samples with subsequent HPLC analysis. It was possible to detect benzo[*a*]pyrene-*r*-7, *t*-8, *t*-9, *c*-10-tetrol in urine of persons who had consumed charbroiled beef [88]. In another study a similar combination of immunoaffinity chromatography and HPLC was used with the aim of analyzing benzo[*a*]pyrene metabolites in urine from workers exposed to high levels of PAH [89]. However, in this case no benzo[*a*]pyrene-tetrols could be detected. This might be due to the fact that PAH metabolites originating from higher PAHs are mainly excreted via the faeces. However, relatively high amounts of 1-hydroxypyrene as well as some hydroxyphenanthrenes could be found in the urine samples.

4.4 REFERENCES

1. D.M. Jerina, H. Yagi, O. Hernandez, P.M. Dansette, A.W. Wood, W. Levin, R.L. Chang, P.G. Wislocki and A.H. Conney, Synthesis and biologic activity of potential benzo[*a*]pyrene metabolites, in: R.I. Freudenthal and P.W. Jones (Eds.), Carcinogenesis, Vol. 1, Polynuclear Aromatic Hydrocarbons: Chemistry, Metabolism and Carcinogenesis, Raven Press, New York, 1976.

2. A. Dipple, in: K. Hemminki, A. Dipple, D.E.G. Shuker, F.F. Kadlubar, D. Segerback and H. Bartsch (Eds.), DNA Adducts Identification and Biological Significance, IARC Scientific Publications No. 125, Lyon, 1994, pp. 107–129.

3. E. Cavalieri and E. Rogan (Eds.), Proceedings of the 14th International Symposium on Polynuclear Aromatic Hydrocarbons, Polycyclic Aromatic Hydrocarbons 5–7, Nos. 1–4, 1994.

4. M.L. Lee, M. Novotny and K.D. Bartle, Analytical Chemistry of Polycyclic Aromatic Compounds, Academic Press, New York, 1981.

5. G. Grimmer, H. Brune, G. Dettbarn, J. Jacob, J. Misfeld, U. Mohr, K.-W. Naujack, J. Timm and R. Wenzel-Hartung, Relevance of polycyclic aromatic hydrocarbons as environmental carcinogens, Fresenius J. Anal. Chem., 339 (1991) 792–795.

6. J. Jacob, The significance of polycyclic aromatic hydrocarbons as environmental carcinogens, Pure Appl. Chem., 68 (1996) 301–308.

7. P. Strickland and D. Kang, Urinary 1-hydroxypyrene and other PAH metabolites as biomarkers of exposure to environmental PAH in air particulate matter, Toxicol. Lett., 108 (1999) 191–199.

8. C. Viau, G. Hakizimana and M. Bouchard, Indoor exposure to polycyclic aromatic hydrocarbons and carbon monoxide in traditional houses in Burundi, Int. Arch. Occup. Environ. Health, 73 (2000) 331–338.

9. B. Schoket, M.C. Poirier, G. Mayer, G. Török, A. Kolozsi-Ringelhann, B. Bognár, W.L. Bigbee and I. Vincze, Biomonitoring of human genotoxicity induced by complex occupational exposures, Mutat. Res., 445 (1999) 193–203.

10. J.O. Levin, M. Rhén and E. Sikström, Occupational PAH exposure: urinary 1-hydroxypyrene levels of coke oven workers, aluminium smelter pot-room workers, road pavers, and occupationally non-exposed persons in Sweden, Sci. Total Environ., 163 (1995) 169–177.

11. J. Angerer, C. Mannschreck and J. Gündel, Biological monitoring and biochemical effect monitoring of exposure to polycyclic aromatic hydrocarbons, Int. Arch. Occup. Environ. Health, 70 (1997) 365–377.

12. J. Jacob, G. Grimmer and G. Dettbarn, Profile of urinary phenanthrene metabolites in smokers and non-smokers, Biomarkers, 4 (1999) 319–327.

13. G. Gabbani, S. Pavanello, B. Nardini, O. Tognato, A. Bordin, C. Veller Fornasa, G. Bezze and E. Clonfero, Influence of metabolic genotype *GSTM1* on levels of urinary mutagens in patients treated topically with coal tar, Mutat. Res., 440 (1999) 27–33.

14. D.H. Phillips, Polycyclic aromatic hydrocarbons in the diet, Mutat. Res., 443 (1999) 139–147.

15. IARC Monographs on the Evaluation of the Carcinogenic Risk of Chemicals to Man, Certain Polycyclic Aromatic Hydrocarbons and Heterocyclic Compounds, Vol. 3, IARC, Lyon, 1973.

16. IARC Monographs on the Evaluation of the Carcinogenic Risk of Chemicals to Humans, Vol. 32–35, Parts 1, 3 and 4, Polynuclear Aromatic Compounds, IARC, Lyon, 1983, 1984 and 1985.

17. G. Grimmer, J. Jacob, G. Dettbarn, K.-W. Naujack and U. Heinrich, Urinary metabolite profile of PAH as a potential mirror of the genetic disposition for cancer, Exp. Toxic. Pathol., 47 (1995) 421–427.

18. T. Vo-Dinh, Chemical Analysis of Polycyclic Aromatic Compounds, Wiley, New York, 1989.

19. US Environmental Protection Agency, Guidelines Establishing Test Procedures for the Analysis of Pollutants, Federal Register, Washington, DC, Vol. 49, 209 (1984) 112–120.

20. D.R. Thakker, H. Yagi, W. Levin, A.W. Wood, A.H. Conny and D.M. Jerina, Polycyclic aromatic hydrocarbons: metabolic activation ultimate carcinogens, in: M.W. Anders (Ed.), Bioactivation of Foreign Compounds, Academic Press, New York, 1985, pp. 177–242.

21. E. Bauer, Z. Guo, Y.F. Ueng, L.C. Bell, D. Zeldin and F.-P. Guengerich, Oxidation of benzo[*a*]pyrene by recombinant human cytochrome P450 enzymes, Chem. Res. Toxicol., 8 (1995) 136–142.

22. C.H. Yun, T. Shimada and F.P. Guengerich, Role of human liver cytochrome P4502C and 3A enzymes in 3-hydroxylation of benzo[*a*]pyrene, Cancer Res., 52 (1992) 1868–1874.

23   P.L. Grover, The metabolic activation of the polycyclic aromatic hydrocarbons, VDI-Ber., 358 (1980) 257–271.

24   F. Oesch, Biochemistry of polycyclic aromatic hydrocarbons, VDI-Ber., 358 (1980) 251–256.

25   J. Jacob, G. Grimmer, M. Emura, G. Raab, J.W. Knebel and M. Aufderheide, Metabolism of polycyclic aromatic hydrocarbons in fetal human, rat and hamster epithelial lung cells, Exp. Toxic. Pathol., 47 (1995) 428–431.

26   D.M. Jerina, H. Yagi, R.E. Lehr, D.R. Thakker, M. Schaefer-Ridder, J.M. Karle, W. Levin, A.W. Wood, R.L. Chang and A.H. Conney, in: H.V. Gelboin and P.O.P. Ts'o (Eds.), Polycyclic Hydrocarbons and Cancer, Academic Press, New York, 1978.

27   S. Amin, D. Desai, W. Dai, R.G. Harvey and S.S. Hecht, Tumorigenicity in newborn mice of fjord region and other sterically hindered diol epoxides of benzo[*g*]chrysene, dibenzo[*a,l*]pyrene (dibenzo[*def,p*]chrysene), 4H-cyclopenta[*def*]chrysene and fluoranthene, Carcinogenesis, 16 (1995) 2813–2817.

28   Encyclopaedia of Analytical Science, Polycyclic Aromatic Hydrocarbons, Academic Press, London, 7 (1995) 4044–4058.

29   B.A. Hatjian, J.W. Edwards, J. Harrison, F.M. Williams and P.G. Blain, Ambient, biological, and biological effect monitoring of exposure to polycyclic aromatic hydrocarbons (PAHs), Toxicol. Lett., 77 (1995) 271–279.

30   R.K. Bentsen, H. Notø, K. Halgard and S. Øvrebø, The effect of dust-protective respirator mask and the relevance of work category on urinary 1-hydroxypyrene concentration in PAH exposed electrode paste plant workers, Ann. Occup. Hyg., 42 (1998) 135–144.

31   F.J. Jongeneelen, Methods for routine biological monitoring of carcinogenic PAH-mixtures, Sci. Total Environ., 199 (1997) 141–149.

32   M. Lafontaine, J.P. Payan, P. Delsaut and Y. Morele, Polycyclic aromatic hydrocarbon exposure in an artificial shooting target factory: assessment of 1-hydroxypyrene urinary excretion as a biological indicator of exposure, Ann. Occup. Hyg., 44 (1999) 89–100.

33   A. Karakaya, B. Yücescoy, A. Turhan, O. Erdem, S. Burgaz and A.E. Karakaya, Investigation of some immunological functions in a group of asphalt workers exposed to polycyclic aromatic hydrocarbons, Toxicology, 135 (1999) 43–47.

34   R.K. Bentsen, K. Halgard, H. Notø, H.L. Daae and S. Øvrebø, Correlation between urinary 1-hydroxypyrene and ambient air pyrene measured with an inhalable aerosol sampler and a total dust sampler in an electrode paste plant, Sci. Total Environ., 212 (1998) 59–67.

35   E. Siwinska, D. Mielzynska, A. Bubak and E. Smolik, The effect of coal stoves and environmental tobacco smoke on the level of urinary 1-hydroxypyrene, Mutat. Res., 445 (1999) 147–153.

36   E. Siwinska, D. Mielzynska, E. Smolik, A. Bubak and J. Kwapulinski, Evaluation of intra- and interindividual variation of urinary 1-hydroxypyrene, a biomarker of exposure to polycyclic aromatic hydrocarbons, Sci. Total Environ., 217 (1998) 175–183.

37   J. Hollender, B. Koch and W. Dott, Biomonitoring of environmental polycyclic aromatic hydrocarbon exposure by simultaneous measurement of urinary phenanthrene, pyrene and bay hydroxides, J. Chromatogr. B, 739 (2000) 225–229.

38   Å.M. Hansen, J.M. Christensen and D. Sherson, Estimation of reference values for urinary 1-hydroxypyrene and α-naphthol in Danish workers, Sci. Total Environ., 163 (1995) 211–219.

39   G. Grimmer, G. Dettbarn, K.-W. Naujack and J. Jacob, Excretion of hydroxy derivatives of polycyclic aromatic hydrocarbons of the masses 178, 202, 228 and 252 in the urine of coke and road workers, Int. J. Environ. Anal. Chem., 43 (1991) 177–186.

40   F.J. Jongeneelen, R.B.M. Anzion and P.Th. Henderson, Determination of hydroxylated metabolites of polycyclic aromatic hydrocarbons in urine, J. Chromatogr., 413 (1987) 227–232.

41   F.J. Jongeneelen, R.B.M. Anzion, P.T.J. Scheepers, R.P. Bos, P.T. Henderson, E.H. Nijehuis, S.J. Veenstra, R.M.E. Brouns and A. Winkes, 1-Hydroxypyrene in urine as a biological indicator of exposure to polycyclic aromatic hydrocarbons in several work environments, Ann. Occup. Hyg., 32 (1988) 35–43.

42   F.J. Jongeneelen, Biological monitoring of environmental exposure to polycyclic aromatic hydrocarbons; 1-hydroxypyrene in urine of people, Toxicol. Lett., 72 (1994) 205–211.

43   L. Pyy, M. Mäkelä, E. Hakala, K. Kakko, T. Lapinlampi, A. Lisko, E. Yrjänheikki and K. Vähäkangas,

Ambient and biological monitoring of exposure to polycyclic aromatic hydrocarbons at a coking plant, Sci. Total Environ., 199 (1997) 151–158.

44   T. Petry, P. Schmid and C. Schlatter, Airborne exposure to polycyclic aromatic hydrocarbons (PAHs) and urinary excretion of 1-hydroxypyrene of carbon anode plant workers, Ann. Occup. Hyg., 40 (1996) 345–357.

45   M. Bouchard, K. Krishnan and C. Viau, Urinary excretion kinetics of 1-hydroxypyrene following intravenous administration of binary and ternary mixtures of polycyclic aromatic hydrocarbons in rat, Arch. Toxicol., 72 (1998) 475–482.

46   F.J. van Schooten, E.J.C. Moonen, L. van der Wal, P. Levels and J.C.S. Kleinjans, Determination of polycyclic aromatic hydrocarbon (PAH) and their metabolites in blood, faeces, and urine of rats orally exposed to PAH contaminated soils, Arch. Environ. Contam. Toxicol., 33 (1997) 317–322.

47   M. Bouchard and C. Viau, Urinary and biliary excretion kinetics of 1-hydroxypyrene following intravenous and oral administration of pyrene in rats, Toxicology, 127 (1998) 69–84.

48   S. Øvrebø, D. Ryberg, A. Haugen and H.L. Leira, Glutathione S-transferase M1 and P1 genotypes and urinary excretion of 1-hydroxypyrene in coke oven workers, Sci. Total Environ., 220 (1998) 25–31.

49   R. Kubiak, J. Belowski, J. Szczeklik, E. Smolik, K. Mielzynska, M. Baj and A. Szczesna, Biomarkers of carcinogenesis in humans exposed to polycyclic aromatic hydrocarbons, Mutat. Res., 445 (1999) 175–180.

50   R.S. Whiton, C.L. Witherspoon and T.J. Buckley, Improved high-performance liquid chromatographic method for the determination of polycyclic aromatic hydrocarbon metabolites in human urine, J. Chromatogr. B, 665 (1995) 390–394.

51   M. Bouchard, C. Dodd and C. Viau, Improved procedure for the high-performance liquid chromatographic determination of monohydroxylated PAH metabolites in urine, J. Anal. Toxicol., 18 (1994) 261–264.

52   J.K. Selkirk, R.G. Croy and H.V. Gelboin, Benzo[*a*]pyrene metabolites: efficient and rapid separation by high-pressure liquid chromatography, Science, 184 (1974) 169–171.

53   E.A. Elnenaey and W.P. Schoor, The separation of the isomeric phenols of benzo[*a*]pyrene by high-performance liquid chromatography, Anal. Biochem., 111 (1981) 393–400.

54   M. Lodovici, V. Akpan, L. Giovannini, F. Migliani and P. Dolara, Benzo[*a*]pyrene diol-epoxide DNA adducts and levels of polycyclic aromatic hydrocarbons in autoptic samples from human lungs, Chemico-Biol. Interact., 116 (1998) 199–212.

55   S.P. Boyle and J.A. Craft, Gender specific metabolism of benz[*a*]anthracene in hepatic microsomes from Long–Evans and Hooded Lister rats, Chemico-Biol. Interact., 125 (2000) 209–220.

56   R. Koeber, R. Niessner and J.M. Bayona, Comparison of liquid chromatography–mass spectrometry interfaces for the analysis of polar metabolites of benzo[*a*]pyrene, Fresenius' J. Anal. Chem., 359 (1997) 267–273.

57   E. Fischer, G. Henze and K.L. Platt, Sensitive and selective determination of metabolically formed *trans*-dihydrodiols and phenols of benzo[*a*]pyrene in water and urine samples by HPLC with amperometric detection, Fresenius' J. Anal. Chem., 360 (1998) 95–99.

58   K.-S. Boos, J. Lintelmann and A. Kettrup, Coupled-column high-performance liquid chromatographic method for the determination of 1-hydroxypyrene in urine of subjects exposed to polycyclic aromatic hydrocarbons, J. Chromatogr., 600 (1992) 189–194.

59   J. Lintelmann, C. Hellemann and A. Kettrup, Coupled-column high-performance liquid chromatographic method for the determination of four metabolites of polycyclic aromatic hydrocarbons, 1-, 4- and 9-hydroxyphenanthrene and 1-hydroxypyrene, in urine, J. Chromatogr. B, 660 (1994) 67–73.

60   J. Gündel, C. Mannschreck, K. Büttner, U. Ewers and J. Angerer, Urinary levels of 1-hydroxypyrene, 1-, 2-, 3-, and 4-hydroxyphenanthrene in females living in an industrial area of Germany, Arch. Environ. Contam. Toxicol., 31 (1996) 585–590.

61   J. Angerer, C. Mannschreck and J. Gündel, Occupational exposure to polycyclic aromatic hydrocarbons in a graphite-electrode producing plant: biological monitoring of 1-hydroxypyrene and monohydroxylated metabolites of phenanthrene, Int. Arch. Occup. Environ. Health, 69 (1997) 323–331.

62   J. Gündel, K.H. Schaller and J. Angerer, Occupational exposure to polycyclic aromatic hydrocarbons in a fireproof stone producing plant: biological monitoring of 1-hydroxypyrene, 1-, 2-, 3- and 4-

hydroxyphenanthrene, 3-hydroxybenzo[*a*]anthracene and 3-hydroxybenzo[*a*]pyrene, Int. Arch. Occup. Environ. Health, 73 (2000) 270–274.

63   J. Gündel and J. Angerer, High-performance liquid chromatographic method with fluorescence detection for the determination of 3-hydroxybenzo[*a*]pyrene and 3-hydroxybenz[*a*]anthracene in the urine of polycyclic aromatic hydrocarbon-exposed workers, J. Chromatogr. B, 738 (2000) 47–55.

64   P. Simon, M. Lafontaine, P. Delsaut, Y. Morele and T. Nicot, Trace determination of urinary 3-hydroxybenzo[*a*]pyrene by automated column-switching high-performance liquid chromatography, J. Chromatogr. B, 748 (2000) 337–348.

65   R. Landsiedel, H. Frank, H. Glatt and A. Seidel, Direct optical resolution of *trans*-dihydrodiol enantiomers of fjord-region polycyclic aromatic hydrocarbons by high-performance liquid chromatography on a modified cellulose phase, J. Chromatogr. A, 822 (1998) 29–35.

66   M. Rozbeh and R.J. Hurtubise, Reversed-phase chromatographic separation with β-cyclodextrin as a mobile phase modifier, J. Liquid Chromatogr., 18 (1995) 17–37.

67   C.-K. Lee, S.-H. Cho, J.-W. Kang, S.-J. Lee, Y.-S. Ju, J. Sung, P.T. Strickland and D. Kang, Comparison of three analytical methods for 1-hydroxypyrene glucuronide in urine after non-occupational exposure to polycyclic aromatic hydrocarbons, Toxicol. Lett., 108 (1999) 209–215.

68   R. Singh, M. Tucek, K. Maxa, J. Tenglerova and E.H. Weyand, A rapid and simple method for the analysis of 1-hydroxypyrene glucuronide: a potential biomarker for polycyclic aromatic hydrocarbon exposure, Carcinogenesis, 16 (1995) 2909–2915.

69   C. Johnson, A. Greenberg and L.C. Sander, Separation of benzo[*a*]pyrene sulfate isomers by reversed-phase liquid chromatography, J. Chromatogr. A, 753 (1996) 201–206.

70   R. Andreoli, P. Manini, E. Bergamaschi, A. Mutti, I. Franchini and W.M.A. Niessen, Determination of naphthalene metabolites in human urine by liquid chromatography–mass spectrometry with electrospray ionisation, J. Chromatogr. A, 847 (1999) 9–17.

71   C. Johnson and A. Greenberg, Extraction and high-performance liquid chromatographic separation of selected pyrene and benzo[*a*]pyrene sulfates and glucuronides: preliminary application to the analysis of smokers' urine, J. Chromatogr. B, 728 (1999) 209–216.

72   J. Jacob and A. Seidel, GC/MS-Analytik von Metaboliten polycyclischer aromatischer Kohlenwasserstoffe (PAK), GIT Fachz. Lab., 44 (2000) 570–573.

73   G. Grimmer, J. Jacob, G. Dettbarn and K.-W. Naujack, Determination of urinary metabolites of polycyclic aromatic hydrocarbons (PAH) for the risk assessment of PAH-exposed workers, Int. Arch. Occup. Environ. Health, 69 (1997) 231–239.

74   A.A. Melikian, P. Sun, B. Prokopczyk, K. El-Bayoumy, D. Hoffmann, X. Wang and S. Waggoner, Identification of benzo[*a*]pyrene metabolites in cervical mucus and DNA adducts in cervical tissues in humans by gas chromatography–mass spectrometry, Cancer Lett., 146 (1999) 127–134.

75   A.A. Melikian, P. Sun, C. Pierpont, S. Coleman and S.S. Hecht, Gas chromatographic–mass spectrometric determination of benzo[*a*]pyrene and chrysene diol epoxide globin adducts in humans, Cancer Epidemiol. Biomarkers Prev., 6 (1997) 833–839.

76   G. Gmeiner, C. Krassnig, E. Schmid and H. Tausch, Fast screening method for the profile analysis of polycyclic aromatic hydrocarbon metabolites in urine using derivatisation–solid-phase microextraction, J. Chromatogr. B, 705 (1998) 132–138.

77   X. Xu and R.J. Hurtubise, Influence of organic solvents in the capillary zone electrophoresis of polycyclic aromatic hydrocarbon metabolites, J. Chromatogr. A, 829 (1999) 289–299.

78   U. Krismann and W. Kleiböhmer, Separation of hydroxylated polycyclic aromatic hydrocarbons by micellar electrokinetic capillary chromatography, J. Chromatogr. A, 774 (1997) 193–201.

79   C.J. Smith, J. Grainger and D.G. Patterson, Jr., Separation of polycyclic aromatic hydrocarbon metabolites by γ-cyclodextrin-modified micellar electrokinetic chromatography with laser-induced fluorescence detection, J. Chromatogr. A, 803 (1998) 241–247.

80   U. Krismann, Entwicklung eines selektiven Bestimmungsverfahrens für enzymatische Abbauprodukte Polyzyklischer Aromatischer Kohlenwasserstoffe, Dissertation 1999.

81   S. Kodama, A. Yamamoto, A. Matsunaga, A. Toriba and K. Hayakawa, Micellar electrokinetic chromatography of monohydroxybenzo[*a*]pyrene positional isomers using γ-cyclodextrin, Analyst, 125 (2000) 1555–1559.

82   D. Knopp, M. Schedl, S. Achatz, A. Kettrup and R. Niessner, Immunochemical test to monitor human

exposure to polycyclic aromatic hydrocarbons: urine as sample source, Anal. Chim. Acta, 399 (1999) 115–126.

83 P.T.J. Scheepers, P.H.S. Fijneman, M.F.M. Beenakkers, A.J.G.M. de Lepper, H.J.T.M. Thuis, D. Stevens, J.G.M. Van Rooij, J. Noordhoek and R.P. Bos, Immunochemical detection of metabolites of parent and nitro polycyclic aromatic hydrocarbons in urine samples from persons occupationally exposed to diesel exhaust, Fresenius' J. Anal. Chem., 351 (1995) 660–669.

84 D.H. Kang, N. Rothman, M.C. Poirier, A. Greenberg, C.H. Hsu, B.S. Schwartz, M.E. Baser, J.D. Groopman, A. Weston and P.T. Strickland, Interindividual differences in the concentration of 1-hydroxypyrene-glucuronide in urine and polycyclic aromatic hydrocarbon-DNA adducts in peripheral white blood cells after charbroiled beef consumption, Carcinogenesis, 16 (1995) 1079–1085.

85 P.T. Strickland, D. Kang, E.D. Bowman, A. Fitzwilliam, T.E. Downing, N. Rothman, J.D. Groopman and A. Weston, Identification of 1-hydroxypyrene glucuronide as a major pyrene metabolite in human urine by synchronous fluorescence spectroscopy and gas chromatography–mass spectrometry, Carcinogenesis, 15 (1994) 483–487.

86 Y.C. Hong, J.H. Leem, H.S. Park, K.H. Lee, S.J. Lee, C.-K. Lee and D. Kang, Variations in urinary 1-hydroxypyrene glucuronide in relation to smoking and the modification effects of GSTM1 and GSTT1, Toxicol. Lett., 108 (1999) 217–223.

87 R.M. Santella, C.D. Lin, W.L. Cleveland and I.B. Weinstein, Monoclonal antibodies to DNA modified by a benzo[*a*]pyrene diol epoxide, Carcinogenesis, 5 (1984) 373–377.

88 A. Weston, E.D. Bowman, P. Carr, N. Rothman and P.T. Strickland, Detection of metabolites of polycyclic aromatic hydrocarbons in human urine, Carcinogenesis, 14 (1993) 1053–1055.

89 R.K. Bentsen-Farmen, I.V. Botnen, H. Notø, J. Jacob and S. Øvrebø, Detection of polycyclic aromatic hydrocarbon metabolites by high-pressure liquid chromatography after purification on immunoaffinity columns in urine from occupationally exposed workers, Int. Arch. Occup. Environ. Health, 72 (1999) 161–168.

W. Kleiböhmer (Ed.), *Environmental Analysis*
Handbook of Analytical Separations, Vol. 3

CHAPTER 5

Pesticides defined by matrix

John R. Dean and Lisa J. Fitzpatrick

*School of Applied and Molecular Sciences, University of Northumbria at Newcastle, Ellison Building,
Newcastle upon Tyne NE1 8ST, UK*

5.1 INTRODUCTION

Pesticide is a generic term applied to a range of compounds with insecticidal, herbicidal, fungicidal and rodenticidal activities. Unlike other chemicals, pesticides are quite unique inasmuch as they are deliberately released into the environment to control pests and disease. However, similarly to other chemicals their very nature makes them susceptible to dispersion in different environmental compartments. Of concern in this work is the presence of pesticides in both the aquatic and soil compartments.

5.1.1 Method validation

The need to validate analytical methods is paramount to establishing adequate methods for analysis of a range of analytes in different matrices. In metal analysis, for example, the number of and variety of solid matrices (environmental, food, industrial, metallurgical) available for which metal content has been certified is extensive. This is contrasted by the limited range of solid environmental matrices available for method validation/quality assurance schemes for organic analyses (Table 5.1). No certified reference materials exist for trace organics in aqueous media (Table 5.1). It is recognised that constraints on preparation, storage and use of CRMs are exacerbated by the difficulty in keeping both matrix and determinants homogenous and stable [2].

The lack of CRMs for trace organics in aqueous matrices is a major disadvantage in quality assurance. In order to overcome these problems, it is common practice to spike water samples in the laboratory at an appropriate level to provide an in-house quality assurance mechanism. In addition, for laboratories to produce data for environmental monitoring programmes, laboratory performance can be assessed by participation in external quality assessment programmes, such as, AQUACHECK in the UK. AQUACHECK is a proficiency testing scheme for potable and waste water operated from the Water Research Centre (WRC) in Medmenham, Marlow, Bucks, UK [2].

References pp. 172–173

TABLE 5.1

ENVIRONMENTAL REFERENCE MATERIALS

Pesticide	Matrix	Reference material [a]	Concentration range
Triazine herbicides (atrazine, simazine, terbuthylazine and trietazine)	Spiked distilled water	LGC1004	25.0–26.7 μg kg^{-1}
Uron herbicides (chlortoluron, diuron and isoproturon)	Spiked distilled water	LGC1005	25.8–30.5 μg kg^{-1}
p,p'-DDD, p,p'-DDE, p,p'-DDT, HCB and heptachlor	Coastal sediment	IAEA-357	1.5–46 μg kg^{-1}
p,p'-DDE, dieldrin, and γ-HCH	C18 SPE cartridge	LGC1007	1.54–1.59 μg

Table abstracted from [1].
[a] LGC = Laboratory of the Government Chemist, UK; IAEA = International Atomic Energy Agency, Austria.

In spite of the presence of a limited number of suitable reference materials for pesticides from solid matrices, it is still common practice to evaluate extraction techniques based on spikes to soil matrices. Two approaches are common: 'spot' spiking and 'slurry' spiking.

Spot spiking is where an aliquot of pesticide in organic solvent is added to a relatively large mass of soil. The small volume of solvent removed by evaporation either by heat or ambient temperature (depending on the solvent) and the sample extracted and analysed. In contrast, in the slurry spiking approach an aliquot of pesticide is added in a relatively large volume of organic solvent to the soil matrix. The solvent is then stirred, typically overnight in a fume cupboard, to remove solvent and provide a suitably homogeneous spiked sample. While both approaches are artificial and unrealistic in nature their usage is extremely common in the scientific literature.

5.2 EXTRACTION FROM WATER

The presence of pesticides in water is due to: (a) accidental spillage or seepage; (b) deliberate addition to control weeds, e.g. addition of a herbicide; (c) rain run-off from agricultural or domestic usage; or (d) malicious malpractice. However, irrespective of the source of the pesticides, their input into natural waters can only have long-term detrimental effects. The types of natural waters into which pesticides may be derived include rainwater, freshwater, drinking water and seawater. In addition, pesticides may also be present in waste water. The physical movement of natural and waste waters e.g. turbulence, flow or currents, leads to dilution of the pesticides thereby minimising the potentially harmful effects of the pesticides. It is this dilution of the pesticides in natural and waste waters that have led scientists to develop various methods of clean-up and/or preconcentration prior to chromatographic separation and detection.

Methods of preconcentration used have been historically based on liquid–liquid extraction (LLE). However, the 1970s saw the introduction of solid phase extraction (SPE). Developments in SPE, since the 1970s, have focused on: sorbent technology

TABLE 5.2

MAIN CLASSES OF PESTICIDES MONITORED IN AQUEOUS MATRICES

Pesticide class	Specific pesticide compounds	Matrix
Organochlorine (OCP)	α-,β- and γ-HCH, p,p'-DDE, p,p'-DDD, p,p'-DDT, o',p'-DDT, aldrin, endrin, isodrin, dieldrin, trifuralin, and endosulphan (I and II)	Seawater and freshwater
Organophosphorus (OPP)	Azinphos-methyl, azinphos-ethyl, fenthion, malathion, parathion, parathion-methyl, fenitrothion, dichlorvos, diazinon, fenchlorophos, coumophos and chlorfenvinphos	Drinking water, freshwater and estuarine water
Triazine herbicides	Simazine and atrazine	Drinking water, freshwater and estuarine water

Table adapted from [2].

(silica and polymeric phases); design of sorbent holders (cartridge and disk format); manifolds for multiple, sequential extractions; laboratory automation based on autosamplers and more recently robotics; and, on-line coupling of SPE with chromatographic separation/detection in remote locations.

The last decade has seen the introduction of the complimentary technique of solid phase microextraction (SPME). SPME is based on the partitioning of analytes with a silica-coated fibre. Desorption occurring in the hot injector of the gas chromatograph (most common approach) or via the mobile phase of a high performance liquid chromatograph. Subsequent chromatographic separation/detection allowing quantitation. This part of the review seeks to highlight the main features of each approach for preconcentration of analytes in aqueous matrices and recommend future developments.

5.2.1 Types of aqueous matrices

Water can be divided into five types: rainwater, freshwater, seawater, drinking water and waste water. Each of these types of water has certain characteristics and these are summarised in Table 5.2 and discussed below [2].

5.2.1.1 Rainwater

It is generally unbuffered and contains low levels of dissolved salts. The initial washout after a dry period may contain a high proportion of particulates, but the overall bulk of the precipitation is relatively free from colloidal or suspended material. The concentration of pesticides will depend on the proximity to industrial or agricultural sources.

5.2.1.2 Freshwater

The quality and content of freshwater is dominated by its source and location of the sampling point in the catchment and the flow of the system at the time. Soft water is low in ionic strength and frequently characterised by a heavy load of humic substances predominantly in high flow conditions. This in turn has a great effect on the amount of suspended solids and the natural and anthropogenic organic loading. The pH of the water will affect the precipitation of the colloidal humic materials and the associated trace organic contaminants through adsorption. Hard water can have a varying load of calcium and magnesium salts which will adsorb trace organic contaminants.

5.2.1.3 Drinking water

This is provided to a common minimum standard under European legislation and will have similar characteristics to its source. Although the suspended soils are removed from surface water supplies and the bacterial activity minimised by the disinfection, it can still contain quantities of colloidal humic substances and other dissolved components. Drinking water reflects the characteristics of the geology and land use of the catchment and the purification treatment.

5.2.1.4 Estuarine water

Estuarine waters have a high load of suspended solids from the riverine inputs. Changes in salinity and pH alone cause coagulation, flocculation and precipitation of metal oxides onto the suspended particulates creating a dynamic heterogeneous exchange between each state. These waters ultimately receive much of the run-off from agricultural inputs and the discharges from industrial waste.

5.2.1.5 Seawater

This is relatively more homogeneous over a wider area, although patchiness from a lack of mixing occurs frequently over and above the concentration gradients away from the estuarine and coastal regions. The high salt content and the relatively low concentrations of trace organics requires different analytical methodologies to those used for freshwater supplies. High volume sampling is often required to obtain a sufficient mass of the determinant. Emulsions are easily formed when extracting seawater with organic solvents. Chlorinated solvents are used to increase extraction efficiency without using a highly water soluble solvent. However, these chlorinated solvents have the added disadvantage since their densities are not too dissimilar to that of the seawater.

5.2.1.6 Waste water

The quality and content of waste water is extremely varied. The parameters which generally cover discharge contents include pH, chemical oxygen demand (COD), biological oxygen demand (BOD), suspended solids and the main anions (NO_3^-, SO_4^{2-}, Cl^- and PO_4^{3-}) and cations (Na^+, K^+, Ca^{2+} and Mg^{2+}). The varied concentration of dissolved organic matter can also produce an emulsion with LLE techniques and reduces extraction efficiency both for LLE and SPE by lowering the value of the octanol–water partition coefficient (k_{ow}). The suspended solids themselves are usually highly organic in nature and will absorb most lipophilic trace organic contaminants.

5.2.2 Main pesticides classes monitored in water

The main pesticides monitored in water are shown in Table 5.3 [2]. The main groups are the organochlorine pesticides, organophosphorus pesticides and triazine herbicides.

DDT [1,1,1-trichloro-2,2-bis(*p*-chlorophenyl)ethane] was banned from use in the 1970s in most Western European countries; however, its use and production continued elsewhere in the world. As a consequence, it is still regularly monitored for. Aldrin, endrin and dieldrin are regularly monitored together as a group. Although the main contaminant is dieldrin. Dieldrin has mainly been used as a mothproofer in the woollen textile industry, as an active ingredient in sheep dip and in timber treatment [2]. Of the other organochlorine pesticides, lindane (γ-HCH) is the most common with extensive use in timber treatment and as a general purpose insecticide. Most organochlorine pesticides (OCPs) are stable and persistent in the environment, only being converted in the environment to another toxic form. The OCPs are also water soluble enough to remain in the aqueous state at low concentrations (<1 μg l^{-1}). This solubility is enhanced by the presence of other soluble and particulate organic matter such as humic acids and detritus [2].

The organophosphorus pesticides (OPPs) were introduced as replacements for the persistent OCPs for use in sheep dip, agricultural, forestry sprays as insecticides and for sea lice treatment of infected salmon [2]. As the original intention was for these pesticides to be target and application specific, a range of different compounds were produced. In addition, as pests became resistant to attack more OCPs were developed. As a result a wide variety of OPPs are monitored.

All of the pesticides listed are environmental contaminants of interest. In order to protect the quality of surface and tap waters against the environmental impact of pesticides priority lists ('black' and 'red') have been published [3]. The pesticides listed in the EEC council directive (76/464) on pollution caused by certain dangerous substances discharged into the aquatic environment of the community (black list) are shown in Table 5.4. The EEC directive on the Quality of Water Intended for Human Consumption sets a maximum admissible concentration (MAC) of 0.1 μg l^{-1} per individual pesticide irrespective of its toxicity [4].

The main problems associated with the analysis of pesticides in natural and waste waters are focused on the following (suggested remedies appear in brackets): low concentrations and consequently large volumes required (need for effective preconcentration methods); adsorption of pesticides on walls of storage container (store samples at 4°C and for the minimum amounts of time); colloidal and suspended solids (filter samples through 0.2 μm filters); bacterial and microbial degradation (filter samples through 0.2 μm filters and store at 4°C); and, decomposition of pesticides by hydrolysis, dissociation etc. (separate pesticides from the water matrix, e.g. store on an SPE cartridge or disk).

5.3 LIQUID–LIQUID EXTRACTION

Liquid–liquid extraction (LLE) has been the most common method of preconcentration for aqueous samples due in part to the availability of pure solvents and low-cost

TABLE 5.3

CHARACTERISTICS OF NATURAL, POTABLE AND WASTE WATERS

Characteristic	Rainwater	Freshwater		Seawater		Drinking water	Waste water
		Low flow	High flow	Estuarine	Open ocean		
pH	3.5–6.8	5.0–8.5	4.2–8.5	7.5–8.5	7.5–8.5	5.0–8.0	3–10
Buffer capacity	Low	Medium	Low	Very high	Very high	Low	Medium–high
Salts	Low $(100 \ \mu g \ l^{-1})$	Medium $(1–100 \ \mu g \ l^{-1})$	Low	Very high $(3.5 \ g \ l^{-1})$	Very high $(3.5 \ g \ l^{-1})$	Low–medium	Low–high $(10–100 \ mg \ l^{-1})$
Total organic carbon (TOC)	Low $(<0.5 \ mg \ l^{-1})$	Low	Medium–high	Medium $(0.5–1 \ mg \ l^{-1})$	Low $(0.5–1 \ mg \ l^{-1})$	Medium–high $(<0.5 \ mg \ l^{-1})$	Low–high $(10–100 \ mg \ l^{-1})$
Total organic halogens (TOX)	Low $(5–10 \ \mu g \ l^{-1})$	Low–medium	Low $(5–10 \ \mu g \ l^{-1})$	Low $(5–10 \ \mu g \ l^{-1})$	Low	Medium–high	Medium–high $(10–100 \ \mu g \ l^{-1})$
Anthropogenic organics	Low $(1:10^{-12})$	Medium $(1:10^{-9})$	Low	Low–medium	Very low $(1:10^{-14})$	Medium	Medium–high $(1:10^{-6})$
Natural organics	Low $(1:10^{-12})$	Low	Medium–high $(1:10^{-6})$	Medium–high $(1:10^{-6})$	Medium–high $(1:10^{-6})$	Low	Medium–high $(1:10^{-6})$
Suspended solids	Low	Low $(0.1–1 \ mg \ l^{-1})$	Medium–high $(1–100 \ mg \ l^{-1})$	Medium–high	Very low $(0.01–1 \ mg \ l^{-1})$	Low	Medium–high
Sample volume	10–100 l	1–10 l	10–50 l	1–10 l	10–100 l	1–5 l	0.1–1 l

Table taken from [2].

TABLE 5.4

PESTICIDES LISTED IN 76/464/EEC COUNCIL DIRECTIVE ON POLLUTION CAUSED BY CER-
TAIN DANGEROUS SUBSTANCES DISCHARGED INTO THE AQUATIC ENVIRONMENT OF THE
COMMUNITY (BLACK LIST)

Aldrin	Disulphoton	Monolinuron
Atrazine	Endosulphan	Omethoate
Azinphos-ethyl	Endrin	Oxydemeton-methyl
Azinphos-methyl	Fenitrothion	Parathion-ethyl
Chlordane	Fenthion	Parathion-methyl
Coumaphos	Heptachlor	Phoxim
2,4-D	Hexachlorobenzene	Propanil
DDT	Linuron	Pyrazon
Demeton	Malathion	Simazine
Dichlorprop	MCPA	2,4,5-T
Dichlorvos	Mecoprop	Triazophos
Dieldrin	Metamidophos	Trichlorfon
Dimethoate	Mevinphos	Trifluralin

Taken from [3].

apparatus. The major disadvantage is the use of solvents, often chlorinated, and both
their initial cost and cost for disposal. Liquid–liquid extraction is based on the principle
that the aqueous sample is distributed or partitioned between two immiscible solvents in
which the pesticide(s) and matrix have different solubilities.

5.3.1 Theory of liquid–liquid extraction

The distribution coefficient is an equilibrium constant that describes the distribution of
the pesticide, A, between two immiscible solvents e.g. the aqueous sample matrix and
an organic phase. The process can be written as an equation:

$$A(\text{aq}) \leftrightarrow A(\text{org}) \tag{5.1}$$

where (aq) and (org) are the aqueous and organic phases, respectively. The ratio of the
activities of A in the two solvents is constant and can be represented by:

$$K_d = \{A\}_{\text{org}} / \{A\}_{\text{aq}} \tag{5.2}$$

where K_d is the distribution coefficient.

For a one-step LLE, K_d must be large i.e. >10, for quantitative recovery ($>99\%$) of
the pesticide in the organic solvent. However, practically two or three repeat extractions
are required with fresh organic solvent to achieve quantitative recovery.

5.3.2 Solvent extraction: procedure

The most common approach for LLE utilises a separating funnel. In this case, the
aqueous sample (1 l) is introduced in to a large separating funnel (up to 2 l capacity

with Teflon stopcock) and then an appropriate volume of organic solvent, e.g. 60 ml of dichloromethane, is added. After sealing, the separating funnel is shaken vigorously for 1–2 min. This shaking process maximises the contact between the two immiscible solvents, thus allowing efficient partitioning to occur. As a safety precaution it is necessary to vent excess pressure build-up generated during the shaking process. After a suitable resting period (minimum 10 min) the organic solvent is collected in a clean volumetric flask. The entire process is then repeated using fresh organic solvent; it is common to repeat the process up to three times in total. The organic extracts are then combined. However, as has already been stated the presence of pesticides in natural waters is often at the trace or ultra-trace level, therefore further preconcentration is required prior to chromatographic separation and detection. This is achieved using solvent evaporation methods. Solvent evaporation methods seek to reduce the organic solvent volume without loss of the pesticide(s). The most common approaches use a combination of reduced pressure, heat or an inert gas to reduce solvent volume. However, in all cases, the evaporation method is slow with the risk of contamination from the solvent, glassware and gas supply high.

Unfortunately the process of LLE is not always as straight-forward as has been described. For example, the formation of emulsions can be particularly troublesome. Remedies include centrifugation, filtration through glass wool, refrigeration, 'salting out' or the addition of a small amount of a different organic solvent, in order to break-up the emulsion. The reader is directed to more specialist texts for specific details.

As LLE uses large quantities of organic solvent, 180 ml of DCM to extract 1 l of sample, attempts have been made to reduce the quantity of organic solvent used. An approach to extract five pesticides from industrial waste water using LLE has been reported (in the paper, LLE is compared to SPE) [5]. The pesticides extracted were EPTC (*N,N*-di-*n*-propyl-*S*-ethylthiocarbamate), propachlor, AD-67 (*N*-dichloroacetyl-1-oxo-4-aspiro-4,5-decane), Aktinit (2-chloro-4-ethylamino-6-isopropylamino-*s*-triazine) and acetachlor (pesticides that are widely used and manufactured in Hungary). To a 250-ml sample of industrial waste water as added 10 ml of a saturated sodium chloride solution and 10 μl internal standard (3-nitrophenol). The sample was extracted, by shaking, with 30 ml DCM in a 500-ml volume separating funnel for 10 min. After resting for 15 min, the sample was further extracted with two further quantities of DCM (10 ml only). (In this way, only 50 ml of DCM is required to extract the pesticides). The three fractions were then quantitatively transferred to a 100-ml flask. This extract was dried over anhydrous sodium sulphate and concentrated to 1 ml by using a Kuderna–Danish evaporator. Analysis of the extracts was via GC–MS. Extraction of diluted standard solutions, in the range 5–100 mg l^{-1}, gave recoveries of between $85 \pm 6.7\%$ for Aktinit and $90 \pm 7.4\%$ for acetochlor ($n = 4$). Similarly, detection limits were estimated to be in the range 5 μg l^{-1} for EPTC and propachlor and 10 μg l^{-1} for AD-67. Levels of the pesticides in industrial waste water samples ranged from 177 μg l^{-1} for acetochlor to <4 μg l^{-1} for propachlor.

Another approach sought to use microextraction as a means to reduce the quantity of organic solvent to extract organophosphorus pesticides from water [6]. Four OCPs (initial work was done on up to 13 OCPs) were selected (ethyl-parathion, fenitrothion, fonofos and phorate) with analysis by GC with a flame photometric or nitrogen–

phosphorus detector. An aqueous sample, 500 ml, was spiked with 0.1 or 10 μg l^{-1} of each individual OCP and 10 g of sodium chloride (with stirring). The solution was extracted with 2×1 ml of *n*-hexane for 10 min with stirring. In addition, a few drops of acetone were added to break emulsions. Average recoveries from environmental water at the 0.1 μg l^{-1} level was $89.8 \pm 6.1\%$ while at the 10 μg l^{-1} level was $91.2 \pm 4.4\%$. The approach was recommended for other pesticides.

5.4 SOLID PHASE EXTRACTION

Solid phase extraction (SPE) is the process whereby pesticides present in an aqueous sample are selectively adsorbed on to the surface of a solid sorbent phase [7,8]. After suitable washing to remove extraneous material, the pesticides are selectively desorbed using a suitable solvent. The solid sorbent phase is usually packed in to small tubes (cartridges) or is available in round, flat sheets (disks) that can be mounted in a filtration apparatus. Whichever design is used the aqueous sample is forced by pressure or vacuum through the sorbent. More recently, SPE sorbents have become available as a 96-well plate, designed for high throughput sample clean-up/preconcentration as part of an automated system.

Solid phase extraction sorbents can be sub-divided in to three classes, normal phase (non-polar liquid phase, polar modified solid phase), reversed phase (polar liquid phase, non-polar modified solid phase) and ion exchange (electrostatic attraction of charged group on compound to a charged group on the sorbent's surface). The most common sorbents are based on silica particles (irregular shaped particles with a particle diameter between 30 and 60 μm) to which functional groups are often bonded e.g. C_{18}. While silica is probably still the most common sorbent, a resurgent interest in macroreticular polymeric phases is evident [9]. These phases are based on poly(styrene-divinylbenzene).

The process of SPE can be divided in to five distinct steps (Fig. 5.1) as follows (using a silica-bonded C_{18} sorbent as an example): wetting of the sorbent to allow the bonded alkyl chains to be solvated; conditioning of the sorbent with a similar solvent or buffer as the sample solution; loading of the sample allows retention of the pesticides; rinsing or washing the sorbent to elute extraneous material; and, finally elution of the analyte of interest. By use of the minimum solvent for elution and a large initial sample volume on to the sorbent a significant preconcentration of the analyte of interest can be affected. Successful SPE obviously requires careful choice of the SPE sorbent, the solvent systems to be used and their influence on the pesticides of interest. Selected methods of analysis for SPE are highlighted in Table 5.5.

5.4.1 Automation and SPE

Automated SPE can offer many advantages for sample preparation in the modern laboratory. In choosing the level of automation it is important to consider all options. The following is a guide to advise users prior to implementation [18].

TABLE 5.5

SELECTED METHODS FOR OFF-LINE SPE

Pesticides	Matrix	Sorbent	Wetting/conditioning	Sample loading	Rinsing	Elution	Comments	Reference
OCPs	Drinking and surface water	C18 cartridge	2×5 ml methanol followed by 2×2 ml of water	1 l at 10 ml min^{-1}	None	2×2.5 ml of n-pentane and dichloromethane $(1:1, v/v)$	Extract combined and evaporated to 0.5 ml prior to GC–ECD analysis	[10]
OCPs	Water	C18 or C8, empore disk	10 ml of methanol followed by 10 ml of distilled water	0.5 l	None	5 ml of ethyl acetate and 5 ml of hexane	Extract evaporated to 0.2 ml prior to GC–ECD analysis	[11]
OPPs, OCPs and other pesticides	Water	C18 cartridge	Consecutive additions of 5 ml of isooctane, 5 ml of ethyl acetate, 5 ml of methanol and 10 ml of deionised water	1 l, pH 6.5 at a flow rate of 10–15 ml min^{-1} under vacuum	10 ml of deionised water and dried by aspirating air for 30 min	3×0.5 ml ethyl acetate and 3×0.5 ml isooctane	Extract dried over anhydrous sodium sulphate and washed with an additional 0.5 ml of each eluting solvent. Combined extracts concentrated to dryness under a stream of nitrogen. Residue dissolved in 1 ml hexane prior to GC–ECD or FPD analysis	[12]
Fenamiphos and metabolites	Mineral water	C18 cartridge	5 ml ethyl acetate and 5 ml methanol	100 ml, pH 7	None	Sorbent dried for 15–20 min and then eluted with 5 ml ethyl acetate	Extract evaporated under a stream of nitrogen prior to GC–MSD analysis	[13]
10 Carbamates	River water	C18 Empore disks	10 ml methanol under vacuum, then 10 ml acetonitrile (after drying). Subsequently, 30 ml water ensuring disk does not become dry prior to addition of sample	2 l	None	After drying for 1 h under vacuum, elution with 2×10 ml acetonitrile	Extract evaporated to dryness and residue diluted in 500 µl methanol prior to analysis by HPLC–MS	[14]

TABLE 5.5 (CONTINUED)

Pesticides	Matrix	Sorbent	Wetting/conditioning	Sample loading	Rinsing	Elution	Comments	Reference
Alachlor	Water	C18 Empore disks	10 ml ethyl acetate followed by drying for 2 min. Then, 10 ml acetone followed by drying for 5 min Afterwards, 15 ml methanol, immediately followed by 10 × 10 ml deionised water. Solvent application rate 0.5 ml s^{-1}.	25–300 ml at a rate of 0.5 ml s^{-1}. Sample allowed to dry for 30 min.	None	3 × 20 ml ethyl acetate	Extract evaporated to 0.1–1.0 ml prior to analysis by GC–ECCD and/or GC–MSD	[15]
8 Polar acid herbicides	Surface water	C18 cartridge	10 ml methanol followed by 10 ml of deionised water at pH 2.4–2.6.	500 ml acidified to pH 2.4–2.6 and containing 5 ml methanol and 1 ml of internal standard at a rate of 15 ml min^{-1}. Cartridge was then air-dried under vacuum for 5 min and centrifuged for 15 min at 500 g	0.75 ml methanol (and then discarded)	2 × 2 ml methanol at a rate of 2 ml min^{-1}.	Extract made up to 10 ml with methanol, derivatised followed by column clean-up prior to GC analysis	[16]
Polar thermally labile pesticides	Water	C18 cartridge	5 ml acetone, 5 ml methanol and 5 ml deionised water	1 l at a rate of 9–10 ml min^{-1}. Cartridge dried for 30 min under a stream of nitrogen	None	5 × 1 ml methanol	Extract evaporated under nitrogen to 1 ml for GC or 0.5 ml for HPLC	[17]

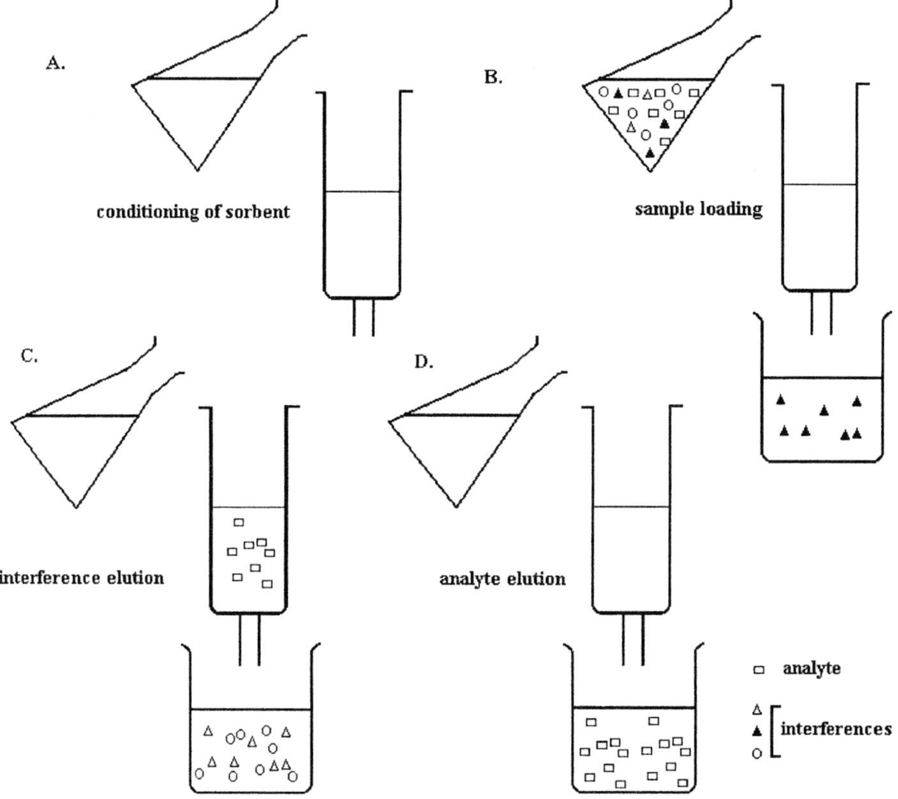

Fig. 5.1. Solid phase extraction: method of operation.

- **Degree of automation**. Five classifications are possible: (1) application specific workstations or modules; (2) on-line instruments such as autosamplers with preparative and liquid addition and transfer capabilities, switching valves, and on-line SPE instruments; (3) xyz liquid handlers; (4) robotic workstations with the ability to pick up and place labware; and (5) full robotic systems.
- **Cost and budget**. Implementation of laboratory automation is both costly in terms of time and capital investment. Careful consideration is required to acquire the best cost-to-benefit ratio for a laboratory.
- **Type and quantity of work**. It is important to consider the suitability of an automated system to the type and quantity of work being undertaken. Typical questions to be considered are: How many samples must be analysed? How many different assays must be accommodated and how frequently will changeovers occur? How similar are the different assays? How challenging are the analytical requirements of the different assays? Will the system be used for method development? What level of process documentation is required and how much of that information should be captured by the automated system?

- **Functionality**. To what extent does the system automate the process? Determine the exact steps that are automated on a particular system. For SPE it is pertinent to consider the following: capping–uncapping samples, weighing, loading a cartridge or block, conditioning, sampling, washing, eluting, vortexing or mixing, evaporating, adding reagents, controlling a vacuum manifold, centrifuging, bar-code labelling and reading, sealing, and injecting.
- **Flexibility and adaptability**. Both the hardware and software architecture determine the versatility of the system. In addition, to the functionality, it is important to consider whether the system can handle other processes and hence broaden its application, extend the useful life of the instrument by allowing adaptation to new technologies, or postpone obsolescence by allowing for upgrades.
- **Space and utilities**. The physical size of the automated SPE system can vary from a small portable unit to a very large system. Good planning involves consideration of the available space for the instrument, whether it will be a dedicated system or require to be shared, and if renovation to the workspace is required. In addition, different equipment requires different utilities such as electrical outlets, computer network and data acquisition ports, compressed air and gases, waste, drains and sinks, and ventilation.
- **Degree of difficulty**. The learning curve associated with operation of an automated sample preparation system depends on staff skill level and equipment complexity. Obviously simpler systems will begin to pay dividends almost immediately, while complex robotic systems may take months to fully operate.
- **Vendor support**. The quality of vendor support is very important in terms of installation, training, maintenance and consumables. Associated with each of these are the additional costs for warranties, servicing, upgrades and consumable costs.
- **Ruggedness**. Difficult to ascertain at the point of purchase unless you are aware of other users. If this is not the case, the use of a demonstration unit for first-hand experience before purchasing is normal. This trial period often provides insight about the quality and availability of technical and service support personnel in your area.
- **Speed, sample capacity and sample throughput**. Depending on the workload, the speed and sample capacity of a system may be important. Sample capacity can vary between 25 and 1000 or more samples. Sample preparation rates for SPE can vary between 3 samples h^{-1} to more than 100 samples h^{-1}, depending on the system. Caution in required in this area, as often speeding up the sample preparation can merely create a bottle neck elsewhere, and has resulted in little impact on the overall productivity of the application.
- **Processing mode**. Automated systems can vary in their mode of sample processing and hence system speed and throughput. For example, differences between serial or parallel sample processing.
- **On-line versus off-line analysis**. Off-line SPE systems offer the most flexibility and minimise coordination problems with other analytical devices. One system can prepare samples for multiple analytical systems. Advantages of operating in an on-line manner include a reduced cycle time for the overall process (analysis concurrent with sample preparation) and a seamless automated process (free of manual intervention steps).

- **Processing flow control**. Consideration should also be given to the means by which solvents are passed through an SPE column or filtration device. Processing flow control can involve using either vacuum, positive pressure, or centrifugation controlled manually or by the automated system.
- **Container format standardisation**. Standardisation of one or several container formats offers benefits such as keeping the automated process and consumable supply process simpler. However, consideration should be given to whether the current system can accommodate or be adapted to handle various sized and shaped SPE formats (e.g. cartridges, cassettes and microplates) and containers. What restrictions are associated with a particular system in terms of sorbent selection and bed volumes, vendor material selection, sample working volumes, and price of consumables? Will consumables continue to be available for a particular system?

A summary of currently available automated systems can be found in Table 5.6. (Note: The information presented is subject to change due to improvements/developments by manufacturers. For up to date information, you are advised to contact the manufacturer directly.)

Selected applications for on-line SPE coupled to HPLC are shown in Table 5.7. Example chromatograms are shown in Fig. 5.2 for 27 pesticides spiked in 100 ml of Amsterdam tap water at 0.1, 0.25 and 1.0 µg l^{-1} levels (1 = Aniline; 2 = Carbendazim; 3 = 1-(3-chloro-4-hydroxyphenol-3,3-dimethylurea; 4 = metamitron; 5 = chloridiazone; 6 = dimethoate; 7 = monomethyl metoxuron; 8 = aldicarb; 9 = bro-

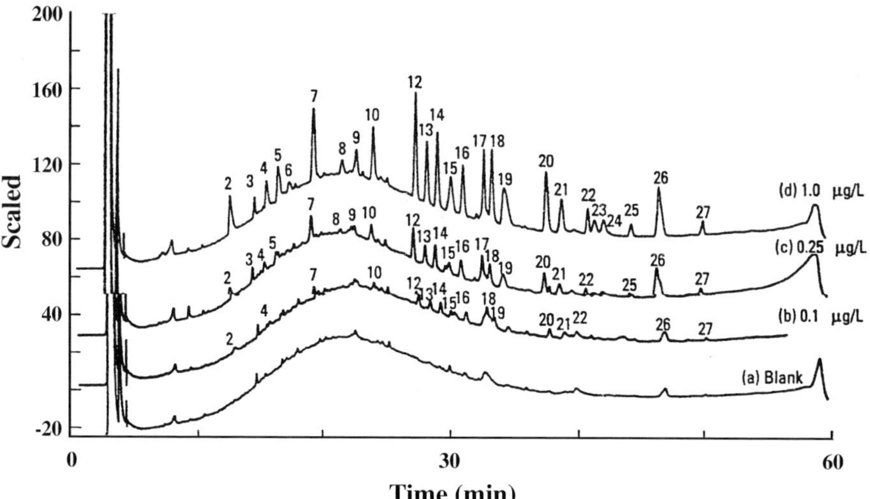

Fig. 5.2. Detection of 27 polar pesticides using SPE–HPLC–diode array detection. Analysis of 100 ml of Amsterdam tap water. (a) Raw tap water spiked at levels of (b) 0.1 µg l^{-1}, (c) 0.25 µg l^{-1}, and (d) 1.0 µg l^{-1}. Peak assignment: 1 = aniline; 2 = carb endazim; 3=1-(3-chloro-4-hydroxyphenyl)-3,3-dimethylurea; 4 = metamitron; 5 = chloridazone; 6 = dimethoate; 7 = monomethyl metoxuron; 8 = aldicar; 9 = bromacil; 10 = cyanazine; 11 = 2-nitrophenol; 12 = chlorotoluron; 13 = atrazine; 14 = diuron; 15 = metobromuron; 16 = metazachlor; 17 = propazine; 18 = warfarin; 19=3,3′-dichlorobenzidine; 20 = barban; 21 = alachlor; 22 = nitralin; 23 = dinoseb; 24 = dinoterb; 25 = phoxim; 26 = nitrofen; and, 27 = trifluralin [19].

TABLE 5.6

COMMERCIAL AUTOMATED SPE SYSTEMS

Product type	Product	Manufacturer[a]	Comments
SPE Application specific workstations or modules	Spe-ed Wiz	Applied Separations	Supports most manufacturers' 1-, 3- and 10-ml barrel-type SPE cartridges; solvent-reagent addition using eight individual lines; positive pressure with flow control; and, easy method development. *Limitations:* interruption for manual sample addition and placement of final elution tubes; 30-sample capacity; serial processing.
SPE Application specific workstations or modules	Rapid Trace	Zymark	Modular systems allows scaleup as required; automated waste segregation; excellent software; easy method development and transfer. *Limitations:* sampling mode (through valve) requires extensive rinsing and daily contamination; cannot be integrated–automated with a robotics system.
On-line instruments	Prospekt	Spark Holland (distributed by Jones Chromatography and Varian)	Fully-automated on-line sample preparation system; high pressure SPE; on-line elution for full mass injection; good method development capacity; column switching capability. *Limitations:* requires significant expertise to run method development and to troubleshoot instrument; works only with proprietary cartridges; eluate must be mobile-phase compatible.
On-line instruments	HP7686 solution-phase synthesizer (formerly known as PrepStation system)	Hewlett-Packard	Fully automated on-line sample preparation system; hydraulic pressure SPE with serial processing in autosampler-vial format; extensive functionality. *Limitations:* limited sample capacity (20–30 samples); format restricted to autosampler-vial shaped SPE cartridges; relatively slow processing speed.
On-line instruments	Asted XL system	Gilson	Similar features to Prospekt system except that it reuses a precolumn for trace enrichment SPE instead of a separate cartridge for each sample processed. *Limitations:* eluate must be mobile phase compatible.
xyz liquid handlers	MicroLab SPE	Hamilton	Proprietary six-port reagent addition nozzle in probe obviates the need to prime between reagent addition steps; positive pressure SPE with flow control; automated sample addition; automated collection tube placement using sliding racks; additional functionality available for evaporation, filtration and injection. *Limitations:* dated software; serial processing.
xyz liquid handlers	Aspec XL (I) and Aspec XL4	Gilson	New parallel processing with XL4 improves speed; great flexibility allows autosampling, dual-valve column switching, conversion to high-capacity vial or plate autoinjector; supports both cartridge and 96-well plate SPE formats. *Limitations:* questionable liquid-level detection feature; flexibility of instrument can be limited when operating from local controller; questionable injection-rinsing port design.

TABLE 5.6 (CONTINUED)

Product type	Product	Manufacturer[a]	Comments
xyz liquid handlers	Multiprobe 104DT and 204DT	Packard Instruments	Four variable spanning probes allow parallel processing and sampling from any size tube; operation in either fixed-probe or disposable-tip mode; designed to use Spec Plus SPE pipette tips; excellent liquid-level detection; automated vacuum manifold flow control. *Limitations:* easy, flexible, but somewhat awkward software; requires special attention to avoid sample dilution by the system liquid when operating in fixed-tip mode; manual intervention to place collection plate during 96-well SPE applications.
xyz liquid handlers	Genesis	Tecan	Eight variable spanning probes allows parallel processing and sampling from any size tube; operation in either fixed-probe or disposable-tip mode; optional pick-and-place capability with robotic arm. *Limitations:* challenging, awkward software; requires special attention to avoid sample dilution by the system liquid when operating in fixed-tip mode.
xyz liquid handlers	Hydra 96	Robbins	96-well pipette station; vacuum manifold that could be adapted for SPE. *Limitations:* requires considerable manual intervention; samples must be presented in 96-well microplate format.
xyz liquid handlers	Quadra	Tomtec Instruments (distributed by Wallac Inc.)	96-disposable-tip pipette station; fastest available SPE using 96-well plates (5–10 min plate^{-1}); capacity for as many as six microplate work positions; custom vacuum manifold for SPE. *Limitations:* samples must be presented in 96-well microplate format; limited deck size requires manual intervention to replenish reagents halfway through SPE process; manual intervention to place collection plate during 96-well SPE applications.
Robotic Workstations	Biomek 2000	Beckman Instruments	Good selection of attachable tools; excellent software; easily integrated with robotics system. *Limitations:* no variable spanning multiple-tip tool.
Robotic Workstations	Bohdan workstation	Bohdan Automation	Customised, application-specific workstations specialising in larger scale volumes and gradient SPE; large variety of functionality can be added to system; parallel processing using vacuum (two at a time) or positive pressure (eight at a time) SPE. *Limitations:* No current 96-well SPE applications.

TABLE 5.6 (CONTINUED)

Product type	Product	Manufacturer[a]	Comments
Robotic Workstations	C-300 workstation	Cyberlab	Customised, application-specific workstations; head of robotic arm can have multiple tools attached simultaneously; fully automated 96-well SPE workstation capability, including 96-tip pipetting; custom vacuum manifold with flow control. *Limitations:* versatile software package can be confusing; cannot automatically attach and park tools (may limit functionality).
Full Robotic systems	Zymate robotics system	Zymark	High degree of flexibility and functionality; customised systems and applications with few format restrictions; very high capacity and throughput potential; automated process documentation. *Limitations:* high level of expertise needed to implement and maintain system; difficult, costly and time consuming to implement; space and utility requirements.
Robotic Workstations	Orca system	Sagian	High degree of flexibility and functionality; customised systems and applications with few format restrictions; very high capacity and throughput potential; automated process documentation. *Limitations:* high level of expertise needed to implement and maintain system; difficult, costly and time consuming to implement; space and utility requirements.
Robotic Workstations	A465 robot system	CRS Robotics	High degree of flexibility and functionality; customised systems and applications with few format restrictions; very high capacity and throughput potential; automated process documentation. *Limitations:* high level of expertise needed to implement and maintain system; difficult, costly and time consuming to implement; space and utility requirements.
Robotic Workstations	Allekto robotics system	SAIC, Laboratory Sensors and Automation	High degree of flexibility and functionality; customised systems and applications with few format restrictions; very high capacity and throughput potential; automated process documentation. *Limitations:* high level of expertise needed to implement and maintain system; difficult, costly and time consuming to implement; space and utility requirements.

Table adapted from [18].

[a] Applied Separations (Lehigh Valley, PA); Beckman Instruments Inc. (Sunnyvale, CA); Bohdan Automation, Inc. (Mundelein, IL); CRS Robotics (Burlington, PQ, Canada); Cyberlab, Inc. (Brookfield, CT); Gilson, Inc. (Middleton, WI); Hamilton Co. (Reno, NV); Hewlett-Packard Co. (Wilmington, DE); Jones Chromatography (Lakewood, CO); Packard Instrument Co. (Meriden, CT); Robbins (Sunnyvale, CA); Sagian, Inc. (a subsidiary of Beckman Instruments Inc., Indianapolis, IN); SAIC, Laboratory Sensors and Automation (Seattle, WA); Spark Holland BV (Emmen (The Netherlands); Tecan (Research Triangle Park, NC); Tomtec Instruments (distributed by Walalc, Inc., Gaithersburg, MD); Varian Instruments (Palo Alto, CA); Zymark Corp., (Hopkinton, MA).

TABLE 5.7

SELECTED APPLICATIONS FOR ON-LINE SPE COUPLED TO HPLC

Pesticide	Matrix	SPE conditions	Comments	Reference
27 Polar pesticides	Amsterdam tap water (100 ml) spike level 0.1, 0.25 and 1.0 μg l^{-1}	C18 (5–10 mm length $\times$ 2–3 mm i.d.) or PLRP-S; rate 5–10 ml min^{-1}	Reversed phase gradient elution using acetonitrile–aqueous phosphate buffer followed by diode array UV detection.	[19]
10 Pesticides	Drinking water (150 ml) spike level 0.3 μg l^{-1}	PRLP-S (10 $\times$ 2.0 mm i.d.)	Reversed phase gradient elution using acetonitrile–aqueous phosphate buffer at pH 7, flow rate 1 ml min^{-1} followed by UV detection.	[20]
12 Triazine herbicides	Drinking water (300 ml) spike level 0.1 μg l^{-1}	PRLP-S (10 $\times$ 2.0 mm i.d.)	Reversed phase gradient elution using acetonitrile–aqueous phosphate buffer at pH 7, flow rate 1 ml min^{-1} followed by UV detection at 220 nm.	[20]
16 Phenyl urea herbicides	Drinking water (300 ml); spike level 0.1 μg l^{-1}.	PRLP-S (10 $\times$ 2.0 mm i.d.)	Reversed phase gradient elution using acetonitrile–aqueous phosphate buffer at pH 7, flow rate 1 ml min^{-1} followed by UV detection at 249 nm.	[20]
Simazine, atrazine, methomyl, oxamyl, MCPA and bentazone	Tap and river water (50, acidified to pH 2.5; 50, 100 and 200 ml volumes used; spike level 2 μg l^{-1}. Addition of 10% Na_2SO_3 to the samples solved problems associated with fulvic and humic acids	Carbopack B 120/400; HYSphere-1; and, Bond Elut PPL (all 10 $\times$ 3 mm i.d.). HYSphere-1 was found to give the highest recoveries	Reversed phase gradient elution using acetonitrile–water (pH 3), flow rate 1 ml min^{-1} followed by UV detection at 240 nm (230 nm for MCPA).	[21]
Phenylurea herbicides (fenuron, methoxuron, chlorotoluron, metobromuron and chloroxuron)	Water, 10 ml at a rate of 1 ml min^{-1}. Spike level 40 μg l^{-1}. Also chromatograph at 40 ng l^{-1}	Polydimethylsiloxane (5 cm $\times$ 4.6 mm i.d.)	Reversed phase LC–UV–MS gradient elution using acetonitrile–water (containing 0.1%, v/v formic acid). Flow rate 40 μl min^{-1} for MS detection.	[22]

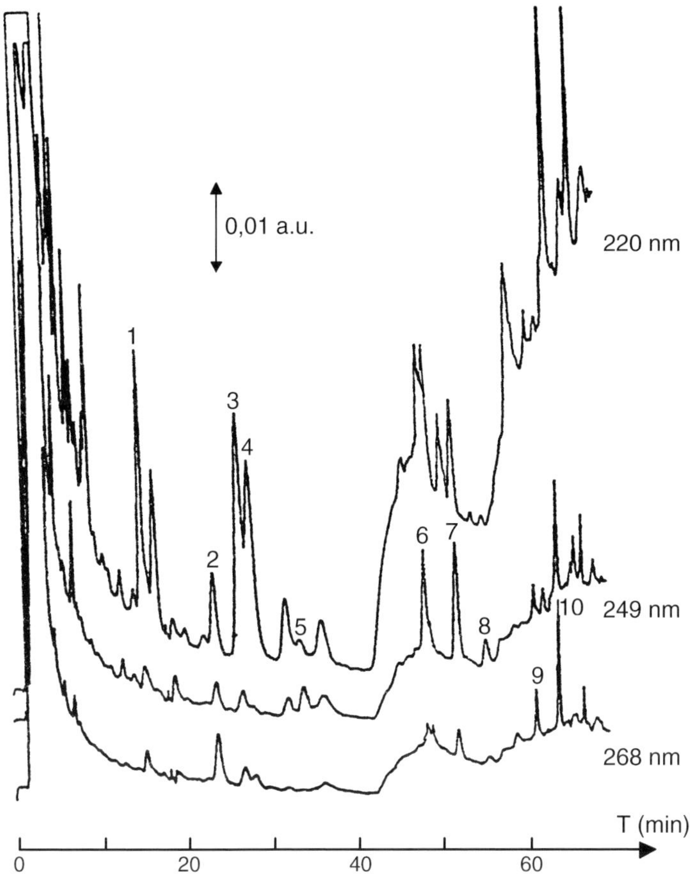

Fig. 5.3. Chromatograms corresponding to the on-line elution of 150 ml of drinking water at different wavelengths; drinking water spiked with 0.3 μg l^{-1} of pesticides. Peak identification: 1 = simazine; 2 = methabenzthiazuron; 3 = atrazine; 4 = carbaryl; 5 = isoproturon; 6 = propanil; 7 = linuron; 8 = fenamiphos; 9 = fenitrothion; and, 10 = parathion. Pre-column, PLRP-S; analytical column, Varian ODS (25 × 0.46 cm i.d.); flow rate, 1 ml min^{-1}; acetonitrile gradient with 0.05 M phosphate buffer at pH 7, gradient 30% acetonitrile from 0 to 38 min; 30–45% from 38 to 44 min; 45–47% from 44 to 52.5 min, 47–100% from 52.5 to 70 min [20].

macil; 10 = cyanazine; 11 = 2-nitrophenol; 12 = chlorotoluron; 13 = atrazine; 14 = diuron; 15 = metobromuron; 16 = metazachlor; 17 = propazine; 18 = warfarin; 19 = 3,3′-dichlorobenzidine; 20 = barban; 21 = alachlor; 22 = nitralin; 23 = dinoseb; 24 = dinoterb; 25 = phoxim; 26 = nitrofen; and, 27 = trifluralin) [19]; Fig. 5.3 for drinking water spiked at the 0.3 μg l^{-1} level with ten pesticides (1 = simazine, 2 = methabenzthiazuron, 3 = atrazine, 4 = carbaryl, 5 = isoproturon, 6 = propanil, 7 = linuron, 8 = fenamiphos, 9 = fenitrathion, 10= parathionethyl) [20]; Fig. 5.4, 12 triazine herbicides spiked at the 0.1 μg l^{-1} into 300 ml of drinking water (1 = de-isopropyltriazine, 2 = hydroxyatrazine, 3 = de-ethylatrazine, 4 = hexainone, 5 = simazine, 6 = cyanazine, 7 =

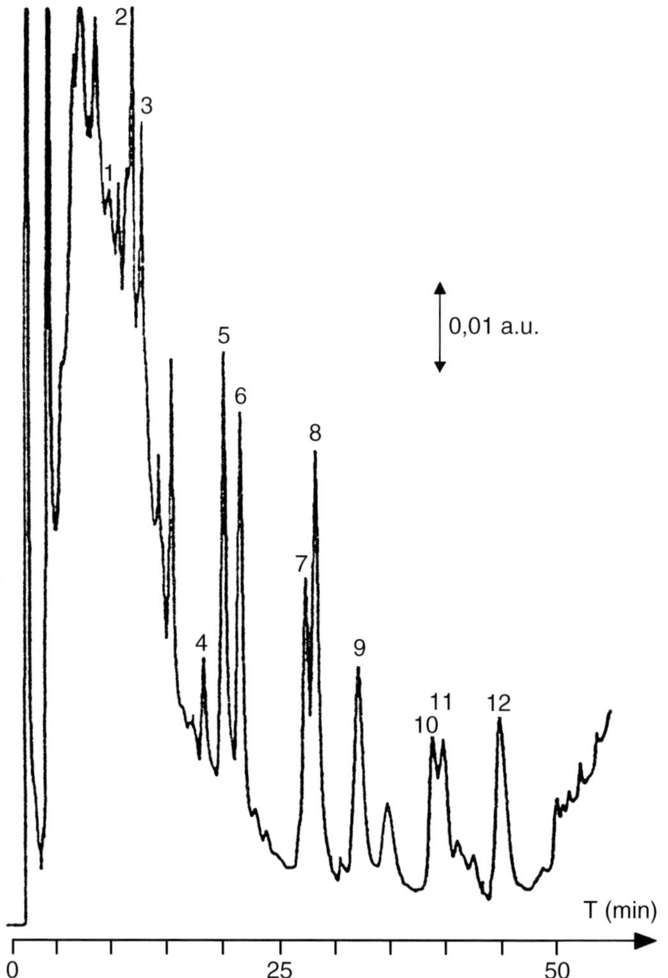

Fig. 5.4. Preconcentration on PLRP-S of 300 ml of drinking water spiked at 0.1 µg l^{-1} with triazines. Peak identification: 1 = deisopropylatrazine; 2 = hydroxyatrazine; 3 = deethylatrazine; 4 = hexazinone; 5 = simazine; 6 = cyanazine; 7 = simetryne; 8 = atrazine; 9 = prometon; 10 = sebutylazine; 11 = propoazine; 12 = terbutylazine. Analytical column, Varian ODS (25 × 0.46 cm i.d.); flow rate, 1 ml min^{-1}; acetonitrile gradient with 0.05 M phosphate buffer at pH 7, gradient 15–30% acetonitrile from 0 to 9 min, 30–34% from 9 to 16 min; 34–40% from 16 to 45 min and 40–60% from 45 to 55 min; detection at 220 nm [20].

symetryne, 8 = de-ethylatrazine, 9 = prometon, 10 = sebutylazine, 11 = propazine, 12 = terbutylazine) [20]; and, Fig. 5.5, 16 phenyl urea herbicides spiked into drinking water at the 0.1 µg l^{-1} (1 = fenuron, 2 = metoxuron, 3 = monuron, 4 = metabenzthiazuran, 5 = chlortoluron, 6 = fluometron, 7 = monolinuron, 8 = isoproturon, 9 = diuron, 10 = difenzoxuron, 11 = buturon, 12 = linuron, 13 = chloroxuron, 14 = chlorbromuron, 15 = difluorbenzuron, 16 = neburon) [20]. It should be noted in all cases that the detection of the analytical peaks after on-line SPE coupled with HPLC with ultraviolet

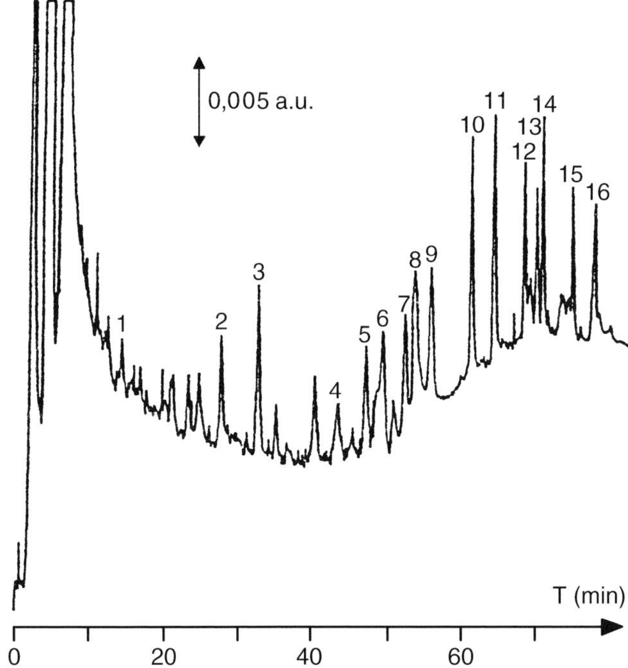

Fig. 5.5. Preconcentration on PLRP-S of 300 ml of drinking water spiked at 0.1 μg l^{-1} with phenylureas. Peak identification: 1 = fenuron; 2 = methoxuron; 3 = monuron; 4 = methabenzthiazuron; 5 = chlortoluron; 6 = fluometuron; 7 = monolinuron; 8 = isoproturon; 9 = diuron; 10 = defenoxuron; 11 = buturon; 12 = linuron; 13 = chloroxuron; 14 = chlorbromuron; 15 diflubenzuron; 16 = neburon. Analytical column, Varian ODS (25 × 0.46 cm i.d.); flow rate, 1 ml min^{-1}; acetonitrile gradient with 0.05 M phosphate buffer at pH 7, gradient 15–30% acetonitrile from 0 to 9 min, 30–34% from 9 to 16 min; 34–40% from 16 to 45 min and 40–60% from 45 to 55 min; detection at 220 nm [20].

detection occurs on top of an elevated broad-based background of ultraviolet absorbing compounds. It is likely that the elevated background is due to the presence of humic substances which have similar chromatographic and ultraviolet characteristics to the pesticides under investigation. The ability to quantify the analytical peaks above the background humic substances is due to the software capabilities of the data handling system.

5.5 SOLID PHASE MICROEXTRACTION

Solid phase microextraction (SPME) is the process whereby an analyte is adsorbed onto the surface of a silica-coated fibre as a method of concentration. This is followed by desorption of the pesticides into a suitable instrument for chromatographic separation and detection. The most important stage of this two-stage process is the adsorption of the pesticide onto a suitably silica-coated fibre or stationary phase. The choice of sorbent is critical, in that it must have a strong affinity for the target pesticides, so that preconcentration can occur from aqueous samples. The most common stationary phases

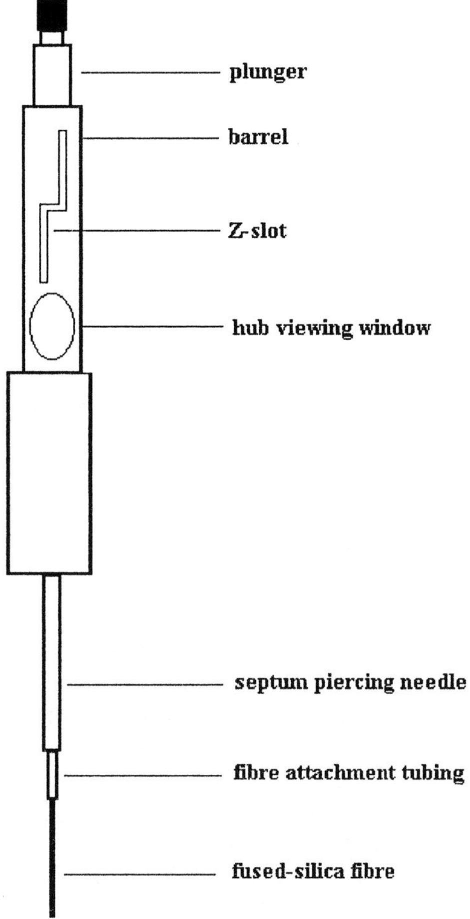

Fig. 5.6. Schematic diagram of a solid phase microextraction device.

for SPME are the non-polar, poly(dimethylsiloxane) and the polar, polyacrylate. The coated fibre is mounted in a syringe-like holder (Fig. 5.6) [23]. The holder has two functions, the first is to provide protection for the fibre during transport, while the second function is to allow piercing of the rubber septum of the gas chromatograph injector via a needle. As the normal method of sample introduction into a gas chromatograph is via a syringe, the use of an alternative syringe-type device (SPME) offers no additional complexity. The majority of applications of SPME have been when coupled to a gas chromatograph (GC), although some recent applications have focused on its use for high performance liquid chromatography (HPLC). In the case of GC, desorption occurs in the hot injector while the latter relies on the mobile phase for desorption.

5.5.1 Theoretical considerations

The partitioning of pesticides between an aqueous sample and a stationary phase is the main principle of operation of SPME. A mathematical relationship for the dynamics of the absorption process was developed by Louch et al. [24]. In this situation, the amount of pesticide absorbed by the silica-coated fibre at equilibrium is directly related to its concentration in the sample, as shown below:

$$n = K V_2 C_o V_1 / K V_2 + V_1$$

where n = number of moles of the analyte absorbed by the stationary phase; K = partition coefficient of pesticide between the stationary phase and the aqueous phase; C_o = initial concentration of pesticide in the aqueous phase; V_1 = volume of the aqueous sample; and V_2 = volume of the stationary phase.

It is unlikely that the values of K are large enough for exhaustive extraction of pesticides from the sample. SPME is therefore an equilibrium method, but provided proper calibration strategies are followed, can provide quantitative data.

5.5.2 Applications of SPME

An extensive study of nitrogen-containing herbicides was reported by Boyd-Bowland and Pawliszyn [25] using a 95-μm polyacrylate SPME fibre. The 22 herbicides chosen were from the following classes: triazines (5), nitroanilines (5), substituted uracils (2), thiocarbamates (6) and miscellaneous (4). A variety of detectors for GC (MS, NPD and FID) were compared for their sensitivity towards the herbicides; MS detection was found to be the most sensitive, followed by the NPD with the FID the least sensitive. The following operating parameters for SPME were investigated: equilibration time (50 min); desorption temperature and time (5 min desorption at 230°C); linearity of the method investigated over the range 0.1–1000 ng ml^{-1} (the majority of pesticides, 21, had correlation coefficients >0.99); precision (7–22% RSD, based on $n = 7$ at a concentration of 10 ng ml^{-1}); and, addition of salt and/or extraction at pH 2 investigated (the results were pesticide class dependent). The calculated limits of detection for the 22 herbicides ranged from 0.01 (metolachlor and oxadiazon) through to 15 ng l^{-1} (propachlor) using MS detection.

The same group extended their approach to the simultaneous determination of 60 pesticides in water using SPME and GC–MS [26]. In this case, the pesticides chosen included representatives from the following classes: organonitrogen, organochlorine and organophosphorus pesticides. Two types of SPME fibre were evaluated (95 μm polyacrylate and 100 μm polydimethylsiloxane). Detection limits for all pesticides and irrespective of fibre ranged from 0.1 to 60 ng l^{-1}. The method was applied for the determination of pesticides in Russian and Canadian Arctic samples (ice and surface snow). The highest reported concentrations in the Canadian Arctic samples were for δ-BHC (491 ng l^{-1} in surface snow), dieldrin (203 ng l^{-1} in surface snow) and endosulfan II (567 ng l^{-1} in ice core). However, these concentrations were relatively low compared to the Russian Arctic samples were concentrations as high as 4000 ng

1^{-1} for *p,p'*-DDE, 4200 ng 1^{-1} for isopropalin, 2200 ng 1^{-1} for trifluralin, 2300 ng 1^{-1} for prothiofos and 2100 ng 1^{-1} for methoxychlor were reported. The precision was estimated to be between 2 and 20% RSD, based on triplicate determinations. The Arctic sample results indicated that pollution by pesticides is a serious environmental concern considering that some of the compounds detected have been banned from use in North America and Western Europe for 15 years.

The results of an inter-laboratory study on the applicability of SPME for the analysis of pesticides in water has been reported [27]. Eleven laboratories in Europe (Italy, Poland, Germany, Belgium) and North America took part in the test. The test sample contained 12 pesticides (dichlorvos, EPTC, ethoprofos, trifluralin, simazine, propazine, diazinon, methyl chlorpyrifos, heptachlor, aldrin, metolachlor and endrin) as representatives of all the main pesticide groups (organochlorine, organonitrogen and organophosphorus). Each laboratory was given a test procedure which included the following steps:

(1) Conditioning of the column according to the manufacturer's specifications and check of the column blank.
(2) Conditioning of the fibre (100 μm polydimethylsiloxane silica coated) according to the manufacturer's specifications and check of the fibre blank.
(3) For quadrupole MS users only: syringe injection of the 10 ppm standard in order to establish the retention times of the pesticides and set up the selected-ion monitoring (SIM) acquisition method.
(4) Preparation of a 30-ppb aqueous standard followed by SPME/GC–MS analysis and subsequent carryover check.
(5) Repeated analysis of a freshly prepared 30-ppb aqueous standard.
(6) Analysis of 10- and 1-ppb aqueous standards in a similar way.
(7) Determination of the calibration curves for all the pesticides.
(8) Preparation of the aqueous solution of the blind sample followed by SPME analysis.
(9) Calculation of the concentration of the analytes in the blind aqueous sample.

The test procedure specified that extractions should be carried out with samples vigorously stirred. The extraction time was set at 45 ± 0.5 min. The chromatographic conditions were as follows: injector and transfer line temperature, 250°C; temperature programme, 40°C held for 5 min, increased at 30°C min^{-1} to 100°C, at 5°C min^{-1} to 250°C and at 50°C min^{-1} to 300°C, held for 1 min. Each participant was required to report the results using the forms included in the test kit and to attach the following: chromatograms of the column blank and fibre blank; chromatogram of the syringe injection and spectra of the pesticides (quadrupole MS only); chromatogram of the 30 ppb standard (all systems) and the spectra of the pesticides (ion trap MS only); chromatograms of the remaining standards (one for each level of quantitation); sample chromatogram of the carryover determination; chromatograms of the unknown sample; and, a copy of the spreadsheet used for the calculation.

A comparison of the confidence interval of the gross average results and the confidence interval of the 'true' values is shown in Fig. 5.7; overlap between the two respective values are due to random factors. The results were characterised by satisfactory repeatability, re-producibility and accuracy in all cases. In contrast to other methods, it was suggested that SPME offers several significant advantages, including complete elimination of solvents,

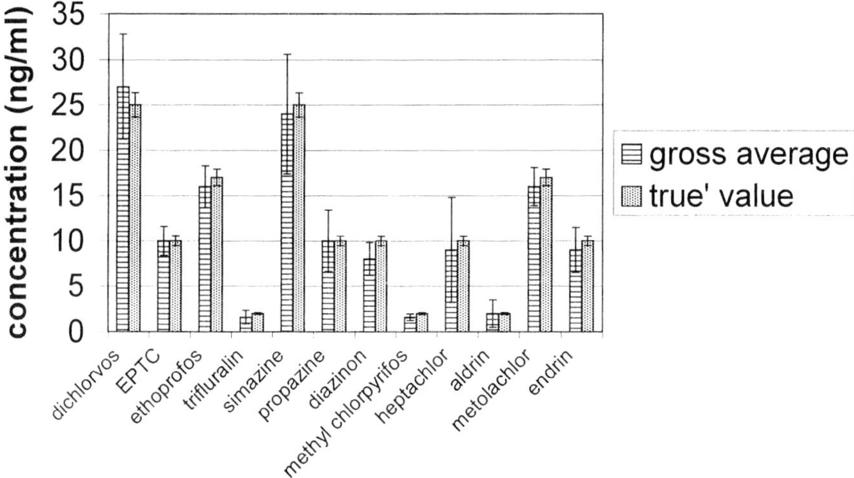

Fig. 5.7. Results of a round robin test.

very low cost related to reusability of the fibre and no requirements for dedicated instrumentation, over-all simplicity and time savings. In addition, the sensitivity of the method is very good, and can be further improved by optimisation of the analytical procedure (coating selection, matrix modification, etc.).

A further inter-laboratory validation of SPME for the determination of triazine herbicides and their degradation products in water samples has been reported [28]. Ten laboratories, all in Italy, took part in the test. The test sample contained nine triazine herbicides (atrazine, ametryn, cyanazine, desethylatrazine, deisopropylatrazine, propazine, prometryn, simazine and terbuthylazine). In addition, to assessing the accuracy and precision of SPME (as done in previous inter-laboratory validations [27]) these workers additionally assessed the ability of SPME to analyse at concentrations near the limit of 0.1 μg l^{-1} for individual pesticides in European drinking water [4] and the analysis of polar metabolites of the triazine herbicides. The optimum conditions used for the inter-laboratory study were: a 65-μm carbowax–divinylbenzene-coated fibre, extraction for 30 min at ambient temperature and neutral pH from a rapidly stirred sample containing 0.3 g ml^{-1} NaCl. Desorption of the fibre was done directly in the injection port of the GC for 5 min at 240°C with a split–splitless injector, were the split was closed immediately before the injection and kept closed during the desorption. The approximate temperature programme was 50°C held for 5 min, rapidly to 150°C and then 5°C min^{-1} to 250°C using a 5% phenylmethylsilicone or similar stationary phase. Detection was achieved with an NPD, ion-trap MS or quadrupole MS. Calibration was achieved by extraction/analysis of spiked mineral water samples containing 50, 100, 200 and 500 ng l^{-1} of each herbicide. Correlation coefficients in all cases were >0.99.

It was noted that two of the ten laboratories who participated encountered severe problems regarding linearity and repeatability in the calibration and hence could not perform the quantitation. This was reported to be due to either detection problems or deposition of salt on the liner of the GC injector. The latter was remedied by washing

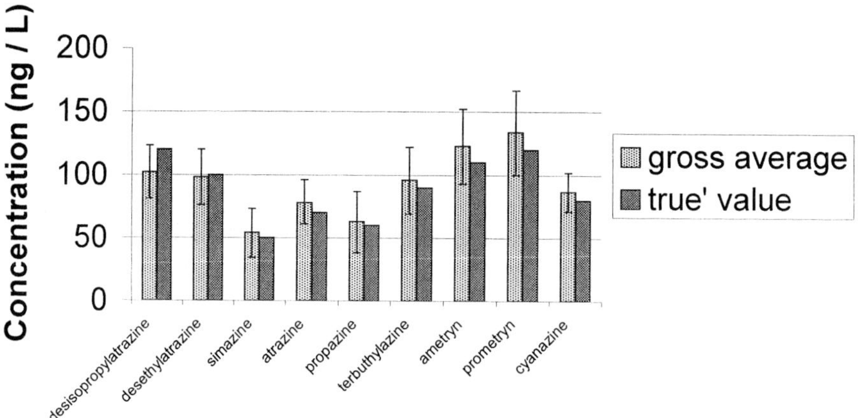

Fig. 5.8. Inter-laboratory validation of SPME for the determination of triazine herbicides.

the fibre with clean water before inserting into the GC injection port for desorption. No losses were noted for the triazine herbicides. It was also noted that not all laboratories reported values for the polar metabolites (desethylatrazine and deisopropylatrazine) and cyanazine. In the case of deisopropylatrazine, only one laboratory value was reported. Detection limits, calculated on the basis of a signal-to-noise ratio of 3 from the 50 ng l^{-1} sample, ranged from 0.1 ng l^{-1} for terbuthylazine to 20 ng l^{-1} for the polar metabolites using the NPD. The relative repeatability standard deviations were between 6 and 14% while the relative reproducibility standard deviations varied from 10 to 17%. Accuracy was assessed by the analysis of an unknown reference sample. The results, mean $\pm$ confidence interval are compared to the 'true' values (Fig. 5.8) and represent good accurate determinations at the ng l^{-1} level. It was concluded that reliable quantitation could be performed at concentrations below the European limit for individual pesticides in drinking water, the precision was satisfactory for most routine analyses, and the accuracy was good in all cases.

5.6 EXTRACTION FROM SOIL

Normally, the analysis of environmental solid materials (soil, sludge or related matter) prior to chromatographic separation and detection requires some form of extraction with an organic solvent (or solvent mixture). This has been traditionally done by either heating or agitation of the organic solvent–solid mixture. The former has utilised such techniques as Soxhlet extraction or Soxtec extraction while the latter utilises sonication or shake-flask extraction. More recently other instrumental extraction techniques have been applied and these include supercritical fluid extraction, microwave-assisted extraction and accelerated solvent extraction. All these approaches are costly in terms of organic solvent usage (and disposal) or equipment costs.

The limited availability of certified reference samples for pesticides from solid matrices (Table 5.1), is detrimental to the development of robust methods of extraction.

Researchers have frequently attempted to circumvent this major limitation by spiking (spot and slurry) soil samples with known quantities of pesticides and then recovering them. This exercise is often limited in value, as it offers no degree of difficulty in terms of extraction capability, as the pesticides have not normally had time to interact with soil constituents (and age). Some researchers have advanced the process further and to provide a more realistic and appropriate sample by attempting to age the spiked soil.

Selected examples are used to highlight the different extraction techniques that have been used for removal of pesticides from soil matrices bearing in mind these limitations. However, it is first pertinent to describe the soil matrix.

5.6.1 What is soil?

Soil is formed through the gradual breakdown of rock, by several mechanisms, including weathering and erosion where the rock is gradually ground down to smaller and smaller particles. Soil can best be defined as a composition of five main components. These are clay minerals, organic matter, air, water and a living component. The quantities of these constituents vary according to the soil type and location [29]. In order to fully understand the effect the soil has on pesticides and the implications for extraction, it is necessary to look at these components in a little more detail.

5.6.1.1 Clay minerals

The effect of soil clay and organic matter is often cited in the literature as having significant effects on the extraction of pesticides [30–34]. The clay minerals, for instance, have been associated with the difficulties in extracting planar (or nearly planar) molecules from soil. There have been several hypotheses as to why this should be true. The main theory is thought to be due to the structure of the clay minerals [35]. Clay minerals are based primarily on silicates and oxides. They typically have a particle size of less than 0.002 mm and form a sticky wet looking mass when wet and clump together when they are dry. There are many types of clay minerals found in soil. They can be based on crystalline structure as in the case of gibbsite, an aluminium oxide, or amorphous (i.e. no regular structure), such as, calcite or Dolomite. But by far the most interesting is the behaviour of certain crystalline silicate minerals. Crystalline clay minerals can be further classified as having chain structure or layer structure. The most interesting are those with a layer structure. These come in two categories: 1 : 1 layers, such as, kaolinite (Fig. 5.9); and 2 : 1 layers, such as, montmorillonite (Fig. 5.10) [36].

The 1 : 1 layer structure is based on a repeating pattern of tetrahedrally (four co-ordinated) silicon and octahedrally (six coordinated) aluminium atoms (Fig. 5.9). The presence of oxygen atom and hydroxyl groups allows hydrogen bonding, and consequently helps to stabilise the crystal structure. The 2 : 1 layer is based on octahedrally coordinated aluminium atom sandwiched between two layers of tetrahedrally coordinated silicon atoms (Fig. 5.10). When water is introduced to either of these systems, the polar water molecules can get in between the layers causing swelling. As the clay dries, the layers return to their original interplanar distance. If the water is associated with pesticides, the pesticides then become retained in the layer structure when the

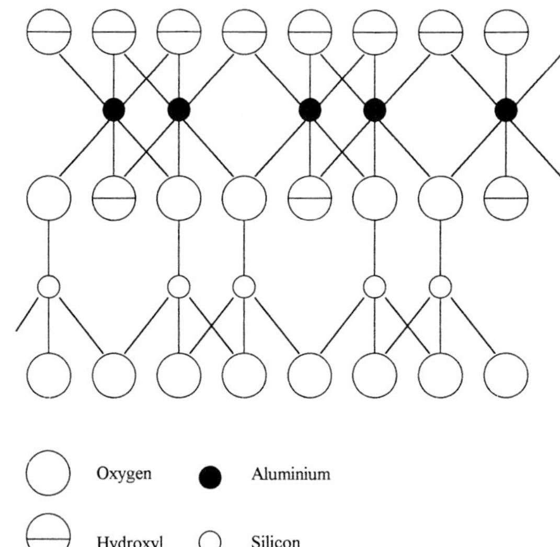

Fig. 5.9. Structure of kaolinite.

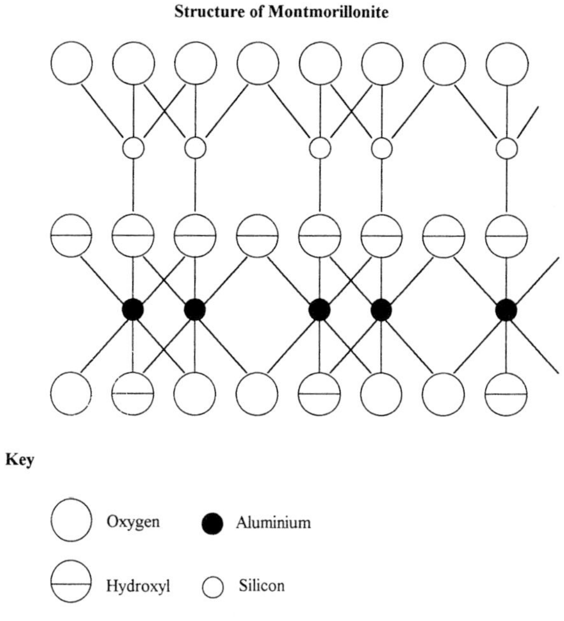

Fig. 5.10. Structure of montmorillonite.

clay is dry. Hence there is an implication for extraction. Since the bonds between the layers are relatively strong, the pesticide is unable to be removed. Extensive studies have been done on the adsorption of pesticides to clay minerals [36–39]. These studies

[36–39] have shown that after the clay has dried, the extraction of the pesticides is very difficult and low extraction efficiencies are common. The cation exchange capacity is also fundamental to the behaviour of the clay minerals. It is a measure of the ability of the clay to substitute metal ions from the lattice to the surrounding environment. These metal ions are not directly bonded to the lattice, but they balance out any charges within the crystal structure [36].

5.6.1.2 Organic matter

The organic matter content of any soil is dependent upon location. For example, a soil in an area with high vegetation, such as in a forest will have a larger amount of organic matter than a soil with very little flora around. The organic matter can be split into three distinct components, humic acid, and humin. Humin is the decomposing remains of both plants and animals. Humic and fulvic acids are not well characterised. They are not classed as single substances, but are a mixture of acids with similar properties. For example, humic acid is that fraction of humus that is soluble in dilute alkali, but is not precipitate upon addition of a mineral acid. Fig. 5.11 summarises the various fractions and their solubilities [40]. Organic matter has been studied by several groups and they have implied that the higher the organic matter content of a soil, the greater the adsorption of the analyte, hence reduced extraction efficiencies are obtained [30–32].

5.6.1.3 Water

Water plays a principal role in the soil environment. Not only does it affect plant growth, it can help to create or destroy soil structure. The relationship between soil and water is very complex, it affects a lot of the physical properties of a soil, for example the

Organic Matter Composition

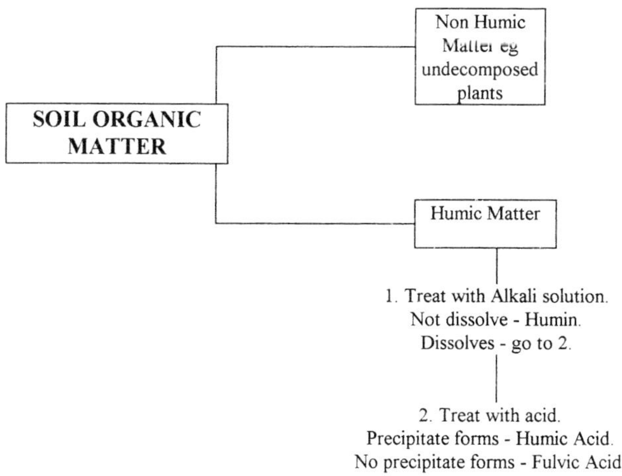

Fig. 5.11. Organic matter composition.

References pp. 172–173

expansion of the clay fraction (see above), and the transport of nutrients through the soil profile. The water content of a soil can vary immensely, from being totally saturated to completely dry. The maximum amount of water a soil can hold without it draining away is known as the field capacity. This is an important quantity that can greatly influence the extraction of pesticides [36].

The living portion of the soil is composed of micro-organisms such as fungi, and bacteria. They are responsible for breaking down dead and decaying matter, they also help to release nutrient into the soil through decomposition [29,41]. The living population can make up around 5% of the soil volume [42]. The micro-organisms present in soil do not directly influence the ease of extraction; frequently, they degrade the parent pesticide to give a wide range of metabolites, that in turn can strongly adsorb to soil, as in the case of atrazine degradation [43].

As stated earlier, the composition of soils varies from location to location. However, a system of classification of soil types has been in use for many years. This classification is based on the sand, clay and silt content of the soil [36,44]. The silt fraction is composed of very fine particles and their size is between 0.002 and 0.05 mm. The silt fraction has a fine texture when it is dry. Sand has a gritty texture and it has a diameter between 0.05 and 2 mm. Sand, clays and silt make up the largest volume of soil (around 45% of the total soil volume). Organic matter is the smallest fraction, only making up around 5% of the soil volume. Water makes up around 25% of the soil volume and carries the nutrients essential for plant life [29,45,46]. The soil textural triangle gives a broad indication of soil type (Fig. 5.12) [36]. Commonly, further information is required about the exact composition of a test soil. Methods exist that allow the calculation of

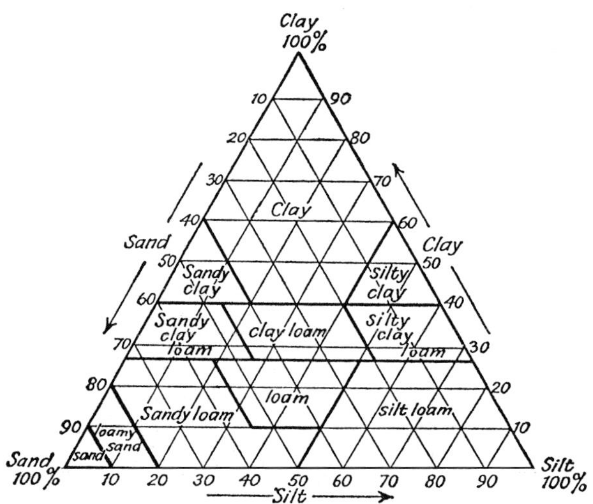

Textural designations according to Mechanical Analysis (American Scale).

Fig. 5.12. Soil composition triangle. Textural designations according to mechanical analysis (American Scale) [36].

several of these parameters, for example, cation exchange capacity (CEC) which is a measure of the ability of a soil to coordinate to multivalent species, such as metal ions. pH is a measure of the soil acidity, which can influence the extraction of ionic pesticides [31]; organic matter and clay type and content an also be established, their importance to the extraction of pesticides is highlighted above.

5.7 MICROWAVE-ASSISTED EXTRACTION

5.7.1 Interaction of microwaves with matter

Microwaves are short wavelength, (1 mm to 1 m) high-frequency electromagnetic radiation. To stop interference with radio transmissions, industrial and domestic microwaves function at a wavelength of around 12.2 cm (1.02×10^{-5} eV). Microwaves are split into two parts, the electric field component and the magnetic field component. Microwaves are made up of two wave components acting perpendicular to each other and the direction of propagation travel and vary sinusoidally; a magnetic field component and an electric field component. Like other electromagnetic energy, microwaves are said to have a dual nature, that is they can act like waves, but also have particulate character (photons). Photons are absorbed by electrons in the lowest energy state (ground state) of a molecule such that the electron is raised to the next energy level. These changes in the energy levels are discrete and do not occur continuously, as electrons occupy definite energy levels. The energy is quantised. The electric field components interacts only with charged (or polar) particles. The dielectric constant of a material determines the ease of polarisation of the molecule. If charged particles (or electrons) present in the molecule are mobile, a current is set up in the material. However, strongly bound electrons undergo a different phenomenon, the particles re-organise themselves so they are in phase with the electric field. This is called dielectric polarisation. Four components have been identified within the total dielectric polarisation. They represent the four main types of charged particles that are found in matter; electrons, nuclei, permanent dipoles and charges at interfaces. An equation lining all four constituents of the total dielectric polarisation is stated in Eq. 5.3 [47].

$$\alpha_1 = \alpha_e + \alpha_a + \alpha_d + \alpha_i \tag{5.3}$$

where α_1 is the total dielectric polarisation, α_e is the electronic, polarisation, α_a is the atomic polarisation, α_d is the dipolar polarisation, and α_i is the interfacial polarisation.

Frequent changes in the orientation of the electric field cause similar changes in the total dielectric polarisation. Changes in the dipolar polarisation result in heating in the material. Interfacial polarisation (the Maxwell–Wagner effect) only has a significant effect on dielectric heating when conducting particles are suspended in a non-conducting material. The other two components have no effect on heating. Therefore, in order to heat a solvent (or mixture of solvents) part of it must be polar. Sensitisers are molecules that preferentially absorb the microwave radiation and pass it on to other molecules [48]. Non-polar solvents such as hexane do not absorb microwave energy, but a mixture of hexane and acetone (1 : 1) does.

References pp. 172–173

5.7.2 Instrumentation for microwave-assisted extraction

A microwave extraction system consists of a microwave generator (magnetron), waveguide, resonant cavity and a power supply. The magnetron consists of a diode in a magnetic field. Indentations in the magnetron act as an anode, causing resonance which acts as a source for the microwave energy. The waveguide then focuses the microwaves onto the sample. Industrial microwave extraction systems are available in two forms: pressurised microwave-assisted extraction (MAE) (Fig. 5.13); and atmospheric MAE (Fig. 5.14). The main difference between the two techniques, is that in the pressurised system, the pressure can be raised up to 200 psi, while no such effect is possible in the atmospheric system. In addition, atmospheric instruments also have a lower power rating; 300 W for atmospheric MAE compared to 950 W for some pressurised systems. Both these parameters have implications for the extraction of pesticides from matrices. In the pressurised MAE up to 12 samples are placed into a carousel in microwave transparent extraction vessels. The vessels are lined with an inert material. The vessels are then irradiated with microwave energy. The temperature and pressure in one of the cells can be monitored and controlled using an infrared sensor and water manometer, respectively. The main parameters that can be controlled are temperature of the extraction, time of extraction, pressure in each vessel as well as the amount of microwave power the vessels receive. Safety features of the microwave include solvent vapour alarm and rupture membranes in each vessel. These will fracture when the pressure exceeds the maximum (typically 200 psi), allowing the contents to syphon into a central container [49].

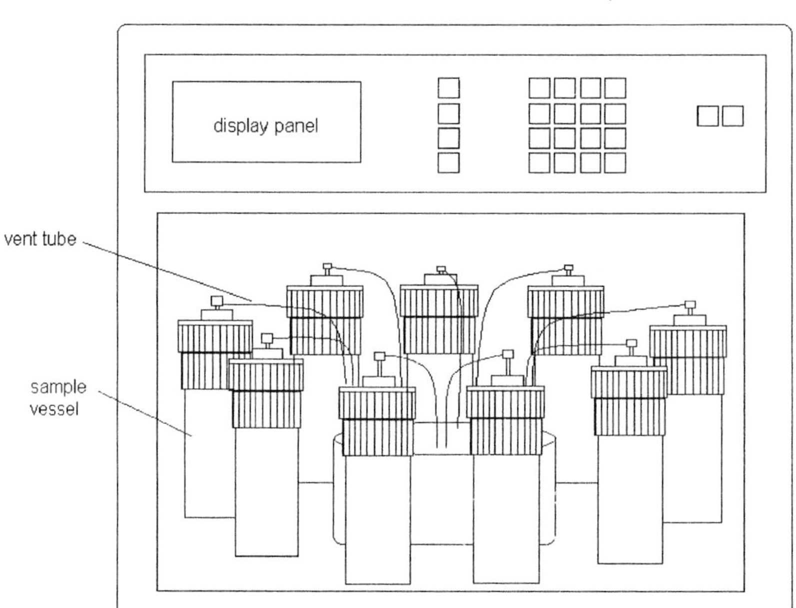

Fig. 5.13. Schematic diagram of a pressurised microwave-assisted extraction system.

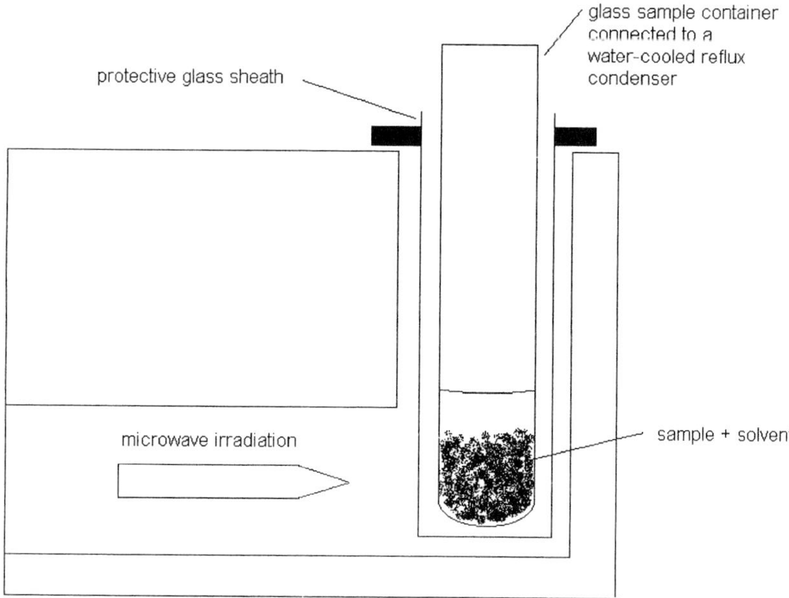

Fig. 5.14. Schematic diagram of an atmospheric microwave-assisted extraction system.

5.7.3 Applications of microwave-assisted extraction

Microwave technology has been used to extract pesticides from spiked and real samples. Steinheimer [43] has used microwave technology to extract the herbicide atrazine and its degradation products, deisopropylatrazine (DIA) and deethylatrazine (DEA) from contaminated soil samples. Briefly, the soil sample was extracted with water and then three successive times with dilute hydrochloric acid. The extracts were combined and analysed using HPLC with UV detection. Two soils of differing composition were investigated. The first was a loamy soil (Nashua) and the second was a silty loam (Treynor). Sample clean-up was required due to the coloured nature of the extract. Solid phase extraction (SPE) and centrifugation was employed for this. The average recoveries of the degradation compounds (DEA and DIA) were between 85 and 95% for the loamy soil and 85 and 115% for the silty loam soil. The recoveries for the parent compound (atrazine) and a surrogate, terbutylazine (TBA) were 65–55% for Nashua soil and 55–50% for Treynor soil. The decrease in the extraction efficiency was thought to be due to the increased basic nature of the degradation products over the parent compound.

Organochlorine pesticides (OCPs) have been extracted by MAE. McMillin et al. [50], have used MAE for the extraction of Arochlor residues from soil samples. They compared this approach with Soxhlet extraction and sonication. Microwave extraction consistently extracted greater amounts of Arochlor for all the 12 samples investigated.

References pp. 172–173

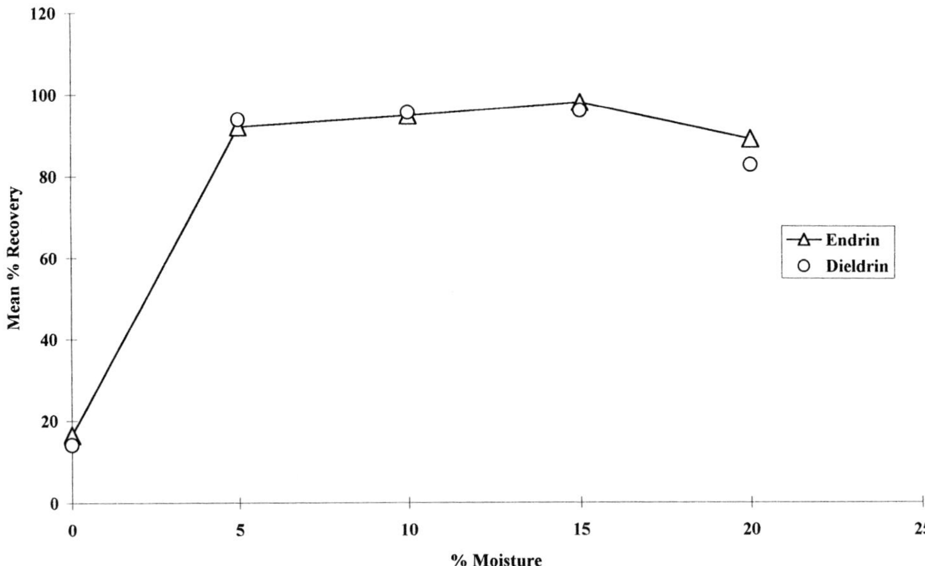

Fig. 5.15. Effect of soil moisture on the microwave-assisted extraction of OCPs.

However they initially expressed concern at the possible loss of solvent; this was later shown to be unfounded with losses of <2% reported. The major problem encountered with the microwave approach was that additional sample clean-up was required compared to the other two techniques. However the high dilution factors utilised negated the necessity of extensive sample clean-up. They also included published data that shows that the bias of microwave extraction is low. Studies by Lopez Avila et al. [51] and Onuska and Terry [52] have shown that extraction time and temperature have no effect on the extraction of 20 OCPs from certified marine sediments and soils. However, the moisture content of the sediment/soil is crucial. Onuska and Terry [52] showed that the presence of moisture is required for quantitative extraction of OCPs (Fig. 5.15) from sediment.

MAE has been shown by Lopez Avila et al. [53] to be a viable alternative to higher solvent consumption techniques. This group extracted a mixture of OCPs from spiked top soil, by both sonication and MAE (Fig. 5.16). MAE was found to extract all the pesticides, with an average recovery of 76%. Sonication; however, only extracted nine of the 20 target pesticides, with an average recovery of 71%. Li et al. [54] extracted spiked and certified soil samples contaminated with organochlorine pesticides by MAE and compared the results with Soxhlet extraction. The spiked samples gave recoveries of between 95 and 155% for DDD and endrin, respectively. MAE did have several advantages over the traditional extraction approach, namely the reduced extraction time and decrease in solvent consumption.

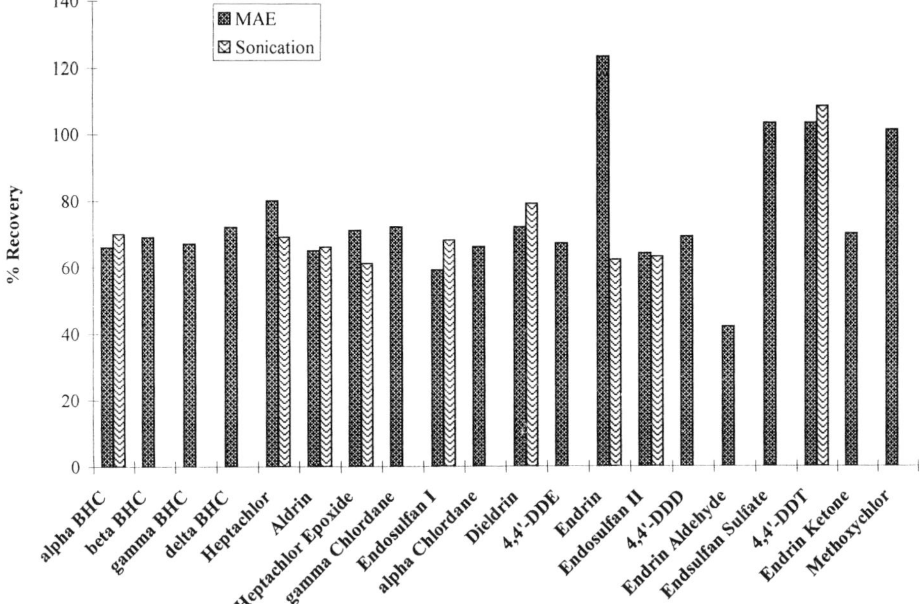

Fig. 5.16. Extraction of OCPs from spiked topsoil.

5.8 ACCELERATED SOLVENT EXTRACTION

5.8.1 Theory

As temperature increases, the physical properties of the solvent are changed, for example viscosity decreases and the surface tension of the solvent is also reduced. Changes also occur with respect to the analyte; it becomes more soluble in the solvent, thus less is required for the quantitative extraction. It has also been shown that for every 10°C rise in temperature the kinetics of a reaction increase, this is seen in accelerated solvent extraction (ASE) as the activation energy of desorption is also more easily overcome aiding extraction. Accelerated solvent extraction (a.k.a. pressurised fluid extraction) has been around since 1995. It is based on a hot pressurised solvent extraction system.

5.8.2 Instrumentation

An automated system is available from Dionex corporation, and consists of a solvent delivery system, oven, carousel and computer controlled software. Fig. 5.17 shows a schematic diagram of the system. Up to 24 samples can be sequentially extracted in stainless steel extraction cells. These consist of two screw thread end caps joined by a cylindrical cell body. In each end cap is a frit to prevent cell blockage. The cells are available in three volumes to allow both wet and dry samples to be extracted efficiently.

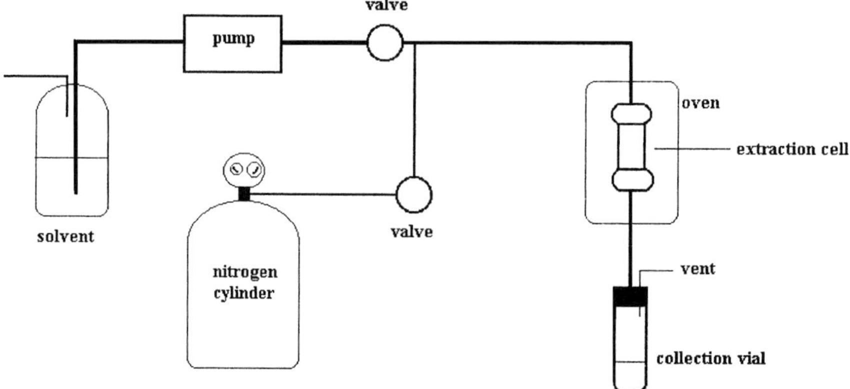

Fig. 5.17. Schematic diagram of an accelerated solvent extraction system.

The sample to be extracted is mixed with an inert matrix to reduce solvent consumption. The sample is quantitatively transferred to a stainless steel extraction cell that has been fitted with a filter to prevent transport of particulates through the cell and into the solvent lines. Hot solvent is pumped into the cell, where it is heated to the temperature and pressure stated in the method (typically between 100 and 200°C). The solvent is then kept in contact with the sample for a static extraction time (between 5 and 20 min). The solvent and analyte(s) are then flushed through the cell to a glass collection vial. A few millilitres (as a % of the cell volume) of solvent is the used to rinse the cell. The lines are then purged with high purity nitrogen to remove the last residues of the solvent and analyte. The extract is then ready for analysis. Automation of the system brings several advantages to the laboratory, up to 24 samples can be run sequentially, allowing a high number to be processed during the working day. The extract is kept separate from the sample, hence little in the way of sample clean-up is required.

5.8.3 Applications of ASE

ASE has been used by several laboratories for the extraction of pesticides from soil and sediments. Conte et al. [55] compared ASE with a traditional method of extraction for the extraction of a herbicide (diflufenican) from freshly spiked soil (Fig. 5.18). There was little difference between the amount extracted for the range of spike levels investigated (0.1–0.4 mg kg^{-1}). Ezzell et al. [56] extracted organophosphorus pesticides (OPPs) and herbicides from spiked soil samples by ASE and compared it with conventional Soxhlet extraction; good agreement between the two sets of results was reported. However, these results are not surprising as spiked samples rarely reflect real aged samples.

In order to fully assess a new technique, a range of aged samples should be extracted and compared with alternative extraction techniques, such as Soxhlet extraction. This was explored by Fisher et al. [57] and Brumley et al. [58]. Fisher et al. [57] extracted soils contaminated with organochlorine pesticides (OCPs) by ASE, Soxhlet extraction,

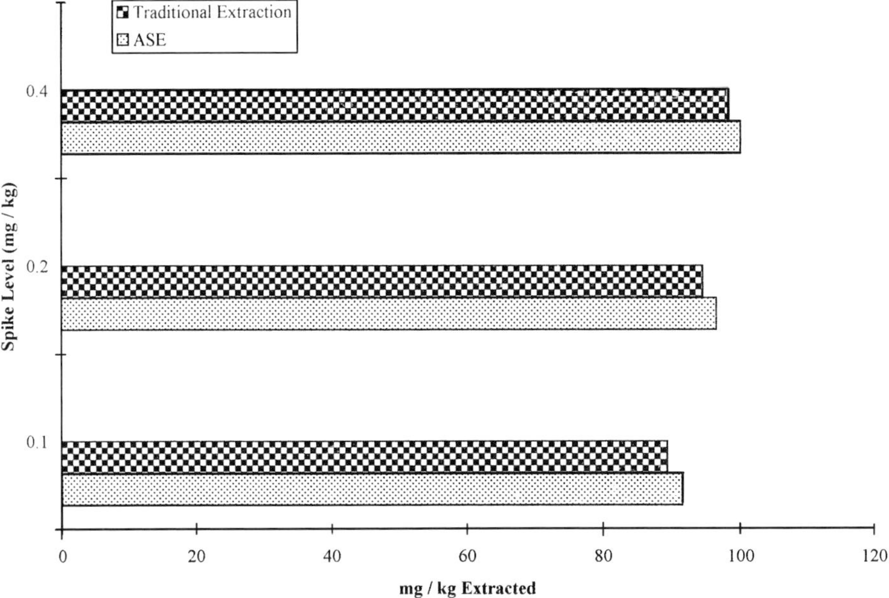

Fig. 5.18. Extraction of diflufenican by ASE and traditional extraction.

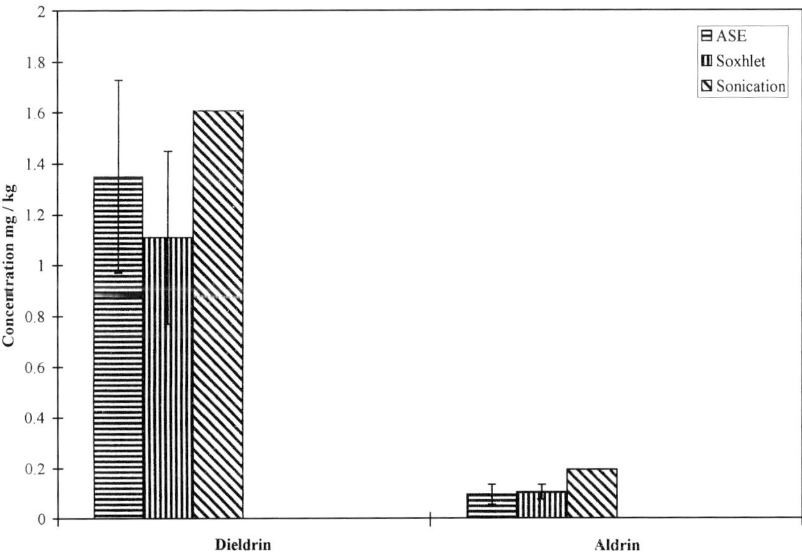

Fig. 5.19. Extraction of OCPs from contaminated soil.

and sonication. The results, presented in Fig. 5.19, show that ASE is at least as good as Soxhlet extraction for the extraction of dieldrin and aldrin, but both methods gave poorer recoveries when compared to sonication. Although the precision of ASE and

Soxhlet extraction were very good (0.04–0.38% for ASE compared with 0.03 to 0.34% for Soxhlet), it was noted that variation in the amount extracted was probably due to the heterogeneity of the soil sample. The sonication method was optimised for OPP, whereas the other two techniques were used as screening techniques, this could also account for the apparent reduced efficiency of OPP extraction by ASE and Soxhlet extraction.

Brumley et al. [58] compared the extraction of chlordane from a spiked soil. Chlordane is a mixture of polychlorinated compounds, hence more than one peak was obtained in the chromatographs. The total amount of chlordane was found by determining the amount for each peak and then summing all the peak. The three techniques they chose to study were ASE and supercritical fluid extraction (SFE) and to compare with results obtained by Soxhlet extraction. Three spiked levels were chosen, 2.0, 0.2 and 0.02 μg g^{-1}. This was done in order to determine the sensitivity of two GC detection techniques, GC–ECD and GC/EC NIMS. The recoveries using the 2 μg g^{-1} spike level, showed that both ASE and SFE were comparable to Soxhlet extraction. ASE gave average recoveries of 85% for all the chlordane component peaks, Soxhlet gave a mean recovery of 82% and SFE gave an average of 125%. It was also noted that the ASE extracts did not require further treatment, a distinct advantage, as the other two techniques required SPE clean-up before the analysis could be performed.

Popp et al. [59] extracted two soils from flood plains in Germany, contaminated with organic compounds, mainly OCPs. This group compared ASE with Soxhlet and Soxtec. Fig. 5.20 shows that ASE was equivalent to Soxhlet for 18 h, or Soxtec for 6 h. They also completed a solvent study that showed, unlike microwave extraction, ASE does not seem to be solvent dependent (Fig. 5.21).

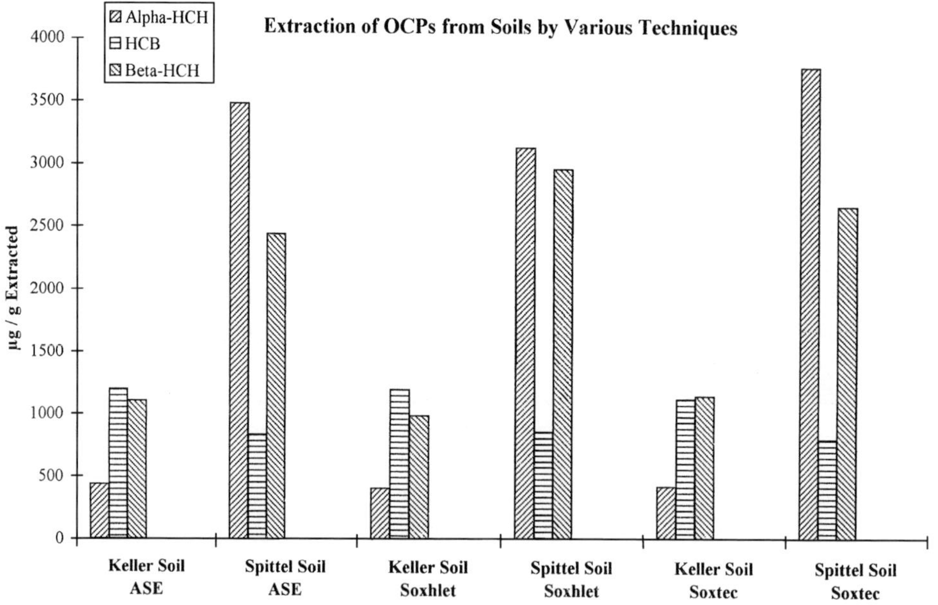

Fig. 5.20. Extraction of OCPs from soils by various techniques.

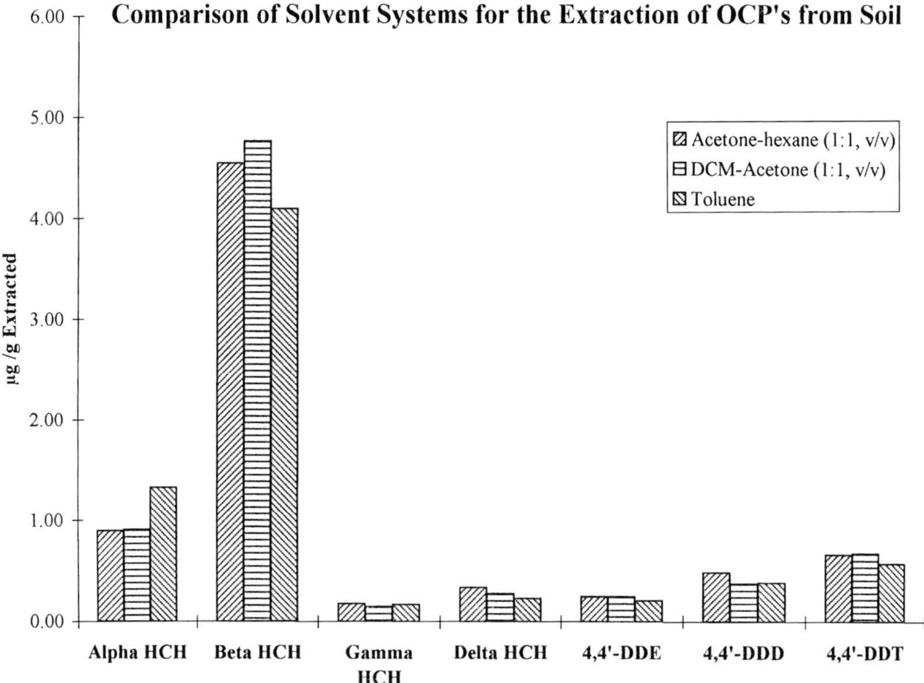

Fig. 5.21. Comparison of solvent systems for the extraction of OCPs from soil.

5.9 SUPERCRITICAL FLUID EXTRACTION

5.9.1 Introduction

The use of supercritical fluids for extraction accelerated around the early 1980s. As early as 1879 [60], however, the enhanced extraction abilities of supercritical fluids were noted. However, it was not until the 1960s that supercritical fluids were used to extract caffeine from coffee on a commercial scale [61]. Since then, further developments in supercritical fluid extraction have enabled the use of SFE on an analytical scale. Several solvents can be used in their supercritical state to extract analytes from matrices, these include N_2O, pentane, CO_2 and NH_3. However, all except CO_2 have safety problems, such as high reactivity and flammability [62]. The move to supercritical fluids as extraction solvents was prompted by environmental concerns, as typically, the extraction procedure uses environmentally innocuous compounds, such as water and carbon dioxide [63]. Environmental incentives for using supercritical fluids include the fact that it is inert to most materials and biologically non-toxic, on the commercial side, it is also inexpensive and gives minimal solvent residues [64].

5.9.2 Theory

Gases have temperature and pressure thresholds over which they become fluids. The temperature and pressure at which this occurs are termed the critical temperature and pressure. The fluids are known as supercritical fluids. The physical properties of supercritical fluids give them several advantages over liquids. The viscosity of the supercritical fluid is lower than that of an analogous liquid, this aids penetration of the solvent into the matrix, thus assisting extraction [65].

5.9.3 Instrumentation

SFE systems (Fig. 5.22) consist of two pumps, one each for high purity organic modifier and high purity solvent (usually carbon dioxide). The extraction cell is held in an oven, and the carbon dioxide pressure is controlled by use of a restrictor. The extract is collected in a suitable vessel. CO_2 is the usual supercritical fluid that is used as the solvent; however, this is non-polar liquid. In order to extract more polar pesticides, such as triazines herbicides, small quantities of organic modifiers are employed. The role of the organic modifier (typically methanol) is to increase the polarity of the extraction solvent. The liquid carbon dioxide is pumped into the extraction cell where it is raised to its supercritical temperature and pressure. The sample is mixed with a drying agent (for example anhydrous sodium sulphate) and placed in the stainless steel extraction cell. The supercritical fluid is passed through the sample and dissolves the target analytes. Then the extract is pumped out of the cell and passes through a restrictor into a collection vessel. While in the restrictor, the fluid cools and returns to its gaseous state, hence in order to quantitatively collect the analyte, the collection vial usually contains a few millilitres of organic solvent (typically the organic modifier). The role of the restrictor is to allow fluid flow while maintaining supercritical fluid conditions in the extraction cell. The extract can then undergo further preparative treatment prior

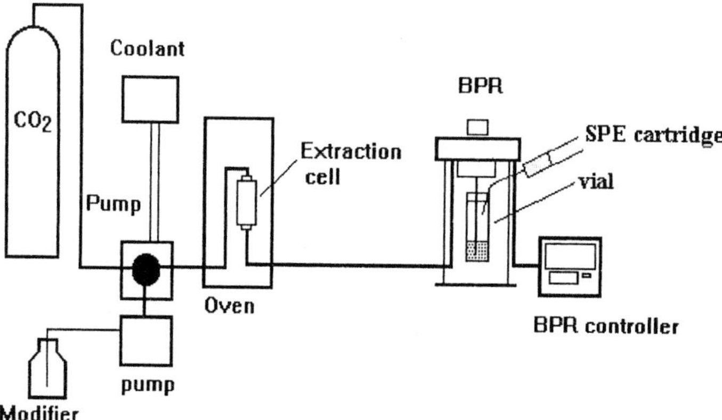

Fig. 5.22. Schematic diagram of a supercritical fluid extraction system.

to analysis, such as filtration. The entire extraction process typically takes less than an hour [49,62,66,67].

5.9.4 Applications of SFE

Extensive literature is available dealing with the extraction of pesticides from spiked inert matrices such as Celite [68], cleaned glass wool [69] and C_{18} silica [70]. All three studies showed that SFE could quantitatively extract pesticides after both spot and slurry spiking. Fig. 5.23 shows the extraction of OCPs from spiked Celite by SFE [68]. Two of the three papers noted that although spiking to inert matrices does not represent real samples, it can be an invaluable tool for the initial assessment of SFE conditions. Spiked soil samples have also been extracted by SFE [70–75] and many more. Wuchner et al. [69] compared two types of spiking, spot and slurry spiking (see above for a description of the spot spiking approach). The latter should better represent the situation in the environment. Two matrices were investigated, sand and soil. This study, on the extraction of organophosphorus pesticides, concluded that the analytes were harder to extract after slurry spiking, indicating some pesticides matrix interactions were occurring; for example spot spiking gave recoveries of around 70%, whereas slurry spiked soil gave recoveries of around 40% for dimethoate.

CO_2 is a non-polar molecule and so often organic solvents (modifiers) are added to the carbon dioxide to increase its polarity. Wuchner et al. [69] studied the effect of various modifiers (methanol, acetone and ethylacetate) on the extraction efficiency of OPP from spiked soil (Fig. 5.24). The addition of a modifier increased the extraction efficiency of all the pesticides by at least 10% with methanol giving the highest recoveries. Similar results were also reported by Janda et al. [74] who extracted

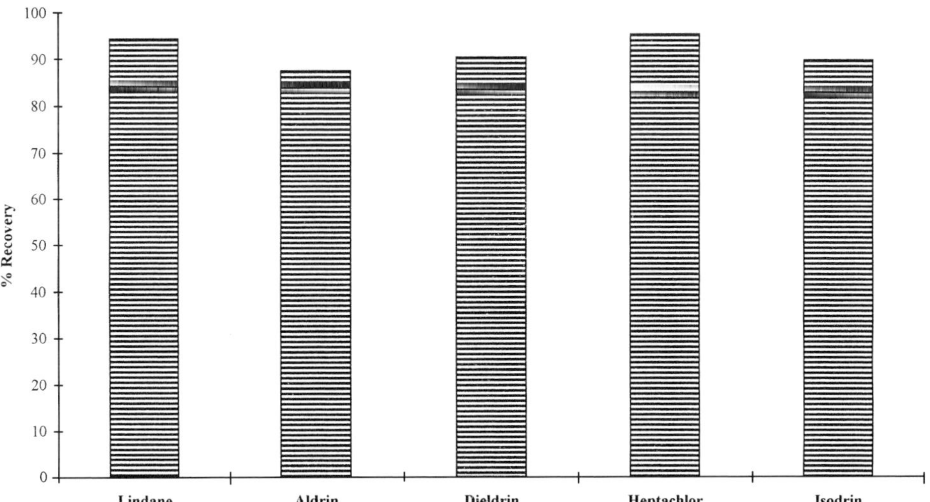

Fig. 5.23. SFE of OCPs from Celite.

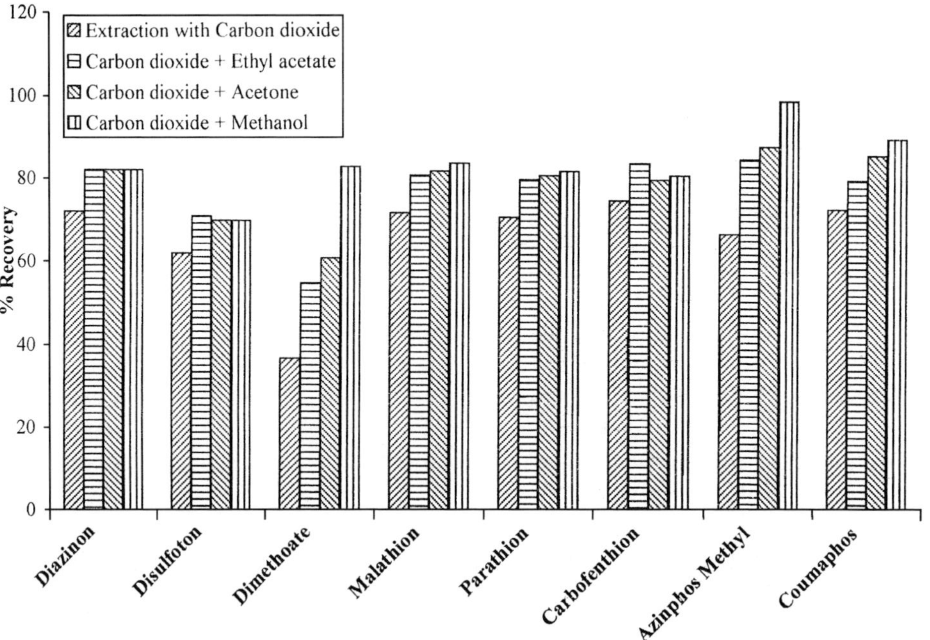

Fig. 5.24. Cumulative OPP recoveries using various organic modifiers.

five triazine herbicides from spiked sediment using unmodified supercritical CO_2 and methanol-modified supercritical CO_2. Recoveries of all the herbicides except simazine were greater than 82% (42.5% for simazine) for unmodified supercritical CO_2 and greater than 90% for the methanol-modified supercritical CO_2 (Fig. 5.25).

Hawthorne and Miller have investigated the effect of temperature on the extraction of triazine herbicides and OPPs in real site contaminated soil samples with unmodified supercritical CO_2 [76]. The soil samples were from various sources: railroad bed soil, industrial site soil, agricultural soil and diesel. Highest recoveries from agricultural soil were obtained at 200 vs. 50°C. A further increase in temperature to 350°C reduced the recoveries significantly, however (Fig. 5.26). SFE at 200°C gave comparable results to Soxhlet extraction with similar precision.

Dean et al. [30] also investigated the effect of interactions between the soil and the pesticide. Soils were characterised primarily for their organic matter content prior to spiking with a mixture of OPPs and OCPs and three urea herbicides. Organic matter content of the soil had a negative effect on the extraction of OCPs, as the amount of organic matter increased, the recovery by methanol-modified supercritical CO_2 decreased significantly (Fig. 5.27). High organic matter soil was also detrimental to the recovery of OPPs when the content was >35%. Extraction of the three urea herbicides used in this study was independent of soil organic matter.

Snyder et al. [71] compared SFE with Soxhlet and sonication for the extraction of a mixture of OCPs and OPPs. Four soil-based matrices were spiked with a mixture of 12 OCPs and OPP and extracted by all three methods. The recoveries were greater than 85%

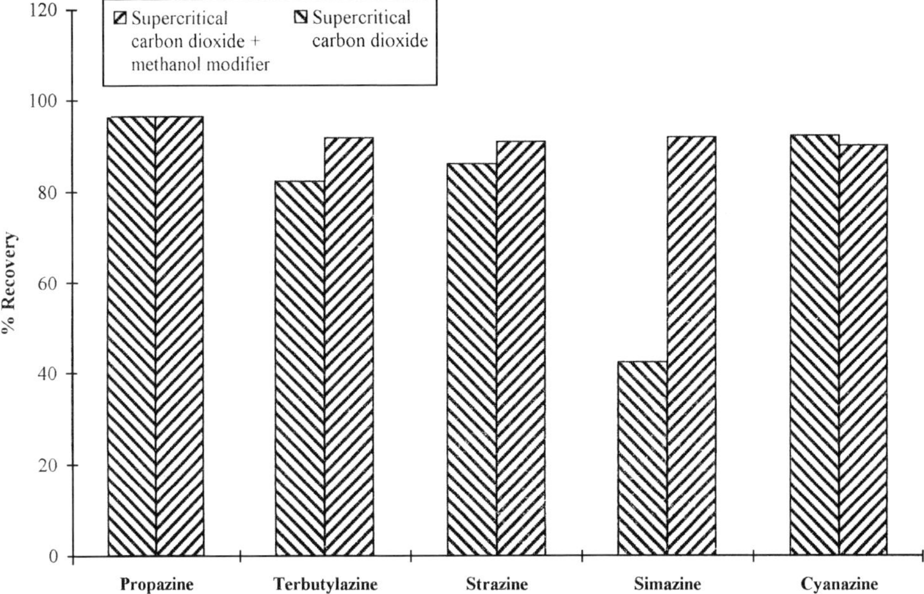

Fig. 5.25. Recovery of *s*-triazines by methanol modified and unmodified supercritical carbon dioxide extraction from sediment.

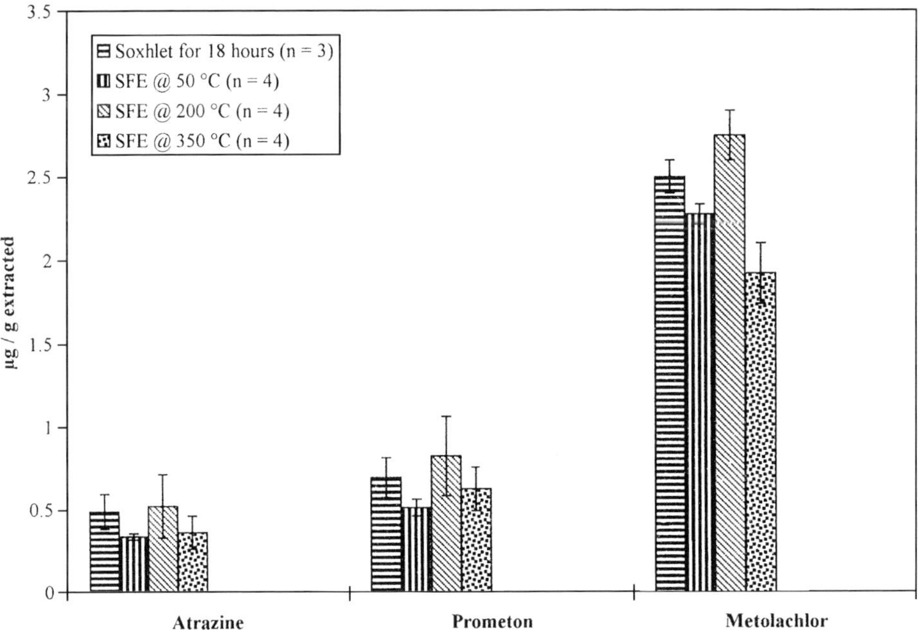

Fig. 5.26. Extraction of pesticides from agricultural soil.

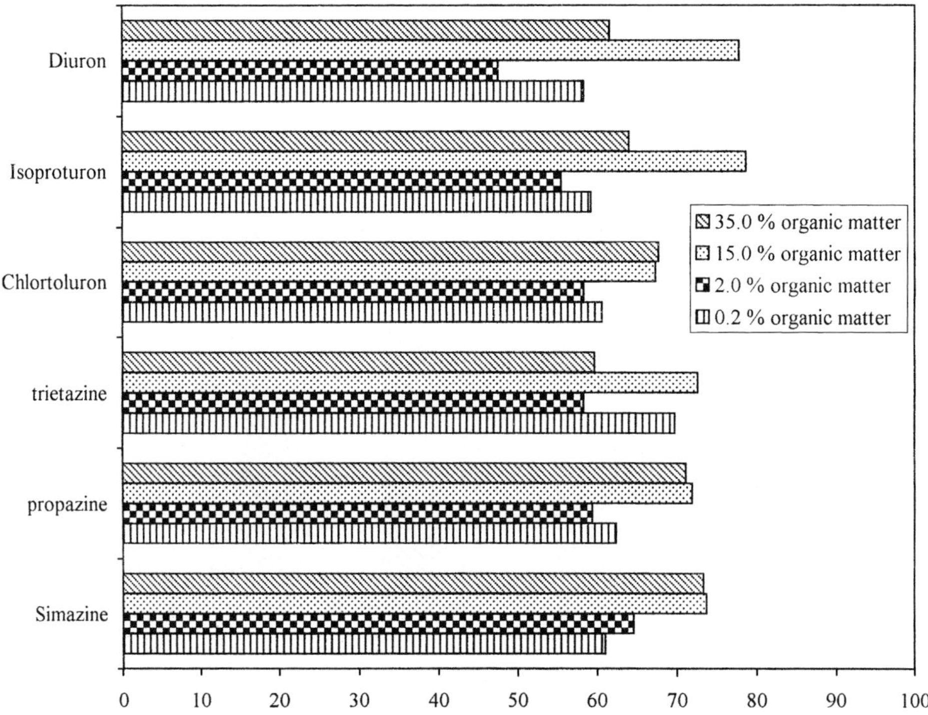

Fig. 5.27. Effect of organic matter content on the recovery of pesticides by SFE.

for all the pesticides for all the matrices. Native soils contaminated with the pesticides were also extracted. Ten grams of each soil both spiked and real soil were mixed with an equivalent amount of sodium sulphate and extracted with hexane: acetone for 20 h in a Soxhlet extraction. Sonic probe extraction was performed as the EPA method i.e. 3×40 ml portions of DCM and acetone were used to extract 10 g portions of soil for 3 min. The extracts were filtered and analysed by GC. The results were compared with methanol-modified supercritical CO_2. SFE extraction gave good recovery and precision for all 12 compounds independent of the matrix (Fig. 5.28).

Native contaminated soils have also been investigated although examples pertaining to extraction of pesticides from native soils and sediments are harder to find. Snyder et al. [71] and Naude et al. [73] have extracted residues from weathered soils and sediments. Both groups came to similar conclusions, that SFE could act as a potential replacement for older high solvent consumption techniques. Snyder et al. [71] extracted three native soils. The first was a dark top soil and was found to be contaminated with DDT and its metabolites. There was no significant difference between the amount extracted by either sonication or SFE. SFE did give better precision (average for SFE 6.9% RSD as compared to 13.5% RSD for sonication, $n = 3$). The other two soils were both found to be contaminated with other organochlorine pesticides, including endrin and endrin ketone. One was a sandy loam soil and the other was a sandy soil. Again, there was very little difference in the amount of each pesticide extracted by both

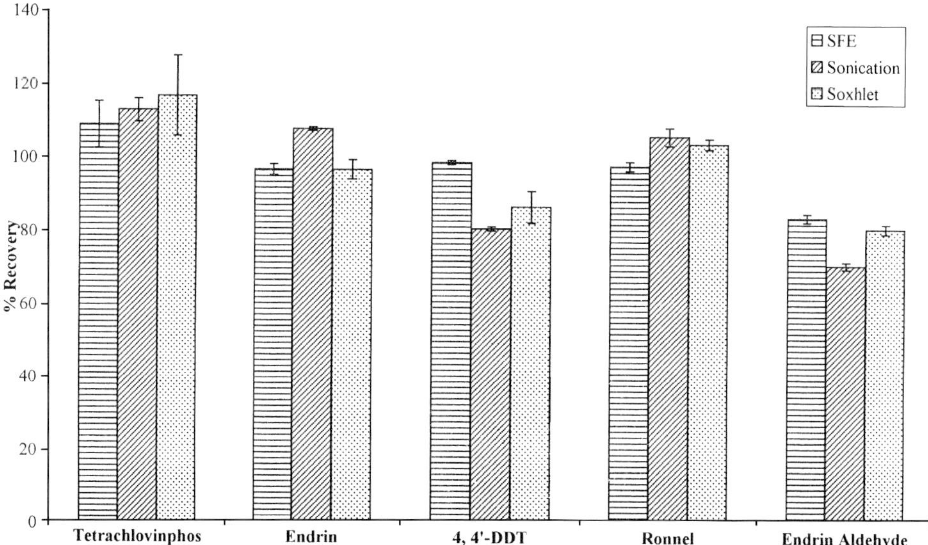

Fig. 5.28. Extraction of OCPs and OPP from spiked soil.

SFE and sonication at the ng g^{-1} level. However, for the sandy loam soil, sonication gave better precision than SFE for all the analytes, with the exception of endosulfan II. The main conclusions of this work was that sonication did not yield significantly different results than SFE, and also that SFE gave the best overall precision for the 12 pesticides investigated. Naude et al. [73] collected native samples with a grab sampler from various areas of the Pongolo flood plain in KwaZulu-Natal, South Africa. The samples were freeze dried and sieved. For each of the four area samples, Soxhlet gave lower results than SFE for DDT and DDD, but comparable results for DDE. However, further statistical analysis showed that the differences between the amount extracted by both Soxhlet and SFE were not significantly different, indicating that SFE could potentially replace Soxhlet (Fig. 5.29).

Frost et al. [32] compared four extraction techniques, SFE, ASE, MAE and Soxhlet for the extraction of weathered hexaconazole residues from two fully characterised soils. Extraction of the soil after 52 weeks application showed that the amount of extractable material had reduced significantly. An extraction time study using SFE was performed but showed no difference between 20, 40 and 60 min. There were differences between the two soils types, however; the soil with the lower organic matter gave comparable recoveries for all four techniques, whereas the results for the soil with the higher organic matter gave more varied results. For example, recoveries by MAE and SFE were half of the target Soxhlet value. ASE, however, gave comparable results to Soxhlet. Using this data, it was tentatively suggested that ASE is matrix independent (Fig. 5.30).

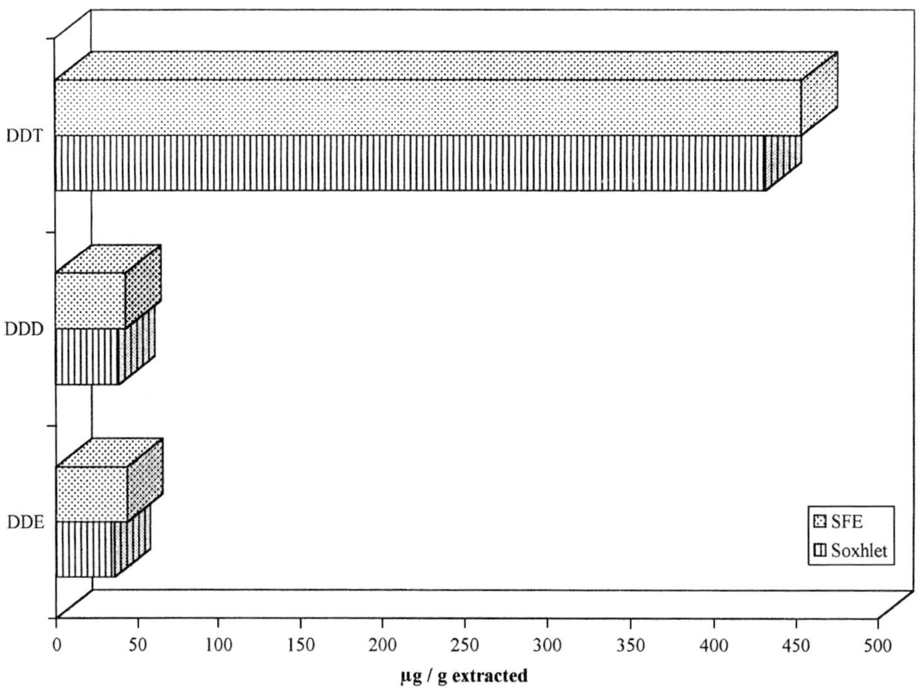

Fig. 5.29. Extraction of DDT and its degradation products from sediment.

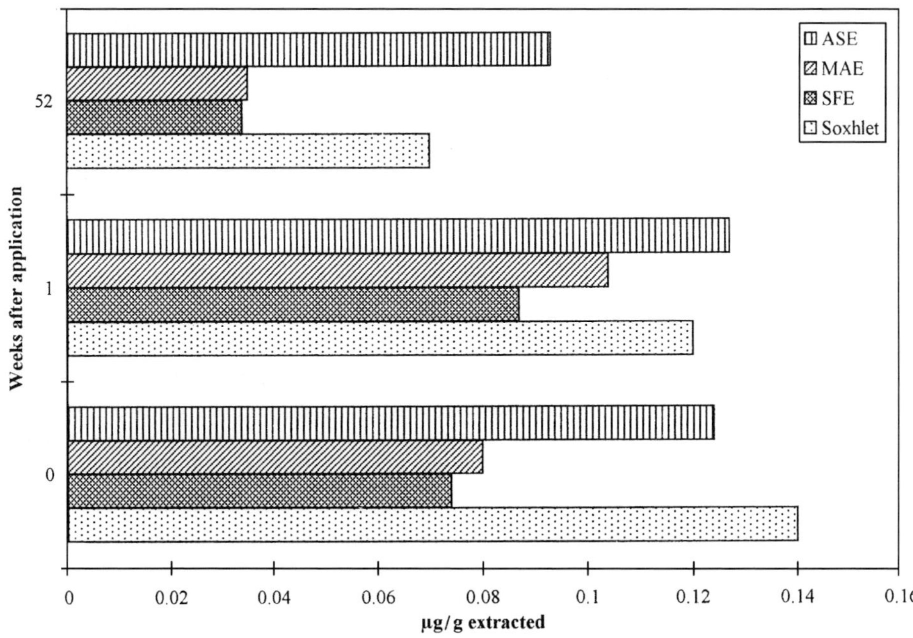

Fig. 5.30. Effect of weathering on the extraction of hexaconazole.

5.10 OTHER SOLID/LIQUID TECHNIQUES

5.10.1 Soxhlet extraction

Soxhlet extraction is used as the benchmark against which any new extraction technique is compared. The basic Soxhlet extraction apparatus consists of a solvent reservoir, extraction body, a heat source (e.g. isomantle) and a water cooled reflux condenser (Fig. 5.31). Soxhlet uses a range of organic solvents to remove organic compounds, primarily from solid matrices. The sample is combined with a drying agent, e.g. anhydrous sodium sulphate. The mixture is then placed in a porous extraction thimble, and extracted under reflux conditions. During the extraction process, the solvent is boiled and the vapour passes though the water-cooled apparatus and is condensed. The liquid solvent then passes through the sample, removing the analytes as it does so. The extract then passes into the boiling solvent and the whole process is repeated. As the boiling point of the analyte and solvent mixture is higher than that of the solvent alone, fresh solvent is continually circulating through the sample matrix. This process

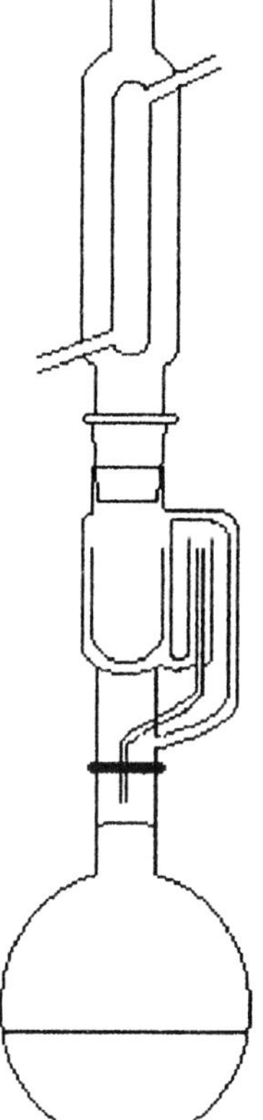

Fig. 5.31. Schematic diagram of a Soxhlet apparatus.

is continued for extended time intervals, e.g. 16 to 24 h. If required, the extract is then exchanged into a solvent that is compatible with the analysis technique [77].

Studies have been performed comparing Soxhlet extraction with several other extraction techniques for the extraction of pesticides from various spiked and real samples and these have been referred to in the discussion above.

5.10.2 Sonication

Sonic extractions can be achieved by the placement of the sample in a solvent in a sonic bath, or via the insertion of a sonic probe into the sample solvent system. This technique has been taken over by newer automated techniques, but has been used in the past to compare various solvent systems for the extraction of four herbicides from aged, spiked soil samples [78]. The technique was compared with shake-flask extraction (see below). The soils investigated were of various composition. Soils were sampled and extracted after both 6 months and 17 months of weathering. Not surprisingly, in all cases the amount of each herbicide that was extracted decreased the longer the soil was weathered. After 6 months of weathering, all the herbicides (no data for triallate) were quantitatively extracted from all the matrices. After 17 months of weathering, comparison of sonication and shake flask for the extraction of both nitrofen and profluralin was performed, using acetonitrile and water as the solvent. The results were not significantly different between the two extraction techniques. An average recovery of 87% by both shake flask and sonication for the extraction of profluralin; and 81% by sonication, compared with 87% recovery by shake flask (recovered on two soils only), for nitrofen. After 17 months of weathering all the herbicides (no data for benzoylprop-ethyl) were extracted from all the matrices with at least 85% recovery.

5.10.3 Shake flask

Solid samples, e.g. soil, sediment etc. are placed in a stoppered flask along with the extraction solvent. The entire system is then shaken using a mechanical shaker, for a set period of time, usually for around an hour. Repeat extractions can be performed to quantitatively remove the analyte from the sample and then the extracts are combined before analysis. As the sample is in contact with the solid, sample clean-up via SPE or similar is require prior to analysis. Although this technique is rarely used as a definitive method, variations have been used extensively in the past and also when assessing the efficiency of new techniques, such as, SFE and ASE.

Cotterill [79] has published work that uses the technique to assess the efficiency of solvent systems for the extraction of weathered herbicide residues from soil. Samples of soil of different compositions spiked with herbicides from a range of families were used. The herbicides included linuron, simazine and propyzamide. The soils were extracted 3 months after herbicide application. Several solvent systems were evaluated by two different techniques, Soxhlet extraction and shake flask. Methanol : water (4 : 1) consistently gave higher recoveries of selected herbicides from all the soil types using shake-flask extraction. A comparison of acetonitrile : water (9 : 1) as the extraction solvent for both Soxhlet extraction and shake flask showed that for soil with a high sand content and low organic carbon, there was very little difference between the recovery of the herbicides. Aqueous methanol extraction of herbicides from a soil with high sand content and pH 7.0 showed that recoveries of greater than 73% were possible for the pesticides.

5.10.4 Solid phase microextraction

An alternative approach to extraction of pesticides from solid matrices has been proposed using SPME. In this case, SPME is used to extract analytes from soil slurries. This is achieved by stirring a known quantity of soil with a solvent (e.g. water) and then exposing the SPME fibre directly to the resultant slurry. A polar poly(acrylate) fibre was utilised by Boyd-Boland and Pawliszyn [25] for the qualitative extraction of nitrogen-containing herbicides from an aqueous slurry. The soil sample (garden lawn) had been previously treated with benfluralin; a GC–MS chromatogram indicated the presence of the herbicide. The same group [26] was able to obtain quantitative data for the assay of metachlor, a pesticide, on soil. The soil sample was prepared by stirring 0.5 g in 4 ml of water and then exposing the SPME fibre directly to the resultant slurry for 50 min. It was found that the level of metachlor in soil was 1.84 mg kg^{-1} and 1.85 mg kg^{-1} by SPME and Soxhlet extraction, respectively. This approach demonstrated the applicability of SPME for the analysis of analytes from solid matrices. However, some concern was expressed by the authors in terms of the limited applicability of the approach as metachlor is a relatively water-soluble compound (solubility $= 530$ mg l^{-1}) and this is not likely to be the case for other compounds.

5.11 FUTURE PROSPECTS

It is sure that the environmental laboratory of the future will be based on automation and techniques that maximise sample throughput without sacrificing data quality Current trends to eliminate tedious and solvent intensive sample preparation techniques, such as Soxhlet and LLE, in favour of solid phase extraction, solid phase microextraction, accelerated solvent extraction, microwave-assisted extraction and supercritical fluid extraction are evident.

5.12 RECOMMENDED FURTHER READING

J.R. Dean, Extraction Methods for Environmental Analysis, John Wiley and Sons, Chichester, 1998.

E.D. Ramsey (Ed.), Analytical Supercritical Fluid Extraction Techniques, Kluwer Academic Publishers, London, 1998.

A.J. Handley (Ed.), Extraction Methods in Organic Analysis, Sheffield Academic Press, Sheffield, 1999.

References pp. 172–173

E.M. Thurman and M.S. Mills, Solid Phase Extraction: Principles and Practice, Wiley-Interscience, New York, 1998.

J. Pawliszyn, Solid Phase Microextraction: Theory and Practice, Wiley-VCH, New York, 1997.

J. Pawliszyn (Ed.), Applications of Solid Phase Microextraction, Royal Society of Chemistry, Cambridge, 1999.

5.13 REFERENCES

1   The Office of Reference Materials, Laboratory of the Government Chemist, CRM Catalogue, issue number 3.

2   D.E. Wells, Analyst, 123 (1998) 983.

3   D. Barcelo, J. Chromatogr., 643 (1993) 117.

4   Fielding, M., Barcelo, D., Helweg, A., Galassi, S., Torstenson, L., van Zoonen, P., Wolter, R. and Angeletti, G., in: Pesticides in Ground and Drinking Water (Water Pollution Research Report, 27), Commission of the European Communities, Brussels, 1992, pp. 1–136.

5   B.M. Mahara, J. Borossay and K. Torkos, Microchem. J., 58 (1998) 31.

6   D. Bourgeois, J. Gaudet, P. Deveau and V.N. Mallet, Bull. Environ. Contam. Toxicol., 50 (1993) 433.

7   M. Moors, D.L. Massart and R.D. McDowall, Pure Appl. Chem., 66 (1994) 277.

8   E.M. Thurman and M.S. Mills, Solid Phase Extraction: Principles and Practice, Wiley-Interscience, 1998.

9   E.S.P. Bouvier, P.C. Iraneta, U.D. Neue, P.D. McDonald, D.J. Phillips, M. Capparella and Y.F. Cheng, LC–GC International (Supplement), September (1998) 35.

10 M. Biziuk, J. Namiesnik, J. Czerwinski, D. Gorlo, B. Makuch, W. Janicki, Z. Polkowska and L. Wolska, J. Chromatogr., 733 (1996) 171.

11 E. Viana, M.J. Redondo, G. Font and J.C. Molto, J. Chromatogr., 733 (1996) 267.

12 C. de la Colina, A. Pena, M.D. Mingorance and F. Sanchez Rasero, J. Chromatogr., 733 (1996) 275.

13 M. Terreni, E. Benfenati, M. Pregnolato, A. Bellini, D. Giavini, S.F. Bavetta, C. Molina and D. Barcelo, J. Chromatogr., 754 (1996) 207.

14 M. Honing, J. Riu, D. Barcelo, B.L.M. van Baar and U.A.Th. Brinkman, J. Chromatogr., 733 (1996) 283.

15 G.A. Penuela and D. Barcelo, J. Chromatogr., 754 (1996) 187.

16 M. Vink and J.M. van der Poll, J. Chromatogr., 733 (1996) 361.

17 R. Eisert, K. Levsen and G. Wunsch, Intern. J. Environ. Anal. Chem., 58 (1995) 103.

18 G.A. Smith and T.L. Lloyd, LC-GC, May (1998) S22.

19 U.A.Th. Brinkman, Env. Sci. Technol., 29 (1994) 80A.

20 V. Pichon and M.C. Hennion, J. Chromatogr., 665 (1994) 269.

21 N. Masque, R.M. Marce and F. Borrull, J. Chromatogr., 793A (1998) 257.

22 E. Baltussen, H. Snijders, H.-G. Janssen, P. Sandra and C.A. Cramers, J. Chromatogr., 802A (1998) 285.

23 Z. Zhang, M.J. Yang and J. Pawliszyn, Anal. Chem., 66 (1994) 844A.

24 D. Louch, S. Motlagh and J. Pawliszyn, Anal. Chem., 64 (1992) 1187.

25 A.A. Boyd-Bowland and J. Pawliszyn, J. Chromatogr., 704 (1995) 163.

26 A.A. Boyd-Bowland, S. Magdic and J.B. Pawliszyn, Analyst, 121 (1996) 929.

27 T. Gorecki, R. Mindrup and J. Pawliszyn, Analyst, 121 (1996) 1381.

28 R. Ferrari, T. Nilsson, R. Arena, P. Arlati, G. Bartolucci, R. Basla, F. Cioni, G. Del Carlo, P. Dellavedova, E. Fattore, M. Fungi, C. Grote, M. Guidotti, S. Morgillo, L. Muller and M. Volante, J. Chromatogr., 795A (1998) 371.

29 M. Alexander, Introduction to Soil Microbiology, Wiley, Chichester, 1964.

30 J.R. Dean, I.J. Barnabas and S.P. Owen, Analyst, 121 (1996) 465.

31 J.L. Snyder, R.L. Grob, M.E. McNally and T.S. Oostdyk, J. Chromatogr. Sci., 31 (1993) 183.

32 S.P. Frost, J.R. Dean, K.P. Evans, K. Harradine, C. Cary and M.H.I. Comber, Analyst, 122 (1997) 895.

33 R.G. Gerritse, J. Beltran and F. Hernandez, Aust. J. Soil Res., 34 (1996) 599.

34 J.M. Bollag, C.J. Myers and R.D. Minard, Sci. Total Environ., 123 (1992) 205.
35 C.A. Goring and J.W. Hamaker (Eds.), Organic Chemicals in the Soil Environment, Vol. 1, Marcel Dekker, 1972.
36 W.N. Townsend, An Introduction to the Scientific Study of the Soil, 5th edn., Edward Arnold, 1973.
37 E. Barriuso, D.A. Laird, W.C. Koskinen and R.H. Dowdy, Soil Sci. Am. J., 58 (1994) 1632.
38 L. Calamai, O. Pantani, A. Pusino, C. Gessa and P. Fusi, Clays Clay Miner., 45 (1997) 23.
39 S.B. Haderlein, K.W. Weissmahr and P.R. Schwarzenbach, Environ. Sci. Technol., 30 (1996) 612.
40 S. Ross, Soil Processes: A Systematic Approach, Routledge, London, 1989.
41 http://www.silsoe.cranfield.ac.uk/sslrc/soil/info/level5.htm
42 http://ww.soils.rr.ualberta.ca/pedosphere/content/section01/page02.cfm
43 T.R. Steinheimer, J. Agric. Food. Chem., 41 (1993) 588.
44 http://www.silsoe.cranfield.ac.uk/sslrc/soil/info/level4.htm
45 http://www.crop.crinz/curresea/soil/vegsoil.htm
46 http://www.silsoe.cranfield.ac.uk/sslrc/soil/info/level1.htm
47 J. Jacob and F.Y.C. Boey, J. Mater. Sci., 30 (1995) 5321.
48 S. Caddick, Tetrahedron, 51 (1995) 10403.
49 J.R. Dean, Extraction Methods for Environmental Analysis, Wiley, 1998.
50 R. McMillin, L.C. Miner and L. Hurst, Spectroscopy, 13 (1997) 41.
51 V. Lopez-Avila, R. Young and W.F. Beckert, Anal. Chem., 66 (1994) 1097.
52 F.I. Onuska and K.A. Terry, Chromatographia, 36 (1993) 191.
53 V. Lopez-Avila, R. Young, R. Kim and W.F. Beckert, J. Chromatogr. Sci., 33 (1995) 481.
54 K. Li, J.M.R. Belanger, M.P. Llompart, R.D. Turpin, R. Singhvi and J.R. Pare, Spectroscopy, 13 (1996) 1.
55 E. Conte, R. Milani, G. Morali and F. Abballe, J. Chromatogr., 765A (1997) 121.
56 J.L. Ezzell, B.E. Richter, W.D. Felix, S.R. Black and J.E. Miekle, LC–GC, 13 (1995) 390.
57 J.A. Fisher, M.J. Scarlett and A.D. Stott, Environ. Sci. Technol., 31 (1997) 1120.
58 W.C. Brumley, E. Latorre, V. Kelliher, A. Marcus and D.E. Knowles, J. Liq. Chrom. Rel. Technol., 21 (1998) 1199.
59 P. Popp, P. Keil, M. Moder, A. Paschke and U. Thuss, J. Chromatogr., 774A (1997) 203.
60 J.B. Hannay and J. Hogarth, Proc. Roy. Soc. Lond., 29 (1879) 324.
61 K. Zosel, US Patent 3969 196 (1976), Chem. Abstr., 63 (1995) 11045b.
62 L.R. Snyder, J.J. Kirkland and J.L. Glajch, Practical HPLC Method Development, 2nd edn., Wiley, Chichester, 1997.
63 http://ww.nd.edu/ēnviro/supercritical.html
64 http://www.simssfe.com/index_main.html
65 S.B. Hawthorne, D.J. Miller and M.S. Kreiger, Fresenius Z. Anal. Chem., 330 (1988) 211.
66 http://www.pnl.gov/WEBTECH/voc/sfe.html#desc
67 C.L. Phelps, N.G. Smart and C.M. Wai, J. Chem. Ed., 73 (1996) 1163.
68 J.R. Dean, I.J. Barnabas and S.P. Owen, Analyst, 121 (1996) 465.
69 K. Wuchner, R.T. Ghijsen, U.A.Th. Brinkman, R. Grob and J. Mathieu, Analyst, 118 (1993) 11.
70 S. Papilloud and W. Haerdi, Chromatographia, 38 (1994) 514.
71 J.L. Snyder, R.L. Grob, M.E. McNally and T.S. Ostdyk, Anal. Chem., 64 (1992) 1940.
72 E.G. Van der velde, M.R. Ramlal, A.C. Van Beuzekom and R. Hoogerbrugge, J. Chromatogr., 683A (1994) 125.
73 Y. Naude, W.H.J. de Beer, S. Jooste, L. Van der Merwe and S.J. van Rensburg, Water SA, 24 (1998) 205.
74 V. Janda, G. Steenbeke and P. Sandra, J. Chromatogr., 479 (1989) 200.
75 A.M. Robertson and J.N. Lester, Environ. Sci. Technol., 28 (1994) 346.
76 S.B. Hawthorne and D.J. Miller, Anal. Chem., 66 (1994) 4005.
77 http://www.pnl.gov/WEBTECH/voc/soxhlet.html#desc
78 A.E. Smith, Pestic. Sci., 9 (1978) 7.
79 E.G. Cotteril, Pestic. Sci., 11 (1980) 23.

CHAPTER 6

Phenols

Olga Jáuregui and M. Teresa Galceran

Department of Analytical Chemistry, University of Barcelona, Diagonal 647, 08028 Barcelona, Spain

6.1 INTRODUCTION

Phenolic compounds are ubiquitous in the environment, they occur as natural constituents and they mediate in many industrial processes, for example in the manufacture of plastics, dyes, drugs and antioxidants. Phenols are also breakdown products from natural organic compounds such as humic substances, lignins and tannins, which are widespread in the environment. In addition, the pulp industry uses a hydrolysis step to extract cellulose from wood in large-scale processes whereby a liquid fraction, lignocellulose, is formed as a by-product containing high levels of phenolic compounds and their derivatives which may be released into the environment (rivers, lakes, sea). Chlorophenols (CPs) from the bleaching process have traditionally attracted most attention in analysis of industrial waste because of their high toxicity. However, CPs are released to the environment from other sources such as the degradation of certain insecticides and herbicides, especially 2,4-D and 2,4,5-T in soil and biological materials, or by application as fungicides, and chlorination of drinking water. Among the different chlorophenols, pentachlorophenol (PCP) has been extensively used as wood preservative. Other chlorinated phenols such as 2,4,6-trichlorophenol and 2,3,4,6-tetrachlorophenol have also been used in mixtures to prevent fungal/microbial infection in wood and to preserve manufactured goods. 2,4-Dichlorophenol, 2,4,6-trichlorophenol and 2,4,5-trichlorophenol have been used as industrial intermediates in the manufacture of the chlorophenoxy herbicides 2,4-D and 2,4,5-T [1]. Nitrophenols (NPs) are also present in polluted groundwaters near former ammunition plants and as decomposed products of carbamate and phosphorus pesticides and are found in polluted groundwater and river water [2,3]. Additionally, these compounds, as products of photochemical reaction between nitric oxides and aromatic hydrocarbons, can be formed in the atmosphere and in consequence have been found in rain, snow and fog in surprisingly high concentrations [4,5]. Alkylphenols from industrial processes in the petroleum industry and in systems receiving runoff from asphalt roadways are also found in the environment. The need to determine phenols in crude oils and fuels lies in that, upon storage, phenols and other polar compounds promote deposit formation, which has a deleterious effect on fuel quality and stability

References pp. 231–236

TABLE 6.1

PHYSICO-CHEMICAL CHARACTERISTICS OF SOME PHENOLIC COMPOUNDS [7,12]

Compound	pK_a	$\log P_{ow}$	S (mg 1^{-1})
Phenol	9.99	1.46	82000
2,4-Dimethylphenol	10.50	2.42	4200
2-Chlorophenol	8.52	2.29	23250
3-Chlorophenol	8.97	2.64	22200
4-Chlorophenol	9.37	2.53	2600
2,3-Dichlorophenol	7.71	3.26	8200
2,4-Dichlorophenol	7.90	3.20	5550
2,5-Dichlorophenol	7.51	3.36	–
2,6-Dichlorophenol	6.80	2.92	2650
3,4-Dichlorophenol	8.60	–	9250
3,5-Dichlorophenol	8.25	3.60	7400
2,3,4-Trichlorophenol	7.00	–	915
2,3,5-Trichlorophenol	6.43	3.85	770
2,3,6-Trichlorophenol	5.80	–	590
2,4,5-Trichlorophenol	6.72	4.02	650
2,4,6-Trichlorophenol	6.00	3.67	710
3,4,5-Trichlorophenol	7.55	–	–
2,3,4,5-Tetrachlorophenol	5.64	–	165
2,3,4,6-Tetrachlorophenol	5.22	4.24	180
2,3,5,6-Tetrachlorophenol	5.02	5.02	100
Pentachlorophenol	4.74	5.85	18
2-Nitrophenol	7.21	1.78	2100
3-Nitrophenol	8.27	–	–
4-Nitrophenol	7.16	1.90	16000
2,4-Dinitrophenol	4.09	1.53	5600
2-Methyl-4,6-dinitrophenol	4.34	2.12	100
4-Chloro-3-methylphenol	9.55	3.10	3850

[6]. Moreover, it is known that combustion and pyrolysis of building materials containing phenolic resin produce methylphenols, dimethylphenols and trimethylphenols [1]. Another group of phenolic compounds, out of the scope of this chapter, is the nonylphenols that have been found in many aquatic environments as degradation products of polyethoxylate nonylphenols widely used as non-ionic surfactants.

Phenols are a very heterogeneous group of compounds, as can be observed in Table 6.1 which gives some physicochemical properties of phenols. Although they are classified as acids, mainly the higher chlorinated ones ($pK_a \approx 5$) and dinitrophenols ($pK_a \approx 4$), some of them have a pK_a considerably high, e.g. phenol and methylphenols around 10. Water solubility (S mg 1^{-1}), which is one of the most significant factors affecting the fate and transport of chemicals in the environment, also varies greatly depending on the phenolic compound. For instance, water solubility for pentachlorophenol is 18 mg 1^{-1} whereas for phenol it is 82,000 mg 1^{-1}. For chlorophenols the solubility in water decreases when the degree of chlorination increases. Another interesting parameter is the octanol–water partition coefficient which is related to the tendency of phenols to be present in the different environmental reservoirs and has also been recognized as a good indicator of the transfer of chemicals to the site of action in vivo. In Table 6.1 there are the log P_{ow} values for certain phenols. For instance, phenols with higher log P_{ow} (chlorophenols) values are mainly found present in soils, sediments or sludges, whereas phenols with lower values are mainly found in aquatic reservoirs. In addition, bioconcentration and bioaccumulation are related to this parameter as is demonstrated by the quantitative relationship found between the lipophilicity, expressed as the octanol–water partition coefficient, of a wide range of organic compounds and their bioconcentration factor in human adipose tissue [8].

Chlorophenols are well-known for causing taste and odor problems in water, at levels below the toxic (<1 μg l^{-1}). The environmental half-lives of most chlorophenols are short, rarely as long as a month: once discharge ceases, levels drop rapidly due to bacterial breakdown. Apart from water problems, chlorophenols can be adsorbed into the soil. High substituted phenols such as trichlorophenols and pentachlorophenol have limited transport in water, are more likely to be adsorbed in soil organic matter because of their log P_{ow} value and persist a long time. The adsorption and desorption of chlorophenols from soils depend on the number of chlorine substituents and also on their position on the phenolic ring. In fact, phenols have been detected in soil, sediments and sludges mainly near wood or paper treatment plants with concentrations of up to 500 mg kg^{-1} or even higher [9].

Biological samples have also been studied to find the contamination of humans and animals by chlorophenols. Accumulation of PCP in the human population presumably results from contact with wood products such as paper and cardboard, ingestion of contaminated food-stuffs or water supplies, and contact with materials used in the construction of dwellings for human habitation. Because of its widespread use, animals and humans are exposed to significant amounts of these compounds and detectable levels are found in most people in the industrialized world. For instance the presence of chlorophenols in urine has been shown to be a sensitive indicator of human exposure not only to chlorophenols but also to other organochlorine compounds metabolized to chlorophenols, e.g. hexachlorocyclohexane and chlorobenzenes [10,11]. PCP can be assimilated by humans through inhalation, skin penetration and the digestive system, and may cause toxic damage to the liver and changes in the immunological system [13,14]. Some LC$_{50}$ data in goldfish have been published for chlorophenols with values going from 0.2 μmol g^{-1} for PCP to 2.13 μmol g^{-1} for 2-chlorophenol [15]. However, many commercial batches of technical PCP are known to contain small but possibly biologically significant amounts of highly toxic impurities, e.g. chlorodibenzo-*p*-dioxins (PCDD) and dibenzofurans (PCDF) that seem to be responsible for most of the toxicity in PCP preparations and for the immunosuppression observed in workers exposed to PCP. In fact, PCP and its derivatives represent an important PCDD and PCDF source for sewage sludges [16]. The toxicity of chlorophenols increases as the number of chlorine atoms increases: e.g. pentachlorophenol is about 700 times more toxic than phenol. However, in the chlorophenols with the same number of chlorine atoms, toxicity decreased in the order of non-, mono- and di-*ortho*-chlorophenols, e.g. 4CP > 3CP > 2CP in monochlorophenols, 35DCP > 25DCP > 24DCP > 23DCP > 26DCP in dichlorophenols and 245TCP > 246TCP in trichlorophenols [17].

For these reasons, most phenols are considered high-priority pollutants in water. Eleven phenolic compounds are included in the US Environmental Protection Agency (USEPA) list of priority pollutants [18]: phenol, 2-nitrophenol, 4-nitrophenol, 2,4-di-nitrophenol, 2-methyl-4,6-dinitrophenol, 2,4-dimethylphenol, 4-chloro-3-methylphenol, 2-chlorophenol, 2,4-dichlorophenol, 2,4,6-trichlorophenol and pentachlorophenol. Also, a number of phenolic compounds are listed in the European Community (EC) Directive 76/464/EEC concerning dangerous substances discharged into the aquatic environment [19]: 2-amino-4-chlorophenol, 4-chloro-3-methylphenol, 2-chlorophenol, 3-chlorophen-ol, 4-chlorophenol, 2,4-dichlorophenol, 2,4,5-trichlorophenol, 2,4,6-trichlorophenol and pentachlorophenol.

Under European Community legislation, the maximum admissible concentration of phenols in drinking water should be 0.5 μg l^{-1} for total content and 0.1 μg l^{-1} for individual content [20], whereas in bathing water the maximum admissible level is 5 μg l^{-1} [21]. The level of pentachlorophenol in industrial effluents has been set at a maximum of 1 mg l^{-1}, and in both interior and sea water at 2 μg l^{-1} for superficial waters [22]. The USEPA has set the drinking water standard for pentachlorophenol at 0.1 μg l^{-1} to protect against the risk of cancer or other adverse health effects. Some rulings on other samples have also been found. For instance, the American Conference of Governmental Industrial Hygienists (ACGIH) has fixed for PCP a workplace threshold limit value (TLV) of 0.5 mg m^{-3} in air, and related Biological Exposure Indices fix 2 mg total PCP per gram of creatinine in urine and 5 mg l^{-1} in blood [23].

As the number of phenol derivatives is very large, our attention in this work focused on phenol, methylphenols (RPs), nitrophenols (NPs) and chlorophenols (CPs). Natural phenolic compounds (e.g. vanillic acid, syringic acid, ferulic acid) and nonylphenols are not included. An overview of the current analytical methods for determining phenolic compounds is also given and some criticisms of the numerous methods found in the literature attempt to identify the main advantages and drawbacks of these methods. Moreover, an overview of the EPA methods dedicated to phenols is presented and their advantages and limitations are discussed. Detailed information on methods is not presented within the text but can be extrapolated from the tables.

6.2 EXTRACTION AND PRECONCENTRATION TECHNIQUES

The low concentrations (ng l^{-1} to μg l^{-1}) of pollutants in environmental samples make it necessary to take some enrichment steps before the chromatographic analysis. In this section preconcentration procedures suitable for phenolic compounds from water, soil and biological samples will be discussed.

6.2.1 Water samples

The techniques most frequently used for the preparation of water samples prior to chromatographic analysis are liquid–liquid extraction (LLE) and liquid–solid extraction also called solid-phase extraction (SPE), although recently solid-phase microextraction

(SPME) has appeared as an alternative procedure. The greatest advantages of LLE are its simplicity and the high number of practical verifications and performance evaluations that have been carried out up till now. In fact, the EPA methodology for the analysis of priority pollutant phenols in water, EPA methods 604 and 625 [24,25], involves extraction with dichloromethane after sample acidification at pH 2. To extract phenolic compounds, other organic solvents have also been used such as hexane and diethylether, but dichloromethane is preferred because it allows high recoveries [26–28]. To extract polar pollutants such as phenolic compounds, the formation of ionic pairs, for example with tetrabutylammonium hydroxide in basic medium, can also be used; more apolar compounds than the original phenols are then formed which can be extracted easily with an organic solvent [29,30], although, interferences may also be extracted. Similarly, derivatization of phenolic compounds in the water sample prior to extraction enhances the extractability of these compounds and improves their gas-chromatographic characteristics [31,32]. For instance, the recovery of phenol from water (250 ml) using direct LLE with dichloromethane was found to be 28–41%, whereas the recovery of acetylated phenol was 100%. Although the LLE approach has been thoroughly used, it has some drawbacks such as the large amount of generally toxic and inflammable solvents that are needed. In addition, automation of the method is rather difficult and important losses of analytes during evaporation steps can occur, for which reason a soft evaporation process is recommended after addition of acetonitrile or NaOH (to convert the phenols to phenolates). In addition, dryness should be avoided [33–35]. Some acid–base partition schemes for clean extraction have also been proposed, such as EPA method 3650 [36] but, emulsions which are difficult to eliminate have often been produced and, in some cases, degradation of phenols mainly in basic medium occurred. Moreover, transformation of several contaminants into chlorinated phenols as a result of alkaline pretreatments can happen.

For these reasons, solid-phase extraction has been used as an alternative to LLE in order to extract and preconcentrate phenolic compounds from water samples. This procedure has certain advantages such as its low cost and low solvent consumption (typically a few milliliters), little sample manipulation (so minimizing the risk of sample contamination) and ease of automation. Solid-phase extraction of phenolic compounds from water samples has used both gas chromatography and liquid chromatography to analyze the resultant organic extracts. The SPE system approach can be carried out in the off-line and on-line modes: the advantages and limitations of both modes have been extensively discussed in the literature [37]. At present, SPE coupled on-line is the preferred procedure because the analysis time and the sample manipulation were less than in the off-line approach. Nevertheless, the off-line approach will remain useful for the analysis of complex samples due to its considerable flexibility and the possibility of analyzing the same extract with different separation techniques.

A wide range of sorbent materials for SPE is commercially available mainly in two formats, cartridges and disks. The greater cross-sectional area of the extraction disks makes it possible to percolate water samples at higher flow-rates than with cartridges, so the extraction time can be considerably reduced (15 min for disks vs 1.5 h for cartridges for 250 ml water sample) in an off-line system [34]. In addition, the risk of blockage by particles and channelling in the sorbent bed is greater for cartridges than for disks. However, more sorbents are available in the cartridge format than in the disk format [38]. For the analysis of phenolic compounds in water samples, cartridges and membrane extraction disks in off-line and on-line modes have been employed. A description of the most important methods found in the literature and a discussion of the advantages and limitations of off-line and on-line SPE methods follow.

The first attempts to use off-line SPE methods to extract phenolic compounds from water samples were made in combination with gas chromatography. However, the last 10 years have seen a general tendency towards liquid chromatography because no derivatization steps are needed to separate phenolic compounds. Silica-based materials such as octyl (C_8), octadecyl (C_{18}), phenyl, cyclohexyl (CH) and cyanopropyl (CN) have been used for off-line SPE–GC methods [39–48] and for off-line SPE–LC methods [34,49–58]. Tables 6.2 and 6.3 give a summary of some data found in the recent literature on use of both off-line SPE–GC and off-line SPE–LC. Several authors compared the performance of silica sorbent materials and, generally, concluded that C_{18} silica is the most efficient sorbent because it gives the highest recovery and the best reproducibility [51]. These types of sorbents gave good results for the more apolar phenols (tri-, tetra- and pentachlorophenol). For instance, with LC with electrochemical detection (ED), there were high breakthrough volumes, 500 ml, and low detection limits, 0.03–0.1 μg l^{-1}, for the more apolar phenols. Nevertheless, C_{18} silica failed to retain the more polar compounds (phenol, nitrophenols and monochlorophenols) giving low breakthrough volumes (<20 ml for phenol) and so causing high detection limits (0.25 μg l^{-1} for phenol). Thus, the EC legal levels (0.1 μg l^{-1} for individual concentration, 0.5 μg l^{-1} for total concentration) in tap water were not reached [34]. Other sorbents such as CH [39,40] and phenyl silica sorbents [49] have been used, but in these studies the water samples were spiked at a fairly high concentration of phenols (mg l^{-1}).

The carbonaceous sorbents (GCB, graphitized carbon black and PGC, porous graphitic carbon) have also been employed for the preconcentration of phenolic compounds in combination of both GC [46–48] and LC [54,57,59]. This type of sorbent has been extensively employed by Di Corcia and co-workers [54,60,61] who propose its use because organic compounds are adsorbed more strongly than with C_{18} due to multiple mechanisms. In consequence, very high breakthrough volumes (up to 2 l) and low detection limits (0.1–1 ng l^{-1}) were obtained. However, some problems in the use of these sorbents have been observed. First, carbon-based sorbents initially afford high flow-rates but rapidly undergo compacting of the carbon bed, which increases its resistance to passage of samples thus and so prevents the use of flow-rates above 25 ml min^{-1} even with samples filtered through membranes of 0.45 μm pore size. Moreover, strong adsorption makes it difficult to elute the retained compounds (especially nitrophenols and highly chlorinated phenols), making the use of two distinct solvent mixtures, the elution in the backflush mode or the use of tetramethylammonium hydroxide (TMAOH)-modified eluents necessary to recover retained analytes [58]. Furthermore, some controversies on the use of these type of sorbents have been found in the recent literature. For example, Cela and co-workers [47,48] compared GCB and PS-DVB columns for extracting phenolic compounds at a low level, μg l^{-1}. Their results show similar recoveries for GCB and PS-DVB cartridges, but extraction with PS-DVB has some advantages over GCB because higher sample volumes in shorter times could be

TABLE 6.2

SUMMARY OF OFF-LINE SPE–GC METHODS FOUND IN THE LITERATURE FOR PHENOLIC COMPOUNDS IN WATER SAMPLES

Sorbent	Sample type	Sample volume (ml)	Concentration level	Derivatization	Recovery (%)	LOD (ng l^{-1})	Detector	Ref.
Acetyl–PS–DVB disks	distilled water	20	5 mg l^{-1}	–	87–100	–	FID	[38]
C$_{18}$ disk	tap water	1000	1 μg l^{-1}	acetyl (before extn.)	47–112	2–35	ITDMS	[41]
C$_8$ disk	groundwater	1000				2–35		
	river water	1000				2–50		
C$_{18}$ column	tap/river water	300	0.1–50 μg l^{-1}	acetyl (after extn.)	42–109	5–15	ECD	[43]
C$_{18}$ column	HPLC water	1000	10 μg l^{-1}	–	22–92	–	FID, ECD	[45]
XAD resin		1000	10 μg l^{-1}	–	46–110	–		
GCB column	HPLC water	500	1 μg l^{-1}	acetyl (before extn.)	72–98	–	DD–FTIR, MS, MS–MS	[46]
PS–DVB column	HPLC water	2000	5 μg l^{-1}	acetyl (after extn.)	80–103	50–160 (LOQ)	MIP–AED	[47]
GCB column				acetyl (before extn.)	–			
GCB column	drinking water	1000	6 ng l^{-1}	acetyl (before extn.)	79–107	0.04–0.09	MS–MS (ion-trap)	[48]

TABLE 6.3

SUMMARY OF OFF-LINE SPE–LC METHODS FOUND IN THE LITERATURE FOR PHENOLIC COMPOUNDS IN WATER SAMPLES

Sorbent	Sample type	Sample volume (ml)	Concentration level	Recovery (%)	LOD (μg l^{-1})	Detector	Ref.
C_{18} columns	sea water	250	10 μg l^{-1}	11–100	0.03–0.12 [a]	ED (amperometric)	[34]
C_{18} disks			0.5 μg l^{-1}	41–111 [a]	0.01–0.05 [b]		
PS-DVB disks			0.5 μg l^{-1}	70–108	0.01–0.1		
C_{18} columns	trap water	100	20 μg l^{-1}	53–98 [c]	0.12–2.6	UV	[51]
CH columns				84–94 [c]			
Ph columns				91–99			
C_{18} disks	groundwater	1000	5 μg l^{-1}	20–86	0.1–4	UV	[52]
PS-DVB disks		1000	5 μg l^{-1}	51–81	0.1–4		
GCB columns	HPLC water	2000	0.05–1 μg l^{-1}	92–106	0.1–1 ng l^{-1}	DAD	[54]
PS-DVB disks	HPLC water	500	200 μg l^{-1}	93–102	–	UV	[55]
Acetyl-PS-DVB disks	distilled water	500		94–101		UV	[56]
	tap water	500	low mg l^{-1}	75–108	–		
	river water	500		92–107			
Carbopack columns	HPLC water	100	0.1 mg l^{-1}	54–98	2–6	DAD	[57]
ENVIChrom columns	HPLC water	500	10 μg l^{-1}	77–101	0.06–0.1		
PS-DVB disks	tap water	250	–	–	0.002–0.5	APCI–MS	[67]
		250	–	–	0.004–5		

[a] Phenol not recovered at all ($V > V_b$).

[b] Quantification not possible because of interferences in blank disks coeluted with phenol.

[c] 2-chlorophenol, 2,6-dichlorophenol and 4-chloro-3-methylphenol not recovered at all.

preconcentrated. Puig and Barceló [59] also compared the PGC sorbent to silica-bonded and polymeric sorbents and concluded that breakthrough volumes and detection limits when using PGC sorbents were worse than when using polymeric sorbents, except in the case of aminophenols for which PGC is to be preferred. Moreover, these authors indicate that PGC shows difficulty in recycling the extraction precolumns and lack of selectivity against polar humic substances, whereas Di Corcia and co-workers affirm that after six HPLC water extractions, the recovery of the analytes did not decrease significantly [54]. Therefore, the use of carbon-based sorbents allows high retention of phenols but the elution presents some difficulties making necessary some modifications of the usually applied protocols.

The polymeric sorbents based on copolymers of styrene-divinylbenzene (PS-DVB) have been the materials most widely used to extract phenolic compounds from water samples. They overcome many limitations of bonded silicas and show a broad range of pH stability and great analyte retention mainly for polar phenols because their hydrophobic surface contains a large number of active aromatic sites which allow $\pi-\pi$ interactions with the analyte to be established. Compared with GCB sorbents, the PS-DVB cartridges can preconcentrate larger sample volumes in a shorter time, afford flow-rates four times as high as those afforded by the carbon cartridges, and are not subjected to the compacting problems typically produced in carbon cartridges [47]. This type of sorbent material has been applied in the off-line mode followed by gas chromatography [38,44,47] and also by liquid chromatography [34,52,53,55, 62–64]. In off-line SPE, XAD resins [45,63,64], PLRP-S resin [52,53], PS-DVB cartridges [52,47], PS-DVB disks [34,44,52,55,58], functionalized sorbents [38,56] and styrene-divinylbenzene copolymers with a high degree of cross-linking such as Isolute Env+ or ENVI Chrom P [47,57] have been used for the extraction of phenolic compounds from water samples.

XAD resins have been used with good results [63,64] but some drawbacks including low breakthrough volumes for polar phenolic compounds and in consequence a lack of sensitivity and extensive cleaning before use have been found. Moreover, chemically modified resins have been developed in recent years and some papers contrasting the ability of derivatives (acetyl, carbonyl and hydroxymethyl) of a cross-linked polystyrene material to extract phenolic compounds from water samples showed that the acetyl derivatives gave the highest recoveries [38]. Also, PS-DVB and acetyl PS-DVB resins incorporated into SPE membranes and compared to C_{18} silica disks for the extraction of phenols from water showed the advantage of the derivatized materials. For instance, phenol recovery goes from 36% for the underivatized PS-DVB disks to 91% for acetyl PS-DVB disks [56].

Another approach to increasing the breakthrough volume (V_b) and improving SPE recoveries is the extraction after derivatization of phenolic compounds. Among the derivatization methods for phenols found in the literature, one of the simplest and most common is direct acetylation with acetic anhydride in the presence of carbonate, as first described by Renberg and Lindström [65]. Using this procedure the breakthrough volumes, for example, for monohalogenated and methylphenol acetates are 2 l when the extraction is performed using C_{18} membrane extraction disks [41], whereas a value of a few milliliters (50 to 100 ml) is obtained when no derivatives are formed. The main problem encountered in derivatization with acetic anhydride is that dinitrophenols react incompletely due to their acid character (2,4,6-trinitrophenol does not react at all, pK_a 0.8) [42]. Another way of enhancing extractability is to add an ion-pair reagent (tetrabuthylammonium bromide) to the water sample at pH 9 followed by the extraction of the ionic pairs using PLRP-S cartridges. This enables the breakthrough volume to be increased and the detection limits decreased down to 0.5 μg l^{-1} [53].

From the data in Tables 6.2 and 6.3 it can be deduced that polymeric sorbents are the best choice for off-line SPE of phenols in terms of breakthrough volumes and recoveries. Nevertheless, C_{18} silica and carbonaceous sorbents can be used successfully after a derivatization step of the phenolic compounds in the water sample, with in this case GC being the analytical method of choice. Moreover, better LODs can, in general, be achieved using GC (Table 6.2) especially with ITDMS detection with values between 0.04 and 0.09 ng l^{-1} in drinking water [48]. Nevertheless, it should be taken into account that LC–APCI/MS in SIM mode [67] achieved similar LODs to GC–ECD [43] or GC–ITDMS [41], with the main advantage that no derivatization is needed.

Although off-line SPE methodology has been extensively used to extract organic micropollutants from water, there is a general trend towards on-line methods. Solid-phase extraction coupled with liquid chromatography (on-line SPE–LC) has been used in the last 10 years to analyze phenolic compounds using UV [66,68–71], electrochemical detection [69–75], fluorescence [33,76] and mass spectrometry [77]. Silica-based reversed-phase sorbents (especially C_{18}) [68], carbonaceous sorbents [59,74,78,79] and polymeric sorbents [59,72,74,76,79–81] have been applied for the extraction of phenolic compounds. In fact, SPE can be coupled directly to liquid chromatography using special sample preparation units available under various trademarks such as LiChrograph OSP-2 (Merck, Darmstadt, Germany) and PROSPEKT (Spark Holland, The Netherlands). Another less complex and cheaper approach is to use a coupled column switching system that uses small precolumns or a disk holder connected to a conventional analytical LC column via an electrically or pneumatically driven (six-port) valve. Fig. 6.1 shows the

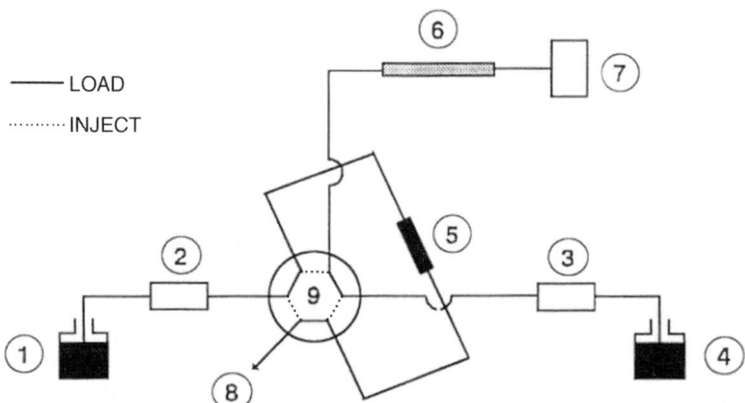

Fig. 6.1. Scheme of an on-line SPE–LC system: (1) water sample; (2) pump delivering sample; (3) mobile phase pump; (4) mobile phase; (5) SPE column; (6) analytical column; (7) detector; (8) waste.

most common system for on-line SPE extractions. A large volume of water is passed through a precolumn placed in the sample-loop position of a six-port liquid sampling valve (LOAD position) using a pumping system, while the mobile phase passes through the analytical HPLC column. The compounds retained by the precolumn are subsequently eluted into the HPLC analytical column (INJECT position). Practical guidelines and operational procedures and considerations can be found in the literature [37].

Table 6.4 summarizes some of the data reported in the recent literature. As in off-line preconcentration, reversed-phase silica-based sorbents have been used, for instance EPA method 555 [86], but the results obtained with polymeric sorbents are better. In the last few years several new polymers with commercial names such as LiChrolut EN, ENVIChrom-P, Styrosorb, Isolute ENV and HYSphere-1 have been used for the preconcentration of phenolic compounds. These sorbents have a high degree of cross-linking and so an open structure which increases their specific surface area and allows interactions between analytes and sorbents. In the recent literature, some comparisons between sorbents for the extraction of phenolic compounds from water samples can be found [59,71,79,80]. For instance, Puig and Barceló compared LiChrolut EN, Isolute ENV, PLRP-S and PGC in terms of breakthrough volume, recoveries and detection limits, and concluded that LiChrolut EN and Isolute ENV are the most suitable sorbents when the whole range of phenolic compounds has to be monitored [59].

It should be mentioned that when choosing polymeric sorbents to perform SPE coupled on-line with reversed-phase LC, the chromatographic profiles show some slight band broadening, which has to be reduced because it produces an increase in detection limits. This effect is due to the poor elution of phenols with eluents of low eluotropic strength. This phenomenon was found to be strongly dependent on the physicochemical characteristics of the sorbent and could be solved, for example, by eluting the precolumn with a mobile phase containing a relatively high percentage of organic solvent (40%) [73] or even with pure organic solvent (methanol or acetonitrile) and then mixing it with the rest of the mobile phase components in a high-pressure chamber [83].

The carbonaceous sorbents (GCB, PGC) have also been employed in on-line SPE. The first attempts to use this type of sorbent combined with LC were made by Weikhoven-Goewic and co workers [78] who, in a study of the preconcentration of chlorophenols from an aqueous solution with a pyrocarbon-modified silica coupled on-line to LC–UV, showed that this type of sorbent could retain phenolic compounds better than C_{18} silicas. However, it has to be coupled to a PGC analytical column to prevent excessive band broadening. The main problem encountered when working with this column is that elution of phenolic compounds with strong resonance capacity substituents such as nitro groups or highly chlorinated phenols is difficult. In addition, an important tailing for certain phenolic compounds was observed [59,84].

When the analysis of pollutants which are widely different in polarity has to be performed, an on-line SPE–LC system with several precolumns can be used. For instance, Brouwer et al. [82] coupled in series a PLRP-S precolumn (used to trap the neutral and non-ionized acidic solutes such as the phenols) and a precolumn packed with the same sorbent and loaded with sodium dodecyl sulfate (SDS) used to trap the positively charged (i.e. basic) solutes. The use of bifunctional extraction disks has also been described: this consisted of the combination of C_{18} and cation exchange (AG 50W-X8) membrane ex-

TABLE 6.4

SUMMARY OF ON-LINE SPE–LC METHODS FOUND IN THE LITERATURE FOR PHENOLIC COMPOUNDS IN WATER SAMPLES

Sorbent	Sample type	Sample volume (ml)	Concentration level (μg l^{-1})	Recovery (%)	LOD (ng l^{-1})	Detector	Ref.
ENVI Chrom P ENVI-18b column	drinking water	200	1	80–103	100	UV–fluorescence	[33]
PLRP-S column	groundwater	100	4	<20–100	10–4000	UV	[59]
LiChrolut EN column				55–103	10–2000		
Isolute ENV column				57–105	10–2000		
PGC column				52–88	1000–2500		
ENVI-Chrom P	tap water	10	1	>80	20–50	UV–ED (amperometric)	[70]
PLRP-S column					20–100		
Carboxybenzoyl resin	tap water	25	–	–	12–35	UV–ED (amperometric)	[71]
	river water	10	–	–	22–77		
LiChrolut EN column	groundwater	5–10	0.5	85–103	0.4–8.5	ED (coulometric)	[72]
PS-DVB disks	HPLC water	250	0.1	80–100	1–10	ED (amperometric)	[73]
	tap water	250	–	–	10–100		
	river water	25	–	–	5–150		
PRP$_1$ column	HPLC water	50	0.4	90	10	fluorescence	[76]
Lichrolut EN column	river water	50	5	62–109 (APCI)	20–100 0.5–40	ES–MS APCI–MS	[77]
PLRP-S column	tap water	10	–	–	100–4000	UV	[82]

traction disks for the simultaneous enrichment of acidic, basic and neutral compounds. In this type of preconcentration using cation exchangers, care should be taken to eliminate calcium ions from surface waters by precipitation with oxalic acid [85].

Finally, matrix effects should be taken into account when optimizing a solid-phase extraction method. Most authors agree that the sample matrix seems to influence detection limits and recoveries, although some authors obtain similar recoveries [56] and breakthrough volumes [82] with different types of water. The influence of the matrix on the recoveries of phenolic compounds was studied by Bao et al. [41] using off-line SPE and three matrixes: tap water, groundwater and river water showing that the recoveries and LODs in tap water and groundwater were similar to those in HPLC water but that LODs for nitrophenols were slightly higher in river water. The increase in the LODs (by between 2 and 5 times) between tap water and seawater was also observed by Jáuregui et al. [67] using LC with a highly selective detector such as API–MS and off-line SPE with PS-DVB membrane extraction disks. In fact, there is general acceptance that humic and fulvic acids affect the extraction of organic micropollutants from natural waters, so extrapolation of results obtained using HPLC water as matrix is not recommended. The breakthrough volumes in natural waters may decrease when humic substances are located on active sites of the adsorbent. Moreover, the adsorption of analytes into humic substances may result in lower recovery since only the dissolved fraction of the analyte will be enriched. This problem can be partly solved by acidifying the samples at low pH which, in addition, prevents partial deprotonation of phenols, but this led to adsorption of humic material and in consequence an enhancement of the peak due to this material in the chromatogram. To reduce the peak at the beginning of the chromatogram when natural waters are analyzed, the use of some chemical reagents such as sodium sulfite, sodium thiosulfate and oxalic acid has been proposed [87]. As an example of the effect of the sample matrix in on-line SPE–LC the chromatograms obtained using liquid chromatography and electrochemical detection (SPE–LC–ED) with PS-DVB membrane extraction disks of 13 phenolic compounds in HPLC-grade water and in tap water are given in Fig. 6.2. In Fig. 6.2B (250 ml of spiked tap water) a large interfering peak shows up, making the identification of phenol and 4-nitrophenol (peaks 1 and 2) impossible. Better results were obtained by decreasing the sample volume (Fig. 6.2C), corresponding to 25 ml of tap water spiked at the same concentration level. For river water with higher amounts of humic substances the matrix peak appearing at the beginning of the chromatogram was greater than for tap water and even with preconcentration of only 2.5 ml of sample, the matrix peak masked the peak of phenol, so preventing its identification and quantification [73].

From data in Table 6.4 it can be deduced that a wide range of LODs can be obtained depending on the sample volume and the detection technique. Nevertheless it can be concluded that electrochemical detection (ED) achieves lower values than UV or fluorescence, with coulometric detection being the most sensitive (LODs between 0.4 and 8.5 ng l^{-1} for 5 to 10 ml of groundwater) [72]. In addition, coupling UV and ED in series has the main advantage of providing the best detection of nitrophenols (UV) and the remaining phenols (ED). If results obtained using off-line SPE–LC (Table 6.3) and on-line SPE–LC (Table 6.4) are compared, in general better detection limits and higher reproducibility values (not shown in tables) are obtained by using on-line precolumn

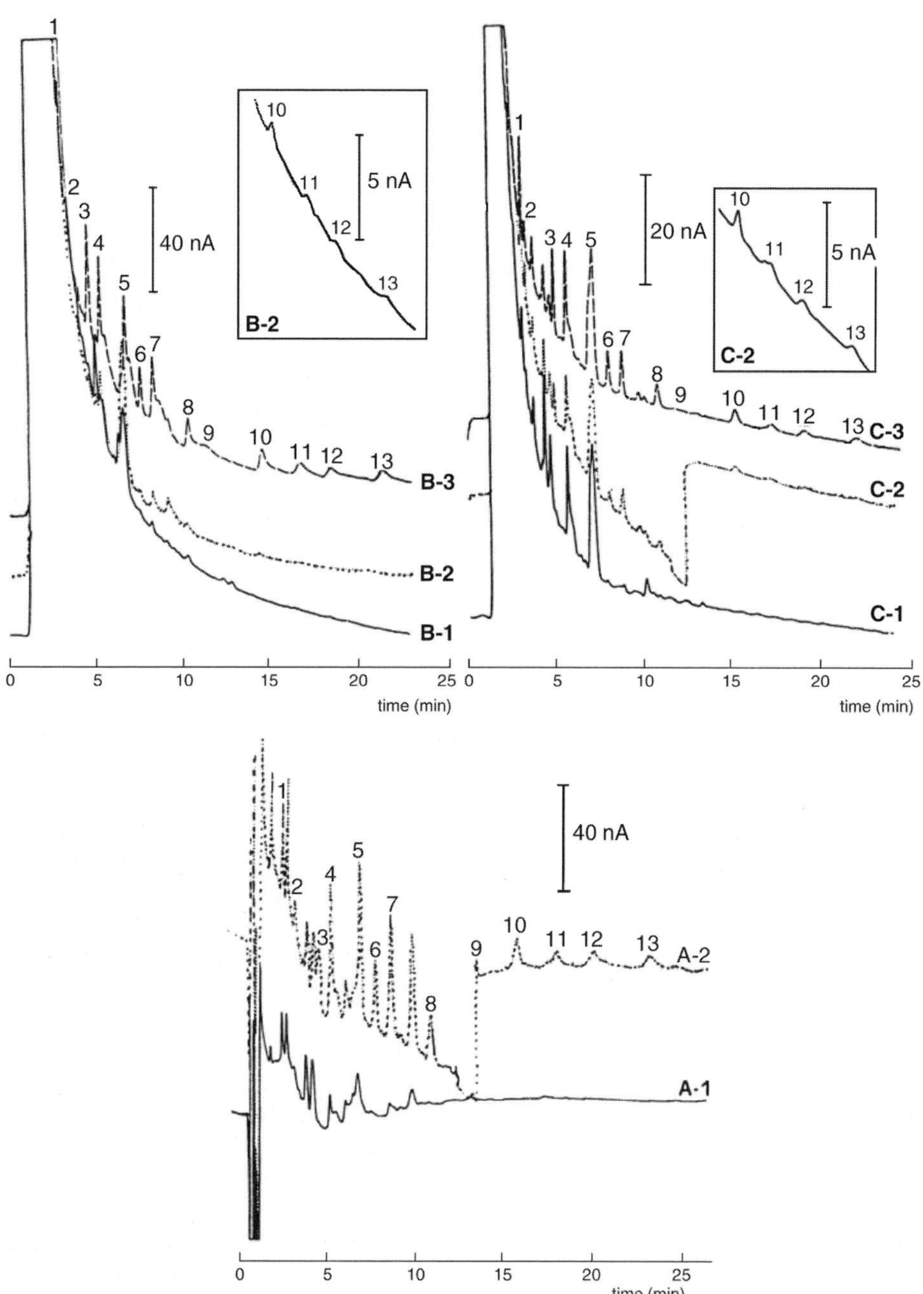

Fig. 6.2. Chromatograms corresponding to the on-line preconcentration through 10–4.6 mm i.d. PS-DVB membrane extraction disks of 250 ml HPLC-grade water at pH 2 non-spiked (A-1) and spiked at 0.01 μg l^{-1} (A-2); 250 ml of tap water at pH 2 non-spiked (B-1), spiked at 0.01 μg l^{-1} (B-2) and at 0.1 μg l^{-1} (B-3); 25 ml of tap water at pH 2 non-spiked (C-1), spiked at 0.05 μg l^{-1} (C-2) and at 0.1 μg l^{-1} (C-3). Peaks: (1) P; (2) 4NP; (3) 2CP; (4) 2NP; (5) 24DMP; (6) 26DCP; (7) 4C3MP; (8) 24DCP; (9) PCP; (10) 24DBP; (11) 246TCP; (12) 234TCP; (13) 245TCP. Tap water samples were treated with Na$_2$SO$_3$. (From [73].)

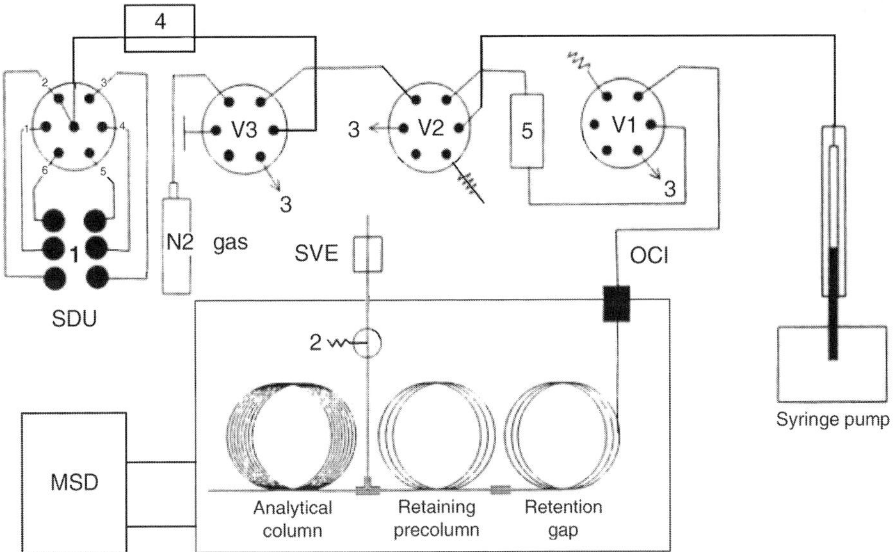

Fig. 6.3. Set-up of a system for automated on-line SPE–GC–MS: (1) solvent channels; (2) purge leak restriction; (3) waste; (4) single-piston LC pump; (5) SPE precolumn; V1–V3, PROSPEKT valves; SDU, solvent delivery unit; SVE, solvent vapor exit; OCI, on-column injector. (From [89].)

technology than off-line SPE because losses during sample manipulation are minimized, even though the enrichment of the low breakthrough volume analytes is a problem due to the restricted size of the precolumns. When using off-line methodologies a larger amount of sorbent can be used and also better LODs for polar phenols are usually achieved [59]. Moreover (see Table 6.3), higher sample volumes can be preconcentrated (100–2000 ml), mainly with membrane extraction disks (which allow flow-rates higher than 25 ml min^{-1}), thus improving the detectability of some compounds.

SPE can also be coupled on-line to GC systems [88,89]. A wholly automated, on-line SPE–GC–MS system including three six-port switching valves, a solvent delivery unit for the automated SPE sequence, and nitrogen to dry the SPE column was developed by Louter et al. [88] and used by some authors to analyze phenolic compounds in water samples [42,90]. In Fig. 6.3 there is a scheme of this type of coupling. This system employs a solvent delivery unit (SDU) to supply the organic solvents and also the water sample. After conditioning of the stationary phase in the precolumn (which usually involves flushing with methanol and, next, with water), a sample volume of 1–10 ml is loaded. The trace-level analytes of interest and also many less desirable sample constituents are preconcentrated on the precolumn; after a clean-up by washing with, usually, HPLC-grade water the analytes are desorbed by a suitable organic solvent and, next, separated on the analytical column. At least three switching valves had to be used, one to switch between aqueous (conditioning and sampling) and organic (desorption) solvents, another to supply drying gas, and the third one to direct the liquid or gaseous stream to waste or into the GC system via a permanently mounted transfer capillary. The solvent vapor exit (SVE) allows to improve the application range of the system and consists of a valve situated after some meters of a so-called retaining pre-column [89]. The main problem with this method arises from the presence of water in the chromatographic column, but this can be overcome by drying the sorbent with nitrogen at room temperature before the elution step. A stream of nitrogen at 1 ml min^{-1} for 3 min was enough to dry a 10 × 1 mm i.d. column. After drying, elution of the retained phenols with derivatization [42] or without [90] was performed using ethyl acetate. Linearity of the response must be taken into account because oversaturation of the GC column can occur. For instance, Jahr and co-workers [42] found a linearity range between 0.02 and 1.0 µg l^{-1}, so samples with phenol concentrations higher than 1.0 µg l^{-1} must be diluted with HPLC-grade water to avoid overloading and memory effects.

Apart from the main goal of extracting traces of the relevant compounds, SPE can also be used to remove the interfering components of the matrix, to change the solvent (e.g. aqueous to organic), and to store and transport analytes. As an example of the last application, Castillo and co-workers [81] showed complete recovery of a series of phenolic compounds after storage for two months at −20°C and after 0.5 months at 4°C, indicating that it is possible to ship SPE cartridges containing phenolic compounds under refrigeration conditions from the sampling site to the laboratory, making it unnecessary to perform the analysis immediately after sampling.

As a general conclusion, on-line SPE–LC using polymeric sorbents (in cartridge or disk format) have been used in preference to the off-line approach because analysis time and sample manipulation are less. In general, on-line SPE–LC provides lower detection limits than off-line SPE–LC but care should be taken with matrix effects when pre-concentrating natural waters, especially for the early-eluted compounds (generally, the more polar ones) because a matrix peak could mask them. So, for the determination of phenolic compounds in tap water and drinking water the use of whole automated on-line SPE–LC systems (mainly with electrochemical detection or MS) can be recommended; for the analysis of phenolic compounds in complex matrixes such as river water or industrial effluents and waste the use of off-line SPE systems is more advisable because more exhaustive clean-up procedures can be performed. Off-line and on-line SPE–GC have also been employed, but a derivatization procedure is necessary, in which case other sorbents different from the polymeric ones such as C$_{18}$ silicas or carbonaceous materials can be used.

Nowadays, a lot of effort is put into developing new selective and sensitive procedures for extracting and isolating components from complex environmental matrixes. Recently, selective elution in reversed-phase extraction and layering sorbents (combination of two membrane extraction disks, a strong anion-exchanger SAX and a C$_{18}$ disk) has been proposed in order to enhance selectivity when analyzing complex environmental samples. Moreover, immunosorbents and more recently molecularly imprinted polymers (MIPs) have been developed to improve the selectivity of the extraction. Immunosorbents (ISs) are solid-phase extraction materials which are based upon molecular recognition with natural antibodies. They allow a high degree of molecular selectivity and have been used for a long time for sample pretreatment in the medical and biological fields. Their application to environmental analysis is relatively recent owing to the difficulties in making selective antibodies for small molecules. This type of adsorbent material can circumvent the problem of coextraction of interferences in SPE, such as the coextraction of humic and fulvic substances with the polar organic compounds, and can be combined with LC in both off-line and on-line modes. Some applications for the enrichment of some organic compounds including triazine and phenylurea pesticides, BTEXs (benzene, toluene, ethylbenzene and xylene isomers), PAHs and benzidine and related azo dyes have also appeared [91]. Molecularly imprinted polymers (MIPs) are being developed to avoid certain disadvantages of ISs such as the instability of biological materials like antibodies. MIPs are stable synthetic polymers possessing selective molecular recognition sites. MIPs have been used as SPE materials in the preparation of complex biological samples but their use in the environmental field is scarce and still being developed. A recent paper has appeared dealing with the extraction of phenols using these sorbents [92].

In 1989 a new extraction method called solid-phase microextraction (SPME) was developed by Belardi and Pawliszyn [93]. Some papers summarizing the main factors affecting this type of extraction have appeared [94–98]. Briefly, SPME is based on the partitioning of the analyte between the extracting phase immobilized on a fused-silica fiber and the matrix, which is mainly water although gaseous and solid samples have also been analyzed. After the equilibrium is reached or after a well-defined period, the absorbed compounds are thermally desorbed by exposing the fiber to the injection port of a gas chromatograph or redissolving it in an organic or aqueous solvent if coupled to liquid chromatography [99] or capillary electrophoresis [100]. When gas chromatography is the analytical technique of choice, SPME completely eliminates the use of organic solvents; for liquid chromatography it greatly reduces their use. In addition, SPME has the advantage of its simplicity, although in general, its detection limits are higher than SPE because the extraction is not exhaustive. The first application of SPME to polar compounds was the determination of phenols in water samples performed by Buchholz and Pawliszyn [101,102]. These authors proposed the use of polyacrylate fibers and the addition of an acid and a salt to the sample to extract the free phenols from water. At pH 2 and saturated salt conditions, limits of detection for phenols using GC–FID are within one order of magnitude of the required EPA method 604 limits, except phenol and nitrophenols, and for GC–MS all the limits of detection exceed the required limits for EPA method 625. Nevertheless, matrix effects may be large. For instance, the authors found significant differences in the recoveries of phenols between a sewage sample and a clean water sample, which was also reported by Möder et al. [103] for the analysis of phenolic compounds in waste waters. It has been recognized that samples with high percentages of suspended matter and high-molecular-weight compounds such as humic acids and proteins present serious problems in SPME because the fiber can be damaged during the extraction or their properties changed due to the irreversible absorbance of the compounds. Humic acids and surfactants in aqueous solution not only obstruct the diffusion of the chlorophenols to the coating, but also inhibit the absorption of chlorophenols onto the fiber. These effects can be combatted by the extension of the extraction time [104] or by using headspace solid-phase microextraction (HS-SPME). In the headspace mode the analytes are transported through the gaseous phase before they can reach the fiber coating. This procedure allows hard sample treatments such as acidification and salting-out and the possibility of analyzing samples with particulate matter. In these cases, the use of HS-SPME permits the lifetime of the fiber coating to be lengthened and more reproducible results to be obtained, although frequently the analysis time increases due to the relatively slow transfer of the analytes from the aqueous to the gaseous phase. For the extraction of phenols by HS-SPME a combination of acidification to below pH 1 and salting-out through saturation with sodium chloride has been proposed. Nevertheless, even at these extreme conditions most analytes need more than 1 hour to reach equilibrium [102,105]. As previously described for SPE–GC, in-situ derivatization of phenols (to acetates) prior to extraction has also been used in SPME [98]. With this method an enhancement of the extraction for most of the EPA phenols can be obtained with a polydimethylsiloxane fiber, in addition, phenol acetates have better peak shapes than free phenols in GC [102,105]. The in-situ derivatization can also be performed directly in the polymeric coating following extraction. For instance, postderivatization to methylation using diazomethane gas has been performed on a polyacrylate fiber to detect various chlorophenols used as herbicides [104].

Solid-phase microextraction has also been coupled to high-performance liquid chromatography and the interface has been commercialized recently. It is based on the initial design of Chen and Pawliszyn [99] and enables the mobile phase to contact the fiber to remove the adsorbed analytes. Till now, only a few applications of SPME–LC, such as the analysis of polyaromatic hydrocarbons [99], alkylphenol ethoxylate surfactants [106] and polar pesticides [95] have been found in the literature. Some chlorinated phenols present in sludges and sediments have been analyzed using SPME and liquid chromatography–electrospray ionization mass spectrometry [107]. Other hydroxyaromatic compounds such as hydroxy-naphthalenes and hydroxyquinones have been analyzed in natural waters using Carbowax-coated SPME fibers and LC–UV [108]. Recently, Whang and Pawliszyn [100] designed an interface to couple SPME to capillary electrophoresis and phenols were selected as test analytes. The extraction was performed with a polyacrylate fiber, as is usual when free phenols have to be extracted. Nevertheless, commercial fibers cannot be used for this coupling, and so this method is not yet available for routine analysis even though one of the advantages of SPME in the micro fiber format could be its ability to remove analytes from very small sample volumes.

SPME is a very effective technique for screening because of its speed and ease of use. Nevertheless, it should be borne in mind that an equilibrium is established and that extraction is rarely quantitative. Therefore, for quantitative analysis, the extraction conditions for the standard solutions and the sample should be kept under strict control. Consequently, for complex matrixes the standard addition method instead of internal or external calibration is recommended. In summary, SPME is an excellent tool for the extraction of derivatized phenols and GC analysis, giving low detection limits thus allowing the analysis of these compounds in complex samples with particulate matter, and can be proposed as a fast screening technique.

6.2.2 Soil samples

The isolation of phenolic compounds from soil samples is not straightforward due to the strong binding of some phenols to soil organic matter. Table 6.5 summarizes

TABLE 6.5

SUMMARY OF SOME METHODS FOUND IN THE LITERATURE FOR THE ANALYSIS OF PHENOLIC COMPOUNDS IN SOIL, SEDIMENT AND SLUDGE SAMPLES

Sample	Extraction	Solvent	Clean-up	Extraction time	Concentration level (μg g^{-1})	Recovery (%)	LOD (ng g^{-1})	Chromatographic analysis	Ref.
Municipal sludge	Soxhlet	CH_2Cl_2	acid–base partition, GPC	24 h	–	–	30–180	LC–ED	[109]
Sewage sludge	Soxhlet	hexane–acetone (2 3) hexane–IPA (1 : 1)	no	–	–	35–103 61–102	–	GC–ECD	[115]
Soil (CRM)	Soxhlet	hexane–acetone (2 : 3)	no	12 h	0.04–1.4	81–107	2–200	LC–ED	[116]
Soil	Shaking	NaOH 0.1 mol l^{-1}	acid–base partition	–	0.01–1	60–83	–	LC–DAD	[120]
Soil (CRM)	Soxhlet	hexane–acetone (2 : 3)	no	12 h	–	–	10–700 [a]	LC–APCI–MS	[123]
Soil (CRM)	SFE	CO_2 (MeOH)	no	1.5 h	5–100	80–100	3–150	LC–ED	[126]
Soil	MAP	hexane–acetone (2 : 3)	no	10 min	2	88–111	–	GC–MS	[127]
	SFE	CO_2 (MeOH)	no	25 min	2		–		
Soil	SFE (in situ derivatization)	CO_2	silica gel column	15 min	11.5	81	–	GC–ECD	[128]
Soil (delta)	MAE	MeOH–H_2O (4 : 1), 2% TEA	SPE (Isolute ENV)	30–40 min	10	38–87	20–100 20–550 0.007–0.4	LC–UV LC–APCI–MS (full-scan) LC–APCI–MS (SIM)	[132]
Sediment	Soxhlet	CH_2Cl_2	no	24 h	15–20	46–81	–	FC–FID	[133]
	SFE	CO_2–MeOH–H_2O			15–20	48–104	–		

[a] LODs for monochlorophenols (300–700 ng g^{-1}) are 10–30 times higher than for other phenols.

some procedures found in the literature. The extraction and preconcentration step is classically performed by mechanical shaking or by Soxhlet extraction using solvents such as dichloromethane, acetone, hexane or mixtures of these [35,109–114], but mostly acetone–hexane has been employed [115,116] despite some controversies in the literature about recoveries with it. For instance, Wild et al. [115] gave a recovery of 30 to 50% for pentachlorophenol and tetrachlorophenols using dichloromethane, whereas Wall and Stratton [117] obtained recoveries higher than 94% for PCP with the same solvent. This showed the huge influence of the sample matrix and the degree of moisture on the extraction of phenolic compounds. Soxhlet extraction, which is very simple and versatile, has been the preferred method for the isolation of phenolic compounds from soil samples and is also the EPA official method (method 3540B) [118]. However, it is time-consuming and costly in the amount of solvent required. In addition, evaporation of the organic extracts must be performed carefully to avoid losses. Some modifications to the classical Soxhlet have appeared such as Soxtec, which uses less solvent and takes less extraction time to provide good results [119]. Mechanical shaking in both alkaline and acid media has also been proposed for the extraction of phenols from solid samples such as soils, sludges and sediments. Alkaline extraction (NaOH 0.1 mol l^{-1}) has been used with good recoveries (>60%) as a more selective method than direct extraction with organic solvents because a large amount of both interfering and less-polar impurities remained in the soil or the sediment [120,121]. The main problems found are the possible degradation of phenols in basic medium, the formation of emulsions in the subsequent reextraction with an organic solvent which are difficult to eliminate, and the possible transformation of other compounds present in the soil into phenols in basic medium. Recently, solid-phase microextraction (SPME) coupled with GC–MS successfully determined chlorophenols in landfill-leaches and soils with an important reduction in extraction time (40 min for SPME vs 8–12 h for Soxhlet), high precision and low LODs as a result of the preconcentration onto the fiber [122]. As can be seen in Table 6.5, several clean-up methods have been used in soil analysis. For instance, acid–basic partition schemes, solid-phase extraction using polymeric sorbents and gel permeation chromatography (GPC) have all been proposed [109]. Comparison of data in Table 6.5 with data in Tables 6.2–6.4 for water analysis shows that a lot of effort has been put into developing clean-up procedures in soil analysis by GC–FID or LC–UV, due to the great complexity of the matrix. When more selective detectors (GC–ECD, LC–ED) or mass spectrometry are used, it seems that the cleaning is not always needed, even in highly polluted sample extracts [116,123–125].

In recent years, supercritical fluid extraction (SFE) has been developed and applied to the extraction of phenolic compounds from soil samples. The SFE technique minimizes sample handling, provides fairly clean extracts, expedites sample preparation and reduces the use (and disposal) of environmentally aggressive solvents. Supercritical CO_2 is by far the most commonly used fluid in analytical-scale SFE. The extraction of polar analytes, such as phenols, requires the addition of an organic modifier such as methanol to enhance extractability due to the poor solvation power of the fluid and an insufficient interaction between the supercritical CO_2 and the matrix [125–127]. Alternatively, the extraction efficiency for phenolic compounds can be boosted by adding a derivatizing reagent to the sample. For instance, Lee et al. [128,129] applied in-situ acetylation to the analysis of phenolic compounds by addition of acetic anhydride in basic medium to the soil before SFE. In this case, triethylamine was used instead of the classical sodium carbonate, because its greater solubility in supercritical CO_2 [128]. One of the advantages of this procedure is the obtaining of phenol derivatives at the extraction step which are also suitable for GC analysis.

In SFE, two factors must be taken into account: the optimization of the experimental conditions and the different behavior of spiked and real samples. The optimization of the operating conditions in SFE is still considered a critical step in the development of a SFE method. Efforts have been made to understand which parameters affect the extraction process and how this can be optimized. Nevertheless, the selection of the operating conditions in SFE is still an area of active research characterized by much trial and error. Most of the reported SFE methods have been optimized by using one variable at a time, assuming no interaction between variables. In order to obtain reliable results in a reasonable time, statistical approaches to SFE have been adopted [126,130]. For instance, Llompart [130] studied in-situ extraction and derivatization for the analysis of phenol in soil samples and considered nine variables (CO_2 flow-rate, fluid density, extraction cell temperature, static extraction time, nozzle and trap temperature, amount of derivatizing reagent, pyridine concentration and time of contact between the derivatization reagent and the sample prior to extraction). Two-level (Packett-Burman) and three-level (central composite) orthogonal factor designs were used. Their results suggest that only the extraction cell temperature and the amount of derivatizing reagent were statistically significant to the overall extraction yield. Other factorial designs have been applied to the SFE of 17 positional isomers of chlorophenols from soils [126], with the conclusion that pressure, percentage of methanol and the first order interaction between these variables were statistically significant. As a result, an optimized method for the analysis of these compounds was developed giving low detection limits (3–150 ng g^{-1}) and reducing the total analysis time from about 24 h using Soxhlet [116] to 1.5 h using the developed SFE method. SFE has also been coupled on-line to LC with the drawback of some operational problems and the major advantage of the enhancement of selectivity, so cleaner chromatograms were obtained than with Soxhlet mainly in highly polluted samples [131].

The second important point to take into account in SFE is that naturally polluted samples may differ from spiked samples in concentrations of pollutants, sites and mechanisms of adsorption and ageing. Unfortunately, the use of spiked samples to evaluate extraction efficiencies can greatly overestimate recoveries because interactions between the sample matrix and both native and spiked analytes may differ and, moreover, the spiking solvent, i.e. acetone, methanol, etc., can act as a modifier and affect the extraction yield. In fact, some of the reported methods for soil analysis have been optimized by using samples spiked with known amounts of analytes prior to extraction. For instance, Soxhlet extraction has been performed after 8–12 h of spiking [116,132]: this contact time seems a reasonable time for reproducing binding effects that occur in environmental soils. However, information in recent literature concerning the extraction of phenolic compounds from soil samples by SFE, has reported ageing periods from 3 weeks to 2 months to ensure the interaction between the analyte and the matrix and to obtain enriched samples close to natural ones [130,133]. In fact,

References pp. 231–236

an interlaboratory study of the Bureau Community of Reference (BCR, Brussels) of chlorobenzenes and chlorophenols that gives information on preparation of candidate reference materials proposes that the soil and the spiking solution must be at least 12 days in contact [35]. In order to avoid such problems, the use of certified reference materials is recommended for the optimization of SFE conditions for the analysis of phenolic compounds in soils.

Recently, microwave assisted extraction (MAE), also called microwave assisted process (MAP), has also been used for the extraction of phenols from soils [127,132], with the main advantage that the simultaneous extractions of four to six samples could be performed quicker (10 min) than Soxhlet extraction, and that similar recoveries to those of SFE were obtained. However, care must be taken when working with flammable solvents or in the case of samples that contain constituents which couple strongly with microwave radiation to cause a rapid rise in temperature and so lead to potentially hazardous situations. Another approach is accelerated solvent extraction (ASE), quite similar to SFE except that a liquid at high temperature and pressure rather than a supercritical fluid is used for a static extraction. A mixture of dichloromethane–acetone (1 : 1) at 100°C and 10 MPa was used to extract some semivolatile EPA priority pollutants (among them phenol and 2,4-dichlorophenol) for 10 min. The results showed good recoveries for both phenols but the GC–MS chromatogram method blank of the ASE contained a number of interfering peaks higher than the Soxhlet, which suggested that reactions involving soil organic matter are greater under the high temperature and pressure conditions of ASE [134].

The analytical separations of phenolic compounds in soil analysis were classically performed by GC with FID or ECD detection or mass spectrometry and, as in the case of water analysis, in recent years LC has been the method of choice due to the main advantage that no derivatization was needed and high sensitivity was achieved with electrochemical detection and API–MS techniques. The detection limits for phenolic compounds in soils are given in Table 6.5. It is difficult to compare these results because values are very dependent on the amount of soil that has been extracted, the volume injected, the sample matrix, and the chromatographic and detection techniques. Nevertheless, the sample matrix seems to affect LOD values greatly. For instance, for 10 g of a clean delta soil spiked with phenols, LODs ranged between 0.007 and 0.4 ng g^{-1} [132], whereas for 1.6 g of a highly contaminated industrial soil LODs increased to 10–50 ng g^{-1} [123], both being in LC–APCI–MS (SIM mode).

In conclusion, the analysis of phenolic compounds in soil samples is dependent on the sample matrix. Therefore, synthetic samples must be prepared carefully. Use of reference materials is recommended. As previously commented, Soxhlet extraction has been the most common technique, but nowadays a lot of effort is being made to introduce new technologies such as microwave, ASE and supercritical fluid extraction in order to minimize time and solvent consumption. For SFE, optimization procedures are tedious and optimum conditions seem to be highly influenced by sample matrix. In contrast to water analysis, several authors have recommended the use of clean-up methods which are especially difficult for non-derivatized phenolic compounds because they are relatively polar. Because of the necessary elution with polar solvents a great many interferences are eluted which can interfere with the subsequent chromatographic analysis. In fact, a clean-up step can be mainly recommended when a derivative is formed during the extraction and, in this case, a non-polar compound is obtained. Then the classical clean-up procedures for non-polar substances using florisil or silica can be applied.

6.2.3 Biological matrices

As described in the introduction to this chapter, chlorophenols may be present in human urine. Monitoring of these compounds is normally used as an indicator of occupational exposure or exposure to environmental contamination [135,136]. Phenols are present in urine in the form of sulfate and/or glucuronic acid conjugates, so the determination of phenols in urine involves first the hydrolysis of the conjugates, then the extraction of hydrolysis products and LC or GC separation and quantification. Traditionally, chlorophenols have been analyzed in human urine by GC–ECD, GC–MS [11] or LC–ED [137,138] after acidic hydrolysis with concentrated H_2SO_4 (100°C) or HCl [10,139] and extraction by LLE [10,137–139] or steam distillation [11]. As described for water and soil analysis, derivatization of chlorophenols is a general trend to improve extraction efficiency and GC performance. Moreover, the complex matrix of urine samples requires some clean-up schemes such as acid base partition [139], SPE with anion exchange resins [138] or C_{18} sorbents [10,11]. In general, good recoveries (>70%) were obtained using the different methods, with limits of detection between 0.4 and 50 µg l^{-1}. Solid-phase microextraction (SPME)–GC–MS has also been applied using polyacrylate fibers. Its optimum conditions were pH 1, salt addition, equilibration times between 20 and 50 min and GC analysis. Lower detection limits (1–40 ng l^{-1}) and shorter analysis time were achieved compared to classical methods [136,140]. As an example, analysis of samples from workers of a sawmill where a sodium chlorophenolate product had been used, gave concentrations of chlorophenols in the range 0.02 to 1.60 µg l^{-1} [140], whereas those from workers exposed to phenol and benzene gave values between 3 and 20 mg l^{-1} [139].

6.3 CHROMATOGRAPHIC ANALYSIS

The present situation in the field of the analysis of phenolic compounds can be characterized as a coexistence between two chromatographic techniques, gas chromatography (GC) and liquid chromatography (LC). Historically, GC has been the method of choice for the analysis of phenols and moreover, GC is the official analytical method of the EPA for the analysis of these compounds in water samples (EPA methods 604, 625 and 8041) [24,25,141]. Nevertheless, in the last ten years a lot of papers have dealt with LC methods because of some advantages they have over GC methods. In this section, the most important GC and LC methods for the analysis of phenolic compounds in environmental samples found in the literature will be presented and the advantages and limitations of both techniques will be discussed. Also, the current USEPA methods for the analysis of phenols are reviewed.

TABLE 6.6

SUMMARY OF SOME GC METHODS FOUND IN THE LITERATURE FOR THE ANALYSIS OF PHENOLIC COMPOUNDS

GC Column	Derivative	Sample type/amount	Extraction	Detector	LOD	Ref.
DB-5	acetate chlorophenols	HPLC water/1000 ml	SPE	ITDMS	$2\text{--}35$ ng l^{-1}	[41]
HP-Ultra-2	acetate	HPLC water/100 ml	on-line SPE	MS	$1\text{--}27$ ng l^{-1}	[42]
WCOT SE-30	acetate	HPLC water/500 ml	SPE	ECD	$5\text{--}13$ ng l^{-1}	[43]
	PFB				$2\text{--}12$ ng l^{-1}	
PTE-5	no	HPLC water/250 ml	SPE	FID/ECD	–	[45]
DB-5	acetate chlorophenols	HPLC water/1000 ml	SPE	MIP–AED	$0.1\text{--}0.3$ μg l^{-1}	[47]
		HPLC water/2000 ml	SPE		$0.05\text{--}0.16$ μg l^{-1}	
DB-5MS	acetate chlorophenols	standard solutions	–	ITDMS	$0.08\text{--}0.19$ μg l^{-1} [*]	[48]
		HPLC water/1000 ml	SPE		$0.04\text{--}0.09$ ng l^{-1}	
HP-1	no	HPLC water/25 ml	on-line SPE	ECD	$2\text{--}8000$ ng l^{-1}	[90]
SPB-5	acetate chlorophenols	sediment/50 g	Soxhlet	MS	$0.2\text{--}2$ ng g^{-1}	[110]
HP-1	free-chlorophenols	water/200 ml	LLE	ECD	$1\text{--}25$ μg l^{-1}	[142]
	acetate chlorophenols				$0.2\text{--}25$ μg l^{-1}	
	PFB chlorophenols				$0.03\text{--}0.05$ μg l^{-1}	
PTE-5	no	groundwater/500 ml	LLE	NPD	$10\text{--}30$ μg l^{-1}	[143]
	PFB chlorophenols	natural water/1000 ml	LLE	ECD	0.1 μg l^{-1}	[146]
DB-17	methylnitrophenols	standard solutions	–	NPD	0.25 μg l^{-1}	[148]
DB-5MS	acetate chlorophenols	standard solutions	–	MIP–AED	$0.18\text{--}0.63$ mg l^{-1} [**]	[149]
				DD–FTIR	$0.11\text{--}0.15$ mg l^{-1}	
				MS	$0.01\text{--}0.02$ mg l^{-1}	
				ITDMS	$0.4\text{--}1.0$ μg l^{-1}	

[*] LOQ, S/N = 6.
[**] LOQ, S/N = 10.

6.3.1 Gas chromatography

Gas chromatography (GC) has been extensively used for the analysis of phenolic compounds in water, sludges and soil samples due to some advantages such as high separation efficiency, resolution and sensitivity as well as its rapidity of analysis. In addition, GC selectivity is improved when highly selective detectors such as electron capture detector (ECD), nitrogen–phosphor detector (NPD) or single-ion monitoring mass spectrometry (SIM/MS) are used. Another advantage of GC is its large amount of retention time data which can help in identification or confirmation. Moreover, in GC many new developments in the sample introduction process have been introduced within the last decade, often reported as LC–GC, on-line SPE–GC or SPME–GC, and have led to a substantial improvement in sensitivity. A summary of some methods found in the literature for the analysis of phenolic compounds in environmental matrixes using GC is given in Table 6.6. As can be seen in this table, phenolic compounds can be directly separated by GC in non-polar stationary phases but although the separation of several phenols can be achieved, irregular peak shapes, especially for tetrachlorophenols and pentachlorophenol, were observed and this tailing became pronounced with increasing age of the column. The use of new capillary columns with highly disactivated stationary phases permits tailing-free chromatograms and avoids adsorption problems. In fact, to analyze underivatized phenols some authors have proposed the use of semipolar GC columns [45,90,142,143], which are especially suited to the analysis of nitrophenols because their derivatization, mainly for dinitrophenols, cannot be performed completely. For example, a PTE-5 column (cross-linked polydimethylsiloxane with 5% diphenyl) has been used by Mußmann et al. [45] and Wennrich et al. [143] with good results, although some tailing was observed for several compounds, as can be seen in Fig. 6.4 which gives the chromatogram for a standard solution. Moreover, these authors claim that after extensive use the column shows tailing for phenols that has to be solved by shortening the column. In addition, GC analysis of underivatized phenols was worse for samples heavily contaminated with substances of high molecular mass: a rapid decrease in column efficiency permitted only a few samples, up to 20, to be analyzed before the performance of the chromatographic system had to be restored [144].

Injection of free phenols can be performed but, in general, GC separation is preceded by a derivatization step to convert phenolics into more apolar compounds in order to improve their chromatographic performance. Derivatization leads primarily to a decrease in the polarity of phenolic compounds, and so lower working temperatures can be used. Derivatization also leads to better separation as the derivatives can differ in their physicochemical properties more than the original phenols and also improves the extractability of these compounds from aqueous solutions if the derivatives are formed directly in the sample before the extraction [145]. Various derivatives have been proposed for the conversion of phenols: methyl or ethyl ethers, trimethylsilyl ethers, acetates or chloroacetates, trifluoroacetates, heptafluorobutyrates, pentafluorobenzoates and others [142]. Of these different derivatives, acetate has been the most commonly used because the procedure for this derivatization is simple and rapid and acetylation can be performed directly in dilute aqueous solutions with an almost quantitative yield.

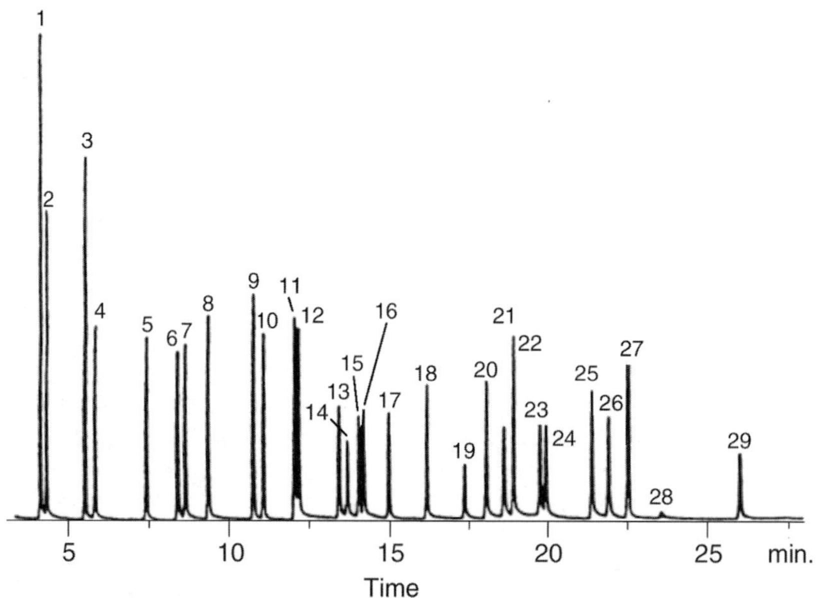

Fig. 6.4. GC–FID separation of underivatized phenols. Column: PTE-5 30 m. Peaks: (1) P; (2) 2CP, (3) *o*-cresol; (4) 2BP; (5) 2NP; (6) 24DCP; (7) 23DCP; (8) 3CP; (9) 4M2NP; (10) 5M2NP; (11) 3BP; (12) 4C3MP; (13) 235TCP; (14) 24DBP; (15) 246TCP; (16) 245TCP; (17) 236TCP; (18) 34DCP; (19) 25DNP; (20) 3NP; (21) 24DNP; (22) 1-naphtol; (23) 4NP; (24) 2356TeCP; (25) 3M4NP; (26) 2M46DNP; (27) 26DM4NP; (28) 4M26DNP; (29) PCP. (Reprinted from [45].)

As an example, in Fig. 6.5 the chromatogram of 21 phenol acetates in an SPB-5 column is given. As can be seen, good separation and peak shapes were obtained [110].

The derivatives containing halogen atoms in their molecule had the added advantage of enhancing detectability using an ECD. For instance, as can be seen in Table 6.6, LODs between 1 and 25 μg l^{-1} for free chlorophenols, between 0.2 and 25 μg l^{-1} for chloroacetates and between 0.03 and 0.05 μg l^{-1} for pentafluorobenzoate ethers of chlorophenols were reported using an apolar GC capillary column (HP-1) and ECD detection after preconcentration of 200 ml of water [142]. Other data found in the literature showed that PFB derivatives improved the detectability of less chlorinated phenols such as mono- and dichlorophenols (13 ng l^{-1} for the acetyl derivative and 2 ng l^{-1} for the PFB one, 500 ml water sample, SPE) whereas this effect is not so important for highly chlorinated phenols [43]. So, PFB derivatization could be the method of choice if simultaneous determination of non- or monochlorinated phenols together with polychlorinated phenols at trace level is required when ECD detection is used, because low detection limits could be achieved. As an example of the separation of the PFB derivatives, in Fig. 6.6 the chromatograms of 22 phenols are given [146]. Nevertheless, the formation of PFB derivatives has the drawback that sterically hindered phenols react slowly and incompletely and, in consequence, these compounds cannot be analyzed as PFB derivatives using GC–ECD. In fact, it has been described that the two dinitrophenols included in the EPA list (2,4-dinitrophenol and 2-methyl-4,6-dinitro-

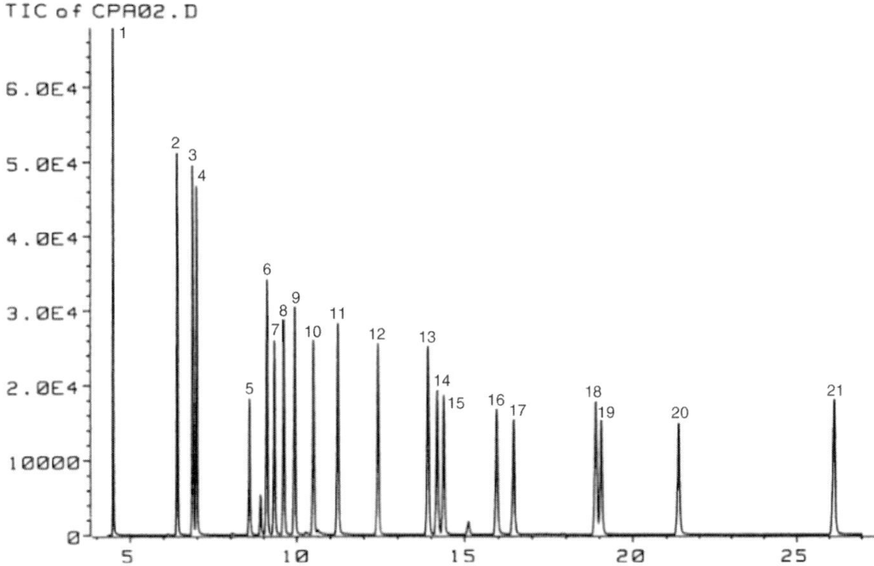

Fig. 6.5. GC–MS separation of acetates of chlorophenols. Column: SPB-5, 30 m. Peaks: (1) P; (2) 2CP; (3) 3CP; (4) 4CP; (5) 2C5MP; (6) 26DCP; (7) 4C3MP; (8) 24DCP; (9) 35DCP; (10) 23DCP; (11) 34DCP; (12) 246TCP; (13) 236TCP; (14) 235TCP; (15) 245TCP; (16) 234TCP; (17) 345TCP; (18) 2356TeCP; (19) 2346TeCP; (20) 2345TeCP; (21) PCP. (Reprinted from [110].)

phenol) do not react with PFBBr in the same way as the other phenols. High reaction times (up to 5 h) must be employed to obtain acceptable recovery yields and the ECD response is considerably lower (50 times) than other phenol derivatives [147]. For this reason, derivatization to methylated phenols instead of to PFB derivatives has been proposed. However, methylation requires the use of diazomethane, which is tedious and has potential hazards associated with its use, as described by Nick and Schöler [148]. Furthermore, a lot of interference appeared when derivatization with diazomethane was performed, probably due to polymerization products formed by excess diazomethane which settled down on the retention gap. Other general drawbacks of derivatization procedures are the additional source of error added to the method, the partial decomposition of derivatives that may occur during their storage and GC analysis, and the toxicity of many reagents used for derivatization which may also be carcinogenic or explosive.

Detection in gas chromatography has been performed using FID, ECD and NPD, but gas chromatography–mass spectrometry (GC–MS) is the most common technique being used today on the vast majority of compounds separable by gas chromatography. Recently, other detectors have been used for the analysis of phenols such as MIP-AED (microwave-induced plasma spectroscopy) and FTIR (Fourier-transform infrared spectrometry) by Rodríguez and Cela [47,149]. The hyphenation GC–MIP-AED provides a good determination technique for a wide variety of organic compounds such as phenols, by selecting the emission line characteristic of an adequate heteroatom in the molecule that for chlorophenols is the emission line of chlorine at 480.19 nm. Nevertheless, relatively high detection limits (0.2–0.6 mg l^{-1}) for standard solutions were obtained

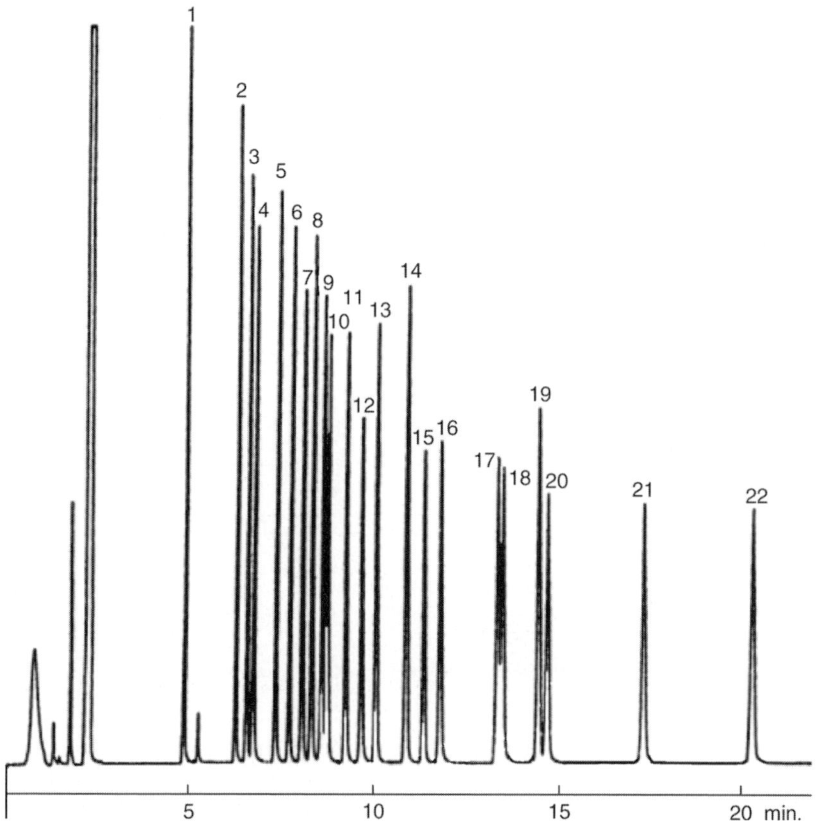

Fig. 6.6. GC–ECD separation of the PFB derivatives of phenols. Column: OV-1, 12 m. Peaks: (1) P; (2) 2CP; (3) 3CP; (4) 4CP; (5) 2C5MP; (6) 26DCP; (7) 4C3MP; (8) 25DCP; (9) 24DCP; (10) 35DCP; (11) 23DCP; (12) 34DCP; (13) 246TCP; (14) 236TCP; (15) 2C4BP; (16) 245TCP; (17) 234TCP; (18) 345TCP; (19) 2356TeCP; (20) 2346TeCP; (21) 2345TeCP; (22) PCP. (Reprinted from [146].)

so, after preconcentration of 500 ml of water sample, a detection limit of only 0.5 μg l^{-1} could be reached. Other drawbacks of this type of detection are that it cannot discriminate between overlapping chromatographic signals produced by chlorophenols and any other chlorinated compounds which have not been separated by the sample handling protocol [149]. Another hyphenated technique used for the analysis of phenols is the GC–DD–FTIR in which peaks are identified according to their corresponding IR spectra, which is very useful for instance for the analysis of positional isomers of chlorophenols. This is an advantage over GC–MS with which it is impossible to distinguish the structural isomers of chlorophenols with the same molecular mass [46,150].

As mentioned above, GC–MS has become a powerful technique for environmental analysis, with the great advantage of the availability of some spectra libraries that allow the unequivocal identification of analytes in complex samples. EPA methods require for qualitative identification the use of retention characteristics in two different columns

TABLE 6.7

CHARACTERISTIC MASSES (m/z) FOR THE POSITIVE IDENTIFICATION OF THE 11 PRIORITY PHENOLIC COMPOUNDS BY GC–MS (EPA METHOD 625)

Compound	Characteristic masses (EI)		
	Primary	Secondary	Secondary
2-Chlorophenol	128	64	130
2-Nitrophenol	139	65	109
Phenol	94	65	66
2,4-Dimethylphenol	122	107	121
2,4-Dichlorophenol	162	164	98
2,4,6-Trichlorophenol	196	198	200
4-Chloro-3-methylphenol	142	107	144
2,4-Dinitrophenol	184	63	154
2-Methyl-4,6-dinitrophenol	198	182	77
Pentachlorophenol	266	264	268
4-Nitrophenol	65	139	109

or the relative abundances of three characteristic masses (m/z) for each compound, whereas for quantitative analysis the addition of an internal standard with a single characteristic m/z is proposed. As an example, the three characteristic masses for the 11 priority free-phenolic compounds proposed by the EPA method 625 are given in Table 6.7.

In general, GC–MS methods found in the literature use the ions recommended by the EPA for quantification of phenols. Nevertheless, depending on the derivative formed, different ions could be monitored. For instance, for acetates of chlorophenols the selected ion is the CH_3CO^+ (m/z 43) that has a relative abundance of 100% for the higher chlorinated phenols (di-, tri-, tetra- and pentachlorophenol), whereas for phenol and monochlorophenols the $(M-42)^+$ ion is the most abundant and the M^+ has only a relative abundance between 7 and 16% for all phenols [110]. When chloroacetylation is the derivatization procedure of choice, the CH_2ClCO^+ (m/z 77) and $(M-77)^+$ ions were the major fragments and the M^+ is about 10% or less as abundant as the $(M-77)^+$ ions [151]. Finally, for the PFB ether derivatives of phenols, the $C_6F_5CH_2^+$ fragment (m/z 181) was always the base peak for all the compounds and the intensities of the M^+ and the corresponding phenoxy ion were much weaker [146]. However, there is a problem not solved by GC–MS and it is that mass spectra of positional isomers are practically identical, and as these compounds appear in the chromatogram with very close retention times, on analyzing real samples the exact nature of some peaks often cannot be accurately established. The solution to the above-mentioned difficulty is the use of tandem mass spectrometry (GC–MS/MS). This technique allows further selectivity because two mass separation steps take place. MS–MS detection can be accomplished by means of multi-quadrupole devices as well as by ion-trap mass spectrometers (ITDMS). A major advantage of the ITDMS is its relatively low cost and easy use for routine analysis. This detector was recently used to determine many classes of environmental organic compounds: its sensitivity and specificity is comparable to those of other instruments. In fact, hyphenation of GC with MS/MS combined with SPE

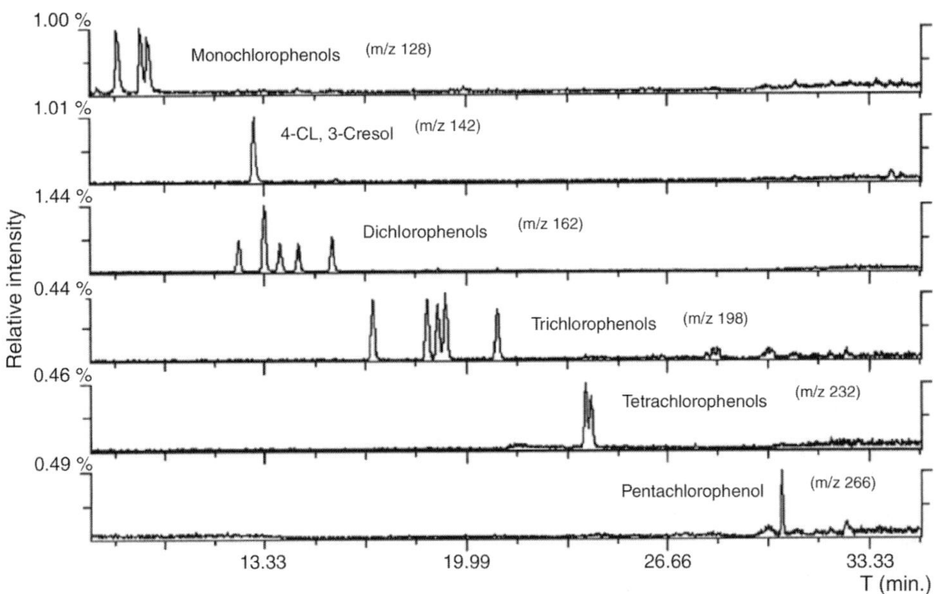

Fig. 6.7. Mass chromatograms for chlorophenol standards by GC–MS/MS. Column: DB-5MS, 30 m. (Reprinted from [48].)

allows the reliable determination of chlorophenols in drinking water at concentration levels 2–3 orders of magnitude below the limits established by current legislation for phenolic compounds (see Table 6.5) [41,48,149]. As an example, Fig. 6.7 shows the mass chromatograms obtained for chlorophenols by GC–MS/MS using an ion-trap (ITDMS) instrument.

Most of the data found in the literature using GC–MS for the analysis of phenolic compounds used electron ionization (EI). Nevertheless, for samples containing an inordinate number of interferences, the use of chemical ionization (CI) mass spectrometry may make identification easier. For instance, in some papers the use of GC–NICI/MS for the analysis of brominated and nitrated phenols in estuarine sediments [152], chlorinated phenols in industrial waste water [32] and nitrophenols in rain and snow [4] is proposed. The NICI spectra of nitrophenols are dominated by the molecular anion and the main fragmentations occur by loss of OH· and NO· [4]. The NICI mass spectra (methane as reagent gas) of the brominated phenols obtained by Tolosa and co-workers [152], exhibit the molecular anion and reliable isotopic distribution according to the halogen composition as the base peak and bromide (m/z 79 and 81) shows lower abundance. In contrast, for highly substituted chlorophenols (e.g. pentachlorophenol), another pattern of fragmentation is observed. In this case, the molecular anion is almost undetectable, and the base peak corresponds to the chloride elimination [152]. Nevertheless, it must be mentioned that source temperature is one of the factors most strongly affecting NICI mass spectra as was demonstrated by Crespín et al. who obtained different fragmentation for chlorophenols at different temperatures. For instance, at 120°C the fragmentation was almost absent; in contrast, above 180°C the fragment ion m/z 35 (corresponding to the chlorine atom) was the base peak in the spectra of all the chlorophenols. At 150°C significant differences in the mass spectra of compounds showing different chlorination degree were observed. For instance, mono- and dichlorophenols form fragment ions corresponding to the methane adduct with loss of HCl, $[M-HCl+CH_4]^-$ and tri-, tetra- and pentachlorophenol form no adducts showing the loss of HCl [153].

Detection limits for phenols found by various authors using GC methods are given in Table 6.6. As can be seen, ECD and MS provide values around the low μg l^{-1} when they are combined with SPE techniques. Some comparisons between various GC detection systems showed that ITDMS gives the lowest values for standard solutions [149] and so, when applied to real water samples, LODs in the low ng l^{-1} range can be achieved [41,48]. Off-line and on-line coupling of SPE with GC provide detection limits in the low ng l^{-1} range [42,90], although not to be overlooked is that on-line coupling suffers from being a sophisticated system and that a tentative knowledge of the concentration in the sample is needed because overloading and memory effects can occur.

6.3.2 EPA methods for analysis of phenolic compounds

As mentioned in the introduction, in 1979 the USEPA included eleven phenols in their lists of priority pollutants. This section reviews the current EPA methods for the analysis of phenolic compounds. It should be pointed out that one of the main advantages of using EPA methods is that detailed instructions for each step in the analytical procedure are given, along with instructions for the preparation of all the solutions needed for the analysis including concentration of standard solutions, solvents, dilution, storage conditions and time to be replaced. Furthermore, the apparatus and equipment are described and safety considerations concerning reagents, standards and consumable materials are included.

Different EPA methods for the analysis of phenolic compounds in drinking water and municipal and industrial waste waters have been proposed. Some of them are intended for the eleven priority phenols, but others include a high variety of phenolics. Table 6.8 gives the phenolic compounds considered in EPA methods 604, 625, 8040, 8041 and 1653 [24,25,154,141,155]. Methods 604 and 8040 are basically the same, although the latter covers a large number of compounds. This method (Table 6.9) proposes the injection of phenols directly onto the column without derivatization after a liquid–liquid extraction at pH 2. Moreover, conditions for derivatization as pentafluorobenzyl ether derivatives using pentafluorobenzyl bromide (PFBBr), a clean-up on silica and GC–ECD to confirm the measurements made by GC–FID are described. Method 625, summarized in Table 6.10, covers the determination of a large number of organic compounds that are partitioned into an organic solvent. A two-step extraction procedure at different pHs is used and phenolics are recovered in the extract obtained at low pH. However, this procedure significantly reduces recovery for phenol, 2-methylphenol and 2,4-dimethylphenol, and increases detection limits. Method 625 is a gas-chromatography mass-spectrometry method that enables compounds to be identified by the relative abundances of three characteristic masses that have been indicated in Table 6.7. In samples that contain a

TABLE 6.8

PHENOLIC COMPOUNDS CONSIDERED IN EPA METHODS

EPA method	Analytes
604	4-chloro-3-methylphenol, 2-chlorophenol, 2,4-dichlorophenol, 2,4-dimethylphenol, 2,4-dinitrophenol, 2-methyl-4,6-dinitrophenol, 2-nitrophenol, 4-nitrophenol, pentachlorophenol, phenol, 2,4,6-trichlorophenol
625	4-chloro-3-methylphenol, 2-chlorophenol, 2,4-dichlorophenol, 2,4-dimethylphenol, 2,4-dinitrophenol, 2-methyl-4,6-dinitrophenol, 2-nitrophenol, 4-nitrophenol, pentachlorophenol, phenol, 2,4,6-trichlorophenol
8040	2-sec-butyl-4,6-dinitrophenol, 4-chloro-3-methylphenol, 2-chlorophenol, cresols (methyl phenols), 2-cyclohexyl-4,6-dinitrophenol, 2,4-dichlorophenol, 2,6-dichlorophenol, 2,4-dimethylphenol, 2,4-dinitrophenol, 2-methyl-4,6-dinitrophenol, 2-nitrophenol, 4-nitrophenol, pentachlorophenol, phenol, tetrachlorophenols, 2,4,6-trichlorophenol
8041	phenol, 2-methylphenol, 4-methylphenol, 2-chlorophenol, 2,6-dichlorophenol, 2,4,5-trichlorophenol, 2,3,4,6-tetrachlorophenol, pentachlorophenol, 4-nitrophenol, dinoseb, 4,6-dinitro-2-methylphenol, 4-chloro-3-methylphenol, 3-methylphenol, 2,4-dimethylphenol, 2,4-dichlorophenol, 2,4,6-trichlorophenol, 2,3,4,5-tetrachlorophenol, 2,3,5,6-tetrachlorophenol, 2-nitrophenol, 2,4-dinitrophenol, 2-cyclohexyl-4,6-dinitrophenol
1653	4-chlorophenol, 2,4-dichlorophenol, 2,6-dichlorophenol, 2,4,5-trichlorophenol, 2,4,6-trichlorophenol, 2,3,4,6-tetrachlorophenol, pentachlorophenol, 4-chloroguaiacol, 3,4-dichloroguaiacol, 4,5-dichloroguaiacol, 4,6-dichloroguaiacol, 3,4,5-trichloroguaiacol, 3,4,6-trichloroguaiacol, 4,5,6-trichloroguaiacol, tetrachloroguaiacol, 4-chlorocatechol, 3,4-dichlorocatechol, 3,6-dichlorocatechol, 4,5-dichlorocatechol, 3,4,5-trichlorocatechol, 3,4,6-trichlorocatechol, tetrachlorocatechol, 5-chlorovanillin, 6-chlorovanillin, 5,6-dichlorovanillin, 2-chlorosyringaldehyde, 2,6-dichlorosyringaldehyde, trichlorosyringol

TABLE 6.9

EPA METHOD 604/8040: PHENOLS

Summary of the method	1 l of water sample is acidified (pH < 2) and extracted with methylene chloride; the solvent is changed to 2-propanol and 2–5 µl are injected onto the GC–FID. Alternatively, derivatization with pentafluorobenzyl bromide, clean-up with silica gel and injection onto the GC could be performed.
GC-columns	Underivatized phenols: 1.8 m × 2 mm i.d. glass, packed with 1% SP-1240DA on Supelcoport (80/100 mesh). Derivatized phenols: 1.8 m × 2 mm i.d. glass, packed with 5% OV-17 on Chromosorb W-AW-DMCS (80/100 mesh).
Detection	Free phenols: FID. Derivatized phenols: ECD.

TABLE 6.10

EPA METHOD 625: BASE/NEUTRALS AND ACIDS

Summary of the method	1 l of water sample is serially extracted with methylene chloride at pH > 11 and again at pH < 2. 2–5 µl are injected into the GC.
GC-columns	1.8 m × 2 mm i.d. glass packed with 1% SP-1240DA on Supelcoport (100/120 mesh).
Detection	MS.

TABLE 6.11

EPA METHOD 8041: PHENOLS

Summary of the method	Water samples are extracted with methylene chloride at pH < 2. Solid samples are extracted using Soxhlet (EPA method 3540) or ultrasonication (EPA method 3550). Acid–base partition clean-up (EPA method 3650) is suggested. Prior to the analysis the solvent is changed to 2-propanol. Alternatively, phenols may also be derivatized with diazomethane or pentafluorobenzylbromide.
GC-columns	Column 1: 30 m $\times$ 0.53 mm i.d. $\times$ 0.8 μm DB-5. RT$_x$-5, SPB-5, or equivalent fused-silica column. Column 2: 30 m $\times$ 0.53 mm i.d. $\times$ 0.8 μm DB-1701, RT$_x$-1701, or equivalent fused-silica column.
Detection	Free phenols and methylated phenols: FID. PFB ethers: ECD.

TABLE 6.12

EPA METHOD 515.1: DETERMINATION OF CHLORINATED ACIDS IN WATER BY GAS CHROMATOGRAPHY WITH AN ELECTRON CAPTURE DETECTOR

Summary of the method	1 l of water sample is adjusted to pH 12 and extracted with methylene chloride. Then the sample is acidified to pH < 2 and extracted with ethyl ether. The chlorinated acids are derivatized using diazomethane or trimethylsilyldiazomethane. Optionally, florisil clean-up could be performed. Injection into the GC.
GC-columns	Column 1 (primary): 30 m $\times$ 0.25 mm i.d. 0.25 μm DB-5 bonded fused-silica column. Column 2 (confirmation): 30 m $\times$ 0.25 mm i.d. 0.25 μm DB-1701 bonded fused-silica column.
Detection	ECD.

lot of interference chemical ionization mass spectrometry can be used. Nevertheless, though this technique is encouraged, it is not required by EPA methods.

Method 8041 (Table 6.11) is a general method that uses gas chromatography to analyze a large number of phenols as free compounds, as pentafluorobenzyl ethers or as methylated derivatives. Methylation is proposed because three phenols, 2,4-dinitrophenol, 2-methyl-4,6-dinitrophenol and Dinoserb, failed to derivatize under the PFBBr method. The extraction of water samples at pH < 2 as in EPA method 604 is proposed, and for solid samples Sohxlet or ultrasonication before an acid–base partition clean-up is recommended (EPA methods 3540, 3550 and 3650) [118,159,36].

In all EPA methods the procedure of cleaning and drying of all glassware is described. Sample collection and storing guidelines are also indicated. For phenols, samples must be collected in amber glass sample bottles (to protect samples from light) and then samples must be iced or refrigerated at 4°C from the time of collection until extraction. If residual chlorine is present (e.g. tap water), dechlorination has to be performed by the addition of a chlorine reductor such as sodium thiosulfate or sodium sulfite. All samples must be extracted within seven days of collection and completely analyzed within 40 days of extraction. The extraction of phenols from the sample, as can be seen in Tables 6.9–6.16, can be performed by liquid–liquid extraction at acid pH, although

TABLE 6.13

EPA METHOD 515.2: DETERMINATION OF CHLORINATED ACIDS IN WATER USING LIQUID–SOLID EXTRACTION AND GAS CHROMATOGRAPHY WITH AN ELECTRON CAPTURE DETECTOR

Summary of the method	250 ml of water sample is adjusted to pH 12 and washed with methylene chloride. The sample is then acidified to pH < 2, extracted with a 47 mm poly(styrene-divinylbenzene) extraction disk. Elution with 10% methanol in methyl-*tert*-butyl ether and derivatization to methyl esters using diazomethane or trimethylsilyldiazomethane. Injection into the GC.
GC-columns	Column 1 (primary): 30 m × 0.25 mm i.d. 0.25 μm DB-5 bonded fused-silica column. Column 2 (confirmation): 30 m × 0.25 mm i.d. 0.25 μm DB-1701 fused-silica column.
Detection	ECD.

TABLE 6.14

EPA METHOD 525.1: DETERMINATION OF ORGANIC COMPOUNDS IN DRINKING WATER BY LIQUID–SOLID EXTRACTION AND CAPILLARY COLUMN GAS CHROMATOGRAPHY/MASS SPECTROMETRY

Summary of the method	1 l of water sample is adjusted to pH < 2 and extracted with C_{18} sorbents (cartridges or disks). Elution with methylene chloride and injection into the GC.
GC-columns	Column: 30 m × 0.25 mm i.d. 0.25 μm DB-5 bonded fused-silica column.
Detection	MS (magnetic sector and ion-trap).

TABLE 6.15

EPA METHOD 555: DETERMINATION OF CHLORINATED ACIDS IN WATER BY HIGH PERFORMANCE LIQUID CHROMATOGRAPHY WITH A PHOTODIODE ARRAY ULTRAVIOLET DETECTOR

Summary of the method	100 ml of water are adjusted to pH 12 with NaOH, shaken and allowed to set for 1 h. Then, the sample is acidified with H_3PO_4 and on-line SPE with C_{18} cartridge or disks is applied. The analytes are separated by LC.
LC-columns	Column 1 (primary): 250 mm × 4.6 mm i.d. ODS-AQ, 5 μm spherical. Column 2 (confirmation): 300 mm × 3.0 mm i.d. Nova-Pak C_{18}, 4 μm spherical.
Mobile phase	0.025 M H_3PO_4–ACN in gradient mode.
Detection	DAD ($\lambda = 310$ nm for 4-nitrophenol; $\lambda = 290$ nm for pentachlorophenol).

TABLE 6.16

EPA METHOD 1653: CHLORINATED PHENOLICS IN WASTE WATER BY IN-SITU ACETYLATION AND GC–MS

Summary of the method	1 l of water sample is adjusted to neutral pH and K_2CO_3 and acetic anhydride are added. Then pH is raised to between 9 and 11.5. Extraction with hexane and injection into the GC were performed.
GC-columns	30 m × 0.25 mm i.d., 0.25 μm, 5% phenyl, 94% methyl, 1% vinyl silicone bonded-phase fused-silica capillary column.
Detection	MS.

some methods also use an acid–base partition scheme, EPA methods 625 and 515.1 [156] (Tables 6.10 and 6.12, respectively). In addition to the extraction, a clean-up step is here taken to some extent. Another clean-up recommended is gel permeation chromatography (GPC, EPA method 3640) [160].

Solid-phase extraction (SPE) procedures are not as frequent as liquid–liquid extraction in EPA methods, but they are proposed for the extraction of chlorinated acids and other organic compounds from drinking and groundwater. One of the methods, EPA method 515.2 [157] (Table 6.13), uses a resin-based polystyrene divinylbenzene disk and another one, EPA method 525.1 [158] (Table 6.14), proposes extraction onto a C_{18} sorbent. Both methods included pentachlorophenol among the compounds to be analyzed. Also, EPA method 555 [86] (Table 6.15) proposes an on-line SPE–LC–DAD method that uses C_{18} cartridges or disks to determine chlorinated acids in water and includes 4-nitrophenol and pentachlorophenol among their analytes of interest. In all EPA methods previously mentioned, phenols are extracted as underivatized compounds. Nevertheless, EPA method 1653 summarized in Table 6.16, involves in-situ acetylation. This method is proposed for the analysis of chlorinated phenolics (see Table 6.8) by acetylation with acetic anhydride, followed by liquid–liquid extraction with hexane and GC–MS.

One requisite of EPA methods is that the analyst should monitor the performance of the extraction and of the clean-up and analytical system and the effectiveness of the method in dealing with each sample matrix, by spiking each sample, standard and water blank with phenolic surrogates. Different compounds such as 2-fluorophenol, 2,4,6-tribromophenol and 2,4-dibromophenol have been proposed as surrogates. When MS detection is used, EPA recommends the use of labeled phenols (2,4-dichlorophenol and pentachlorophenol) as surrogates. Isotope dilution can be applied if labeled compounds are available.

Most of the current official EPA methods for the chromatographic analysis of phenolic compounds use gas chromatography to analyze the organic extracts coming from the extraction procedures. There is only one method, EPA method 555, that uses LC with diode-array detection for the determination of chlorinated acids in water, among them 4-nitrophenol and pentachlorophenol. In GC analysis, although some of the methods recommend the direct injection of free phenols, the derivatization of phenols provides better peak shapes and limits of detection. Three derivatization methods are suggested, one using pentafluorobenzyl bromide (methods 604 and 8040) and more recently one using diazomethane (method 8041). These two methods apply the derivatization step after extraction of phenolics from water. In contrast, the third one (method 1653) uses in-situ acetylation. The use of two chromatographic columns of different polarity, the primary column and a second called the confirmation column, is recommended by the EPA. Packed columns (1.8 m × 2 mm i.d.) are proposed in methods 604, 8040 and 625 (Tables 6.9 and 6.10, respectively). Notwithstanding, GC open-tubular columns (DB-5 and DB-1701) are also recommended to analyze some phenols. For instance, the EPA method 8041 uses a 30 m × 0.53 mm i.d. DB-5 for non-derivatized phenols and a 30 m × 0.53 mm i.d. DB-1701 for derivatives of phenols. More recently, thinner columns of 0.25 mm i.d with the same stationary phases were proposed in EPA methods 515.1, 515.2, 525.1 and 1653. Internal standards must be similar in their analytical behavior to the compounds under analysis and the analyst must

TABLE 6.17

METHOD DETECTION LIMITS (μg l^{-1}) OF EPA METHODS 604/8040 AND 625

Compound	EPA method 604/8040 GC–FID	EPA method 604/8040 GC–ECD	EPA method 625 GC–MS
Phenol	0.14	2.2	1.5
2-Chlorophenol	0.31	0.58	3.3
2-Nitrophenol	0.45	0.77	3.6
2,4-Dimethylphenol	0.32	0.63	2.7
2,4-Dichlorophenol	0.39	0.68	2.7
4-Chloro-3-methylphenol	0.36	1.8	3.0
2,4,6-Trichlorophenol	0.64	0.58	2.7
2,4-Dinitrophenol	13	–	42.0
4-Nitrophenol	2.8	0.70	2.4
2-Methyl-4,6-dinitrophenol	16	–	24.0
Pentachlorophenol	7.4	0.59	3.6

demonstrate that the measurement of the internal standard is not affected by method or matrix interferences. Because of these limitations no internal standard applicable to all samples is suggested by the EPA, but 2,4-dibromophenol and 2,4,6-tribromophenol are, in general, the internal standards recommended. Nevertheless, other compounds such as 2,2′,5,5′-tetrabromobiphenyl (method 8041), 2,2′-difluorobiphenyl (method 1653) and 4,4′-dibromooctafluorobiphenyl (methods 515.1 and 515.2) have also been proposed.

The method detection limits (MDL) of phenolic compounds for EPA methods are difficult to compare because different analytical protocols are employed. As an example, Table 6.17 gives method detection limits obtained with EPA methods 604 and 625 using reagent water spiked at low concentration levels. The higher detection limits obtained with ECD than with FID are due to the dilution of the extract after derivatization to PFB ethers. High method detection limits were also obtained with mass spectrometry (method 625), which can be explained by the effect of the GC–MS interface for packed columns. In fact, when open-tubular columns directly introduced into the MS are used, an important decrease in MLDs occurs. For instance, values ten times lower have been reported for pentachlorophenol with an ion-trap mass spectrometer (method 525.1). Nevertheless, the non-derivatization of phenols usually leads to broad peaks difficult to quantify and, as a result, high detection limits are obtained. For instance, EPA method 525.1 that involves SPE and GC–MS analysis gives much higher detection limits (20 to 30 times) than limits for other organic compounds which are correctly amenable by GC without derivatization. The LC method (EPA method 555) allows to reach a MDL of 1.2 μg l^{-1} for 4-nitrophenol and 1.6 μg l^{-1} for pentachlorophenol with on-line SPE–LC–DAD (20 ml of water). A MDL of 0.3 μg l^{-1} for PCP could be attained if the sample volume was 100 ml. Comparison of different EPA methods reveals a general trend showing that lower detection limits can be obtained for derivatized phenolics and open-tubular columns. For instance, method 1653 (Table 6.16) that involves in-situ acetylation and GC–MS in a capillary column allowed detection limits down to 0.15 μg l^{-1}. Moreover, EPA states that practical quantification limits (PQLs) are highly matrix-dependent and some correction to the method detection limits (MDLs) must be made. For instance, it is indicated that for groundwaters PQLs are 10 times higher than MDLs, and for low-level soils are 670 times higher.

Quality assurance is also a requirement for laboratories using EPA methods. Each laboratory that uses an EPA method must operate a formal quality control program. The minimum requirements of this program consist of an initial demonstration of laboratory capability and an ongoing analysis of spiked samples to evaluate and document data quality. In recognition of advances that are occurring in chromatography, the analyst is permitted certain modifications to improve the separations or to lower the cost of measurements. Each time such a modification is made, the analyst is required to establish the capacity of the modified method to attain acceptable accuracy and precision. To establish this ability quality control (QC) samples have to be prepared in reagent water and analyzed. The average recovery and the standard deviation of the recovery must be determined, and acceptance criteria for precision and accuracy given by the EPA method, fulfilled.

As has been explained, it is clear that USEPA official methods of analysis are still very far behind the latest techniques used in research laboratories. For instance, there are still few EPA methods that use SPE for extracting analytes from water samples, GC capillary columns or MS confirmation and quantification techniques, though these are currently being used in research laboratories. Nevertheless, we believe that EPA methods are very useful as a standard check on the performance of new methods of analysis and also offer a very useful way of obtaining full information about the analysis of phenolic compounds.

6.3.3 Liquid chromatography

As previously described, the usual way of obtaining a good separation of phenols in gas chromatography is by chemical derivatization. Liquid chromatography (LC), however, can be used for direct analysis. LC is a technique with advantages in phenolic compound analysis, in that the polarity of phenols and their relatively low vapor pressure, which complicate GC analysis, have no adverse effect in LC. Table 6.18 summarizes some of the methods found in the literature for LC–UV analysis of phenols in environmental samples including the mobile phase used and the detection limits obtained.

Different stationary phases (silica, amino, cyano, phenyl and octadecyl) have been tested for the separation of phenols but most authors agree that C_{18} is better able to separate substituted phenols. However, some applications use other stationary phases, for instance PRP-1, to separate several chlorophenols at basic pH [161] or diphenyl columns to separate 21 phenols [28]. Binary mixtures of water–methanol or water–acetonitrile at different pHs have been proposed as mobile phases for liquid chromatography. The pH of the mobile phase is known to affect the retention of phenols in the column, depending on the degree of dissociation. In addition, partial dissociation might lead to peak broadening and asymmetric peaks due to co-elution of the acid and appearance of its conjugate base [162]. Acidification of the aqueous mobile phase has a favorable effect on separation, as the dissociation of phenols is suppressed, retention times are shorter and peak asymmetry is improved [29,163]. For this reason, most separations are performed at relatively low pH, mainly with addition of 1% of acetic or phosphoric acid

TABLE 6.18

SUMMARY OF SOME LC–UV METHODS FOUND IN THE LITERATURE FOR THE ANALYSIS OF PHENOLIC COMPOUNDS

Compounds	LC column	Mobile phase	Elution mode	Sample type/volume	Extraction	Detection	LOD	Ref.
16 NFs	C_{18}	phosphate buffer pH 3.25–methanol	G	standard solution HPLC water/1000 ml	– LLE	UV ($\lambda = 316$ nm)	0.4–1.1 ng 0.5 μg l^{-1}	[5]
20 phenols	C_{18} diphenyl	10 mM ammonium acetate pH 4.8–MeOH	I	standard solution	–	UV ($\lambda = 220$ nm)	0.5–1.3 mg l^{-1} 0.5–1.9 mg l^{-1}	[28]
7 CPs	C_{18}	phosphoric acid 7 mM–ACN	I	HPLC water/1000 ml	off-line SPE	UV ($\lambda = 225$ nm)	0.12–2.6 μg l^{-1}	[51]
13 CPs, NPs, RPs	C_{18}	1% acetic acid pH 2.8–MeOH	G	tap water/15 ml	on-line SPE (ion-pair)	UV ($\lambda = 280$ nm)	0.1–2 μg l^{-1}	[68]
13 CPs, NPs, RPs	C_8	1% acetic acid–ACN	G	surface water/500 ml	off-line SPE	UV ($\lambda = 280$; 310 nm)	0.025–0.25 μg l^{-1}	[81]
13 CPs	PRP-1	phosphate buffer pH 9.2–ACN	I	standard solution	–	UV ($\lambda = 254$ nm)	–	[161]
11 EPA	C_{18}	0.1% acetic acid–ACN–MeOH (1:1:1)	I	standard solution	–	UV ($\lambda = 280$ nm)	–	[164]
13 CPs	C_{18}	phosphate buffer pH 7.20–MeOH (1:1)	I	standard solution	–	UV ($\lambda = 254$ nm)	5–45 ng	[165]
11 EPA	C_{18}	phosphate buffer pH 3–ACN	G	surface water/50 ml	on-line SPE	DAD	0.05–1 μg l^{-1}	[166]
15 CPs	C_{18}	10 mM sodium acetate pH 4 + 2 mM sodium EDTA–MeOH	G	standard solution	–	UV ($\lambda = 280$ nm)	0.5–5 ng	[170]

G = gradient; I = isocratic.

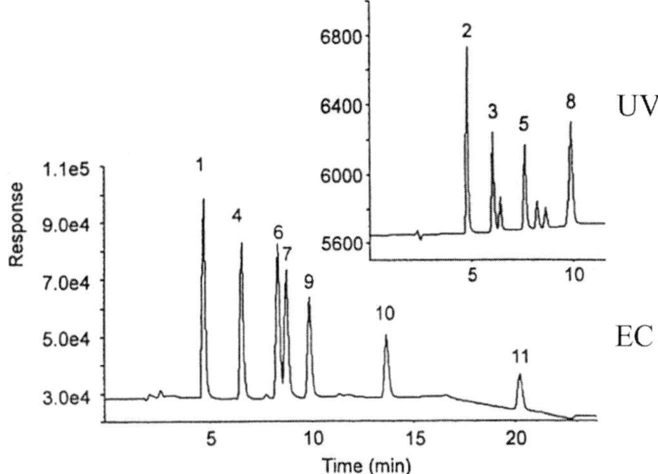

Fig. 6.8. LC–UV and LC–ED separation of 11 EPA phenols. Chromatograms of a standard solution (0.5 mg l^{-1}). Chromatographic conditions: C$_{18}$ column. UV detection at 316 nm for 4NP and 280 nm for the other nitrophenols. Electrochemical detection, amperometric (working potential 1.0 V vs Ag/AgCl). Gradient elution: solvent A, acetic acid 1% +0.05 g l^{-1} KCl, solvent B, acetonitrile. Peaks: (1) P; (2) 4NP; (3) 24DNP; (4) 2CP; (5) 2NP; (6) 24DMP; (7) 4C3MP; (8) 2M46DNP; (9) 24DCP; (10) 246TCP; (11) PCP. (Reprinted from [71].)

to the mobile phase [29,57,81,164]. As an example, Fig. 6.8 shows the LC separation of the 11 priority pollutant phenols listed by the EPA using a binary mobile phase with 1% acetic acid and gradient elution [71]. Nevertheless, some occasional applications used higher pH such as methanol–phosphate buffer at pH 7.2 for the separation of 13 chlorophenols in a C$_{18}$ column [165] or acetonitrile–phosphate buffer pH 9.2 for the separation of 13 chlorophenols in a PRP-1 column [161]. Special attention must be paid to the separation of phenols with LC and coulometric electrochemical detection where a pH around 7 is needed to avoid the contamination of the electrode by the oxidation products of phenolic compounds [30,34] as the main drawback is that at this pH peak tailing for some compounds such as 2,4-dichlorophenol or pentachlorophenol occurred.

The most commonly used detection mode in LC, the UV spectrophotometric method, is also useful for phenolic compounds and has been widely used [5,28,29,57,71,81,161–169]. When diode-array detection (DAD) is used, UV spectra can be recorded and contrasted with the analyte in the sample for positive identification. Selection of the working wavelength is an important point to bear in mind. For most phenolic compounds, the wavelength of maximum absorption is selected and, in general, the most common wavelengths are 254 and 280 nm but other values can enhance the selectivity of the method. For instance, nitrophenols usually show an additional absorption maximum at >300 nm, so at these wavelengths suppression of the non-nitrated phenols is observed and thus selectivity can be enhanced [5,167]. Apart from this advantage of UV detection for nitrophenols, the analysis of water samples using LC–UV suffers from an excessive amount of interference and for this reason some authors have argued for electrochemical and fluorescence detection as more selective and sensitive techniques.

TABLE 6.19

SUMMARY OF SOME LC–ED METHODS FOUND IN THE LITERATURE FOR THE ANALYSIS OF PHENOLIC COMPOUNDS

Compounds	LC column	Mobile phase	Elution mode	Sample type/volume	Extraction	Detection	LOD	Ref.
7 phenols	C_{18}	phosphate buffer pH 7.0–ACN–MeOH (64 : 19 : 17)	I	sea water/250 ml	off-line SPE	coulometric, $+0.75$ V	0.04–0.1 μg l^{-1}	[34]
16 phenols	C_8	acetate buffer pH 4.2–ACN–MeOH	G	groundwater/100 ml	on-line SPE	UV–EC (amperometric) ($\lambda = 280$; 310 nm; $+1.0$ V)	1–10 ng l^{-1}	[69]
11 EPA	C_{18}	acetic acid 1%, 0.05 g l^{-1} KCl–MeOH	G	tap water/25 ml	on-line SPE	amperometric, $+1.0$ V	20–50 ng l^{-1}	[70]
NPs, CPs	C_{18}	phosphate buffer pH 5.2–ACN (75 : 25)[a]	I	groundwater/5–10 ml	on-line SPE	coulometric ($E_1 +0.3$ V; $E_2 +0.65$ V)	2–9 ng l^{-1}	[72]
17 CPs	C_{18}	acetate buffer pH 5.3–ACN–MeOH (60 : 30 : 10)	I	standard solutions / soil samples/1.6 g	– / Soxhlet	amperometric, $+1.1$ V	2–200 pg / 2–200 ng g^{-1}	[116]
Phenols	C_{18}	acetone–acetic acid–KNO$_3$ 0.1 M	G	standard solutions / seawater/1000 ml	– / on-line SPE	amperometric, $+1.25$ V	0.9–6.1 μg l^{-1} / 5–45 ng l^{-1}	[171]
16 NPs	C_{18}	phosphate buffer pH 3–MeOH (49 : 51)	I	standard solutions	–	amperometric, $+1.2$ V, UV ($\lambda = 254$ nm)	3–25 μg l^{-1} / 5–25 μg l^{-1}	[172]
15 RPs	C_{18}	acetic acid 1%–ACN (66 : 34)	I	waste water/100 ml	LLE	amperometric, $+1.1$ V, UV ($\lambda = 280$ nm)	5–170 pg / 4–14 ng	[175]
5 CPs	Phenyl	acetate buffer pH 5–ACN (40 : 60)	I	standard solutions	SPE	dual amperometric ($E_1 +0.60$ V; $E_2 +0.90$ V)	0.4–6 μg l^{-1}	[179]
Phenol	PS-DVB	phosphate pH 9.2–ACN (25 : 75)	I	standard solution	–	coulometric ($E_1 +0.25$ V; $E_2 +0.65$ V)	0.03 μg l^{-1}	[181]

[a] Post-column addition of NaOH 0.1 M.

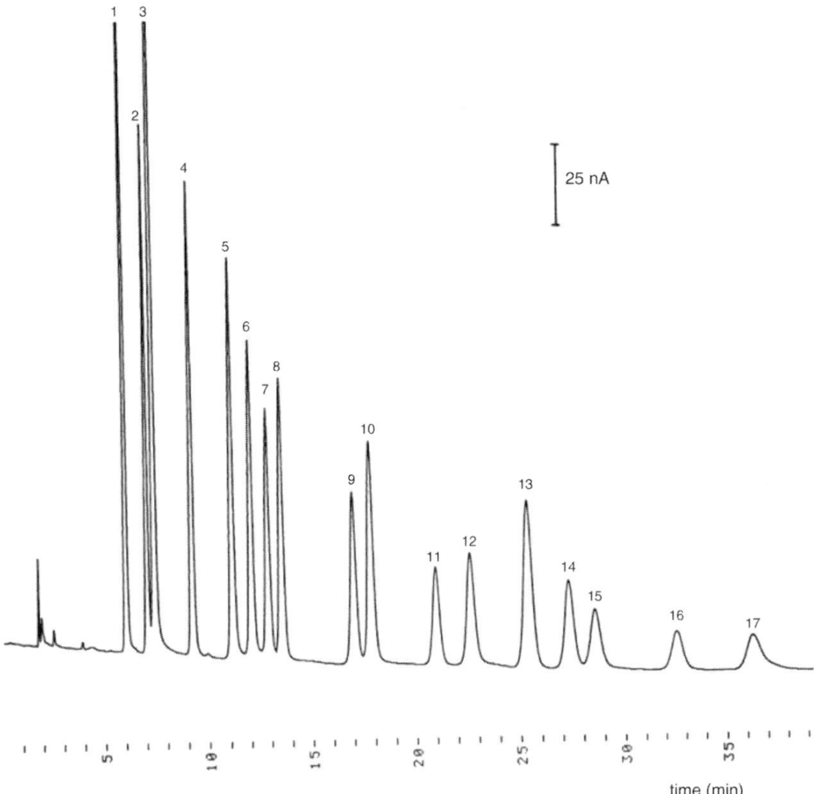

Fig. 6.9. LC–ED separation of 17 positional isomers of chlorophenols. Chromatogram of a standard solution (1 mg l^{-1}). Chromatographic conditions: C$_8$ column, isocratic elution, mobile phase: acetate buffer 30 mM pH 4.5/acetonitrile/methanol (60:30:10), working potential +1100 mV vs Ag/AgCl. Peaks: (1) 2CP; (2) 4CP; (3) 3CP; (4) 26DCP; (5) 23DCP; (6) 25DCP; (7) 24DCP; (8) 34DCP; (9) 236TCP; (10) 35DCP; (11) 234TCP; (12) 234TCP; (13) 245TCP; (14) 235TCP; (15) 2356TeCP; (16) 2346TeCP; (17) PCP. (Reproduced with permission from [116].)

Since most phenols are electrochemically active, LC with electrochemical detection (ED) is a good choice for the analysis of phenolic compounds at trace levels. Table 6.19 summarizes some studies found in the literature on LC–ED. Fig. 6.9 gives the LC–ED separation of 17 positional isomers of chlorophenols using a ternary mobile phase and isocratic elution as an example. Recent years have seen marked growth in the use of this technique, mainly with amperometric detectors [50,69–71,73,74,116,124,171–180], although coulometric detectors have also been successfully used [34,72,181,182] with higher sensitivity than amperometric ones due to the large surface area of the electrode. The main drawback of coulometric detection is that the response decreased in continuous use due to the fouling of the electrode. This is a general problem for electrochemical detection, as several authors have described but, in the case of amperometric detectors cleaning of the surface of the electrodes mechanically, chemically or electrochemically usually restores the response [69,73,116]. In the case of coulometric detectors, however, the porous electrode can only be cleaned in a limited way (by injection of diluted sodium hydroxide solutions or by applying higher potentials) and a decrease in response was observed with the ageing of the electrode that could only be solved by replacing the electrochemical cell [34].

To optimize working potential when using LC–ED, the hydrodynamic voltammograms (HDVs) for each analyte have to be recorded. The half-wave potentials ($E_{1/2}$) and the responses of the phenols mainly depend on the type and position of additional substituents and reflect their electronic effects. Moreover, optimum potential depends on both pH and the composition of the mobile phase, so the HDVs for each analyte must be studied in each chromatographic condition and a compromise between stabilization time, background noise and sensitivity has to be found. For instance, nitrophenols show high oxidation potentials (> +1.2 V) in which noise is very important. Therefore, ED is not a good choice for these compounds and the coupling LC–UV–ED can be recommended for achieving high sensitivities for nitrophenols in the UV detector and for the other phenolic compounds in the ED detector [71]. Another consideration to bear in mind is that the limiting current of the mobile phase is in general lower in basic media than in acid ones, so permitting work at low potentials and, in consequence, a better signal-to-noise ratio. Because of this and since LC separations of phenols are usually performed at acid pH, post-column addition of NaOH 15 mM at 0.1 ml min^{-1} has been proposed in LC with coulometric detection [72].

Gradient elution is not recommended when using electrochemical detection because system stability is affected, causing baseline drift and random noise especially for coulometric detection [34,50,72,178,182]. Nevertheless, some authors performed gradient elution when working with amperometric detection (see Table 6.19), but lower sensitivity was obtained so, the gradient elution could only be used when a relatively high concentration of these compounds had to be analyzed [69,70,171,180]. Although high selectivity can be achieved in LC–ED, the relatively high potentials derived from the HDVs may lead to other matrix components being oxidized, and thus an increase in the background current and a decrease in selectivity, which make the identification of phenols difficult. A solution to the above-mentioned problem is to work with dual electrode detection (E_1 at the first electrode and E_2 at the second electrode, $E_1 < E_2$) where peak identity in samples could be obtained through the comparison of current response ratios at two different potentials (E_1/E_2) previously established in standards [177,179,181]. If a dual electrode detector is not available, two injections of the sample can be performed at two different potentials and the ratio of currents obtained (E_1/E_2) can be compared with that of the standards [50,71,173,174,176]. An interesting application of this procedure is found in the analysis of alkylphenols in which the chromatographic separation of *m*- and *p*-cresol is not possible, as has been described by some authors [175,176]. In this case electrochemical separation is possible because both compounds give different current ratios at +850 mV and +900 mV vs Ag/AgCl [174,177] thus providing a solution to the chromatographic co-elution problem.

Detection limits with LC–ED are considerably lower (up to 100 times) than with LC–UV except for nitrophenols, as mentioned above. For standard solutions LODs in the ng level for UV and in the pg level for ED are obtained for alkylphenols [175]. In water samples, LODs in the μg l^{-1} level for LC–UV and down to 1 ng l^{-1} for LC–ED can be obtained with on-line SPE–LC systems [69,73]. Moreover, improvement by at least one order of magnitude was found when coulometric instead of amperometric detection was used. This improvement in LODs (values ranging from 10 to 50 ng l^{-1} for amperometric and from 0.4 to 2.4 ng l^{-1} for coulometric) should be attributed to the almost complete oxidation of the analytes in the coulometric detector.

Fluorescence detection has also been used in the analysis of phenols because of its high selectivity. In fact, measurement of the native fluorescence is selective and can lead to low mass-detection limits in the higher picogram range, but the main drawback of this type of detection is that most of phenols show only low fluorescence or no fluorescence at all. The only compound which has innate fluorescence is phenol itself, so the lowest detection limits are attained for this compound. For instance, LODs of 0.01 ng for phenol and between 0.3 and 0.7 ng for other chlorinated compounds have been obtained for standard solutions using fluorescence detection [33]. Attempts have therefore been made to solve the detection problem by means of post-column reactions that lead to products with better detection properties for fluorescence detection than the original ones. For instance, photochemical conversion of chlorophenols into fluorescent phenol was achieved by Brinkman and co-workers with good results mainly for mono- and dichlorophenols [183]. Another finding of this group of researchers was the photochemical decomposition by UV irradiation of dansyl derivatives of phenolic compounds in methanol–water mixtures leading to the formation of highly fluorescent dansyl-OH and dansyl-OCH$_3$ thus providing LODs of 200 pg for PCP [184,185]. So, high sensitivity similar to that of electrochemical detection could be reached using fluorescence detection with the main drawback, as in GC, that derivatization of phenols must be performed in order to obtain fluorescent compounds.

Liquid chromatography–mass spectrometry (LC–MS) has also been used to analyze phenols, with the main advantage over other conventional detectors that it is able to provide confirmation or unambiguous identification. Thermospray (TSP) [186–189] and particle beam (PB) [190] interfaces have been reported for the analysis of phenols. LC–PB/MS provides LODs for standard solutions between 4 ng and 51 ng and LC–TSP/MS gives values between 40 and 650 ng in full-scan, going down to 0.4 and 2 ng in single-ion monitoring (SIM) [189]. Nevertheless, in the field of LC–MS there is much current interest in atmospheric pressure ionization methods (API), i.e. electrospray (ES), ionspray (IS) and atmospheric pressure chemical ionization (APCI), due to their being more sensitive than TSP or PB. For instance, the limits of detection in APCI are in general a magnitude lower than in TSP [189]. Some environmental applications for phenols using LC–API/MS methods have been reported. For example, chloronitrophenols in waste water effluents [195] and pentachlorophenol in waters [192] were analyzed by LC–ES/MS, whereas LC–APCI/MS was used to determine some polyphenolic compounds in olive mill waste water [193] and in artichokes [191]. More recently the EPA priority pollutant phenols have been analyzed in environmental waters [67,77,189] and several phenols have been determined in soil samples [123,132], both using LC–API/MS techniques. Fig. 6.10 gives as an example the LC–APCI/MS chromatogram for a polluted soil sample candidate to reference material where some chlorophenols were identified [123].

An important feature to take into account when working with IS or ES is that

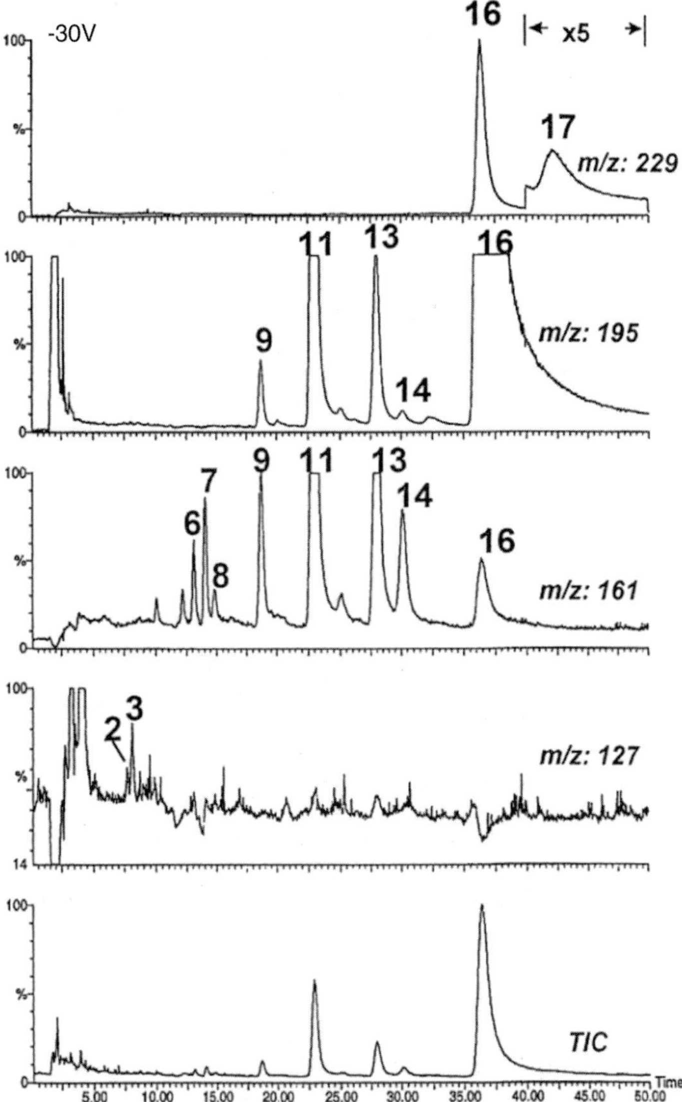

Fig. 6.10. LC–APCI–MS analysis of a soil candidate reference material (CRM-530, M and T, Brussels). Chromatographic conditions: C$_8$ column, mobile phase acetate buffer 5 mM pH 4.5/acetonitrile/methanol (60 : 30 : 10). The lower trace is the TIC obtained by summing all ions above. Peaks: (2) 4CP; (3) 3CP; (6) 25DCP; (7) 24DCP; (8) 34DCP; (9) 236TCP; (11) 234TCP; (13) 245TCP; (14) 235TCP; (16) 2346TeCP; (17) PCP. (Reprinted from [123].)

ionization takes place by ion evaporation, so there is a need for ions to be preformed in solution. Hence, since acid pH is normally required for the LC separation of phenolic compounds, post-column addition of a base is necessary so as to have the ionized species and not affect the chromatographic separation. In the case of APCI/MS, it is rather unlikely that ion evaporation contributes significantly since ionization is a gas-phase process, so there is no need for the addition of a base. Dimethylamine [67], triethylamine [189] or even stronger bases such as KOH [194] have been used to enhance the response of phenols in ES/MS or IS/MS. Nevertheless, the compounds with high pK_a (phenol, pK_a 9.9 and 2,4-dimethylphenol pK_a 10.5) gave no responses even at very high base concentrations or using KOH when the LC mobile phase (acetic acid–ACN or acetic acid–MeOH–ACN) contains a relatively high percentage of water [86,189]. However, it was reported in FIA experiments that phenol was detected in ES/MS with pure methanol as eluent and KOH as additive [194]. On the basis of this study, Barceló and co-workers developed a LC–IS/MS procedure to detect the most polar phenolic compounds such as catechol, 4-chloro-2-aminophenol, phenol and 4-methylphenol using a carbon-based LC column (PGC) and 100% methanol as mobile phase with post-column addition of triethylamine, which allows 0.2–0.4 ng of these compounds to be detected in standard solutions (SIM mode) [77,189]. Fig. 6.11 shows the chromatograms in SIM mode of a spiked river water sample (1.0 μg l^{-1})

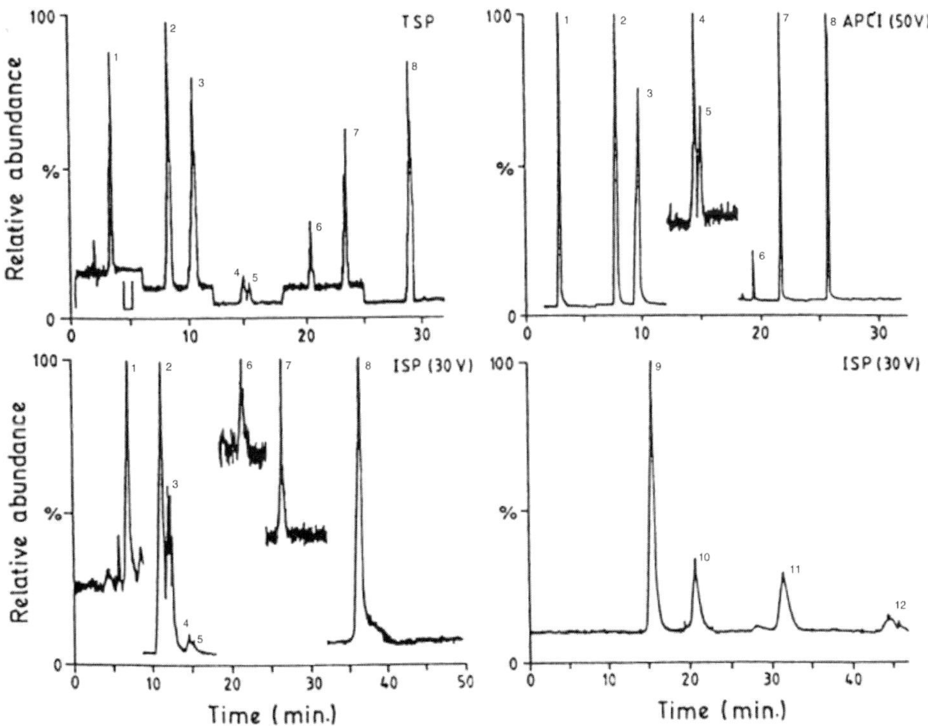

Fig. 6.11. Chromatograms in SIM of spiked river water sample (1 μg l^{-1}, 1000 ml) preconcentrated using SDB-RPS Empore disks using LC–MS with three interfaces (TSP, IS, APCI). Chromatographic conditions: C_{18} column, mobile phase 1% acetic acid and methanol/acetonitrile (1% acetic acid) 1:1 in gradient elution. Ammonium acetate 50 mM in TSP and methanol (0.1 M TEA) in IS were added post-column. Peaks: (1) catechol; (2) 4NP; (3) 24DNP; (4) 4CP; (5) 3CP; (6) 24DCP; (7) 246TCP; (8) PCP; (9) P; (10) 4MP; (11) 2A4CP; (12) 24DMP. (Reprinted from [189]; reproduced with permission.)

TABLE 6.20

LC–API–MS FOR THE ANALYSIS OF PHENOLS IN ENVIRONMENTAL SAMPLES

Compounds	API technique	LC conditions	Sample	Extraction	LOD	Ref.
11 EPA	ES [a]	1% acetic acid–ACN	tap water (250 ml)	off-line SPE	0.02–0.5 μg l^{-1}	[86]
	APCI				0.002–0.5 μg l^{-1}	
18 phenols	IS [b]	1% acetic acid–MeOH–ACN (1% acetic acid)	standard solutions	–	0.04–6 ng	[189]
	APCI			–	0.001–0.08 ng	
	TSP			–	0.4–5 ng	
17 CPs	APCI	ammonium acetate pH 4.5–ACN–MeOH	soil (1 g)	Soxhlet	0.01–0.7 μg g^{-1}	[123]
11 phenols	APCI	1% acetic acid–MeOH–ACN (1% acetic acid)	soil (10 g)	Soxhlet/MAE	0.007–0.4 ng g^{-1}	[132]
18 phenols	IS [b]	1% acetic acid–MeOH–ACN (1% acetic acid)	water (50 ml)	on-line SPE (50 ml)	50–75 ng l^{-1}	[77]
	APCI				0.1–25 ng l^{-1}	
Polyphenols	APCI	formic acid–MeOH–ACN	olive-mill waste water	–	0.03–30 ng	[193]

[a] Post-column addition of dimethylamine 250 mM.

[b] Post-column addition of triethylamine 100 mM.

using LC–MS with three interfaces (TSP, IS and APCI) after preconcentration of 1 l of sample through PS-DVB membrane disks. A comparison of the chromatograms shows that TSP and IS are less sensitive than APCI, as can be seen in this figure which shows a higher signal-to-noise ratio for APCI. Nevertheless, IS is the most suitable technique for the most polar analytes (peaks 9 to 12 in Fig. 6.11) and so the combination of IS and APCI enables all the phenolic compounds at 1–1.5 orders of magnitude lower than TSP to be analyzed [189].

Galceran and co-workers [86,123] and Barceló and co-workers [77,132,189] compared both API techniques, ES and APCI, for the analysis of phenols and concluded that LC–APCI/MS is the method of choice due to its greater capacity for fragmenting the analytes, so allowing their identification, and its greater sensitivity at low potentials. Table 6.20 summarizes the studies found in the literature which use LC–API–MS for the analysis of phenols. As can be seen, APCI was more sensitive than ES or IS and the legislated levels for phenols in water (0.1 μg l^{-1} for individual concentration and 0.5 μg l^{-1} for total content) were reached mainly using on-line SPE–LC–APCI/MS.

In general, API spectra do not produce the degree of unambiguous information that is required to identify the exact compound or to distinguish among isomers but further structural information can be obtained through collision-induced dissociation (CID) in the ion source or by using tandem mass spectrometry. To our knowledge, the MS/MS approach has only been used for chloronitrophenols in a triple quadrupole mass spectrometer with good results [195], whereas the CID technique has been frequently employed with single quadrupole instruments. High extraction voltages led to high fragmentation and a decrease in the intensity of the [M−H]$^-$ ion, so there has to be a compromise between structural information and sensitivity. The differentiation between positional isomers, i.e. chlorophenols, can be accomplished with this approach and LC–APCI/MS. Table 6.21 summarizes abundances of mass spectra fragments

TABLE 6.21

RELATIVE ABUNDANCES (%) OF FRAGMENTS FOR CHLOROPHENOLS USING LC–APCI/MS AT TWO EXTRACTION VOLTAGES [123]

Family of chlorophenols	Fragment	Extraction voltage	
		−30 V	−50 V
Monochlorophenols	[M−H]$^-$	100	100
	[M−H−HCl]$^-$	–	20
Dichlorophenols	[M−H]$^-$	100	67–100
	[M−H−HCl]$^-$	2–8	10–100
Trichlorophenols	[M−H]$^-$	8–100	12–100
	[M−H−HCl]$^-$	100	100
	[M−H−2HCl]$^-$	3–7	45–94
Tetrachlorophenols	[M−H]$^-$	4–25	6–30
	[M−H−HCl]$^-$	100	100
	[M−HCl−Cl]$^-$	–	90
Pentachlorophenol	[M−H]$^-$	1	–
	[M-Cl]$^-$	100	100
	[M−HCl−Cl]$^-$	7	100

for each family of chlorophenols at two extraction voltages (-30 V and -50 V). As can be seen, low chlorinated phenols showed the $[M-H]^-$ ion as the base peak whereas $[M-H-HCl]^-$ or $[M-Cl]^-$ ions are the base peak for highly chlorinated phenols. It should be noted that the spectra at -50 V for all the chlorophenols except monochlorophenols had a pattern which depended on the position of the substituents that can be used to distinguish between positional isomers. For instance, tetrachlorophenols gave a quite different relative abundance for the $[M-H]^-$ ion, which was 25% for 2,3,4,6-tetrachlorophenol and only 4% for 2,3,5,6-tetrachlorophenol. Another example is the abundance of the $[M-H-HCl]^-$ ion at -50 V for the six dichlorophenols: this changes from 10% for 2,6-dichlorophenol to 100% for 3,4-dichlorophenol [123]. Therefore, CID reactions provide information for the confirmation of target analytes by using the abundance ratios of several diagnostic fragment ions.

The loss and multiple loss of HCl and Cl were the only fragmentations for chlorophenols in both LC–ES/MS and LC–APCI/MS. Nitrophenols show greater fragmentation than the other phenolic compounds with losses of NO and NO_2 and, moreover, were the most sensitive phenols, with their LODs even lower than with LC–ED. The literature shows some controversy concerning the identification of fragments in mass spectra, which reveals that mobile phase composition is important in solute ionization, as has been reported by several authors [67,189,196]. For instance, using a single quadrupole mass spectrometer (VG Platform II) all the 11 EPA phenols coincide in the $[M-H]^-$ ion as a main fragment, but abundances for some fragments vary depending on the mobile phase used.

Finally, some controversies on the robustness of LC–API/MS systems have been found in the recent literature. External standard calibration has been used for on-line SPE–LC–APCI/MS and relative standard deviations (RSD) of around 20% have been obtained due to the frequent need to clean the system [77]. Nevertheless, standard deviations can be lowered by the addition of an internal standard. For instance, with 2,4,6-tribromophenol or 2,4-dibromophenol as internal standards and off-line SPE–LC–API/MS, RSD between 7 and 14% were obtained for the determination of phenols in water and soil [67,123]. Internal standard calibration has also been proposed for the analysis of polyphenols in olive mill waste waters using 4-bromophenol as internal standard and RSD lower than 10% were obtained [193]. Therefore, to increase the robustness of the LC–API/MS methods internal standard calibration is needed.

In summary, LC is a helpful tool for analyzing phenols in environmental matrixes, its main advantage over GC being that no derivatization is required, but less separation efficiency was obtained. Among the different detection modes applied for the analysis of these compounds, the electrochemical (amperometric or coulometric) mode has been the most frequently employed due to its high sensitivity and relatively high selectivity, but the passivation of the electrodes and the difficulty of performing gradient elution have become important drawbacks in the use of this technique. Nowadays, steadily falling cost and the general simplification of LC–MS instrumentation is resulting in a rapid take-up of MS detection which allows unambiguous identification and confirmation of the analytes. In addition, relatively low detection limits and the use of gradient elution (not advisable in LC–ED) are other advantages of this coupling. In general, choosing between GC or LC depends on a great number of factors, among them the experience of each laboratory and the instrumentation available and the type of sample to be analyzed but, to our opinion, even if possible, very polluted samples can be better analyzed by GC after a derivatization of phenols because classical clean-up methods (florisil, silica) can be then applied to obtain cleaner extracts. Drinking and surface waters can be analyzed by LC–ED or LC–MS with the main advantage that shorter analysis time were obtained if compared with GC because there is no need to form a derivative of phenols and because of the possibility of using on-line preconcentration techniques.

6.4 CAPILLARY ELECTROPHORESIS

Capillary electrophoresis (CE) is a powerful separation tool which has rapidly developed and matured since its introduction [197,198]. The key advantage of CE lies in its great efficiency, speed, simplicity and economy. Initially introduced as a separation technique for biological macromolecules, CE has also attracted a lot of interest in the environmental field, as is demonstrated by the growing number of publications and reviews [199,200]. Because of their acidity, phenols can be analyzed as anions under capillary zone electrophoresis (CZE) conditions or as either anions or neutrals under micellar electrokinetic chromatography (MEKC). By using an untreated capillary surface, anodic injection and a buffered alkaline solution, phenols are swept to the cathode by the electroosmotic flow (EOF) of the buffer. Size and charge differences between species result in electrophoretic mobility differences (in the opposite direction to EOF) that facilitate their effective separation. This mode of operation is termed counterelectroosmotic CE and has been extensively applied to the analysis of phenols [201–209]. Another approach is the coelectroosmotic one that consists in establishing an anodic EOF which is achieved by positively coating the inner surface of the silica capillary using long-chained alkylammonium ions such as cetyltrimethylammonium (CTAB) or polycations such as hexadimethrine (1,5-dimethyl-1,5-diazaundecamethylene polymethobromide, HDB) and switching the polarity of the power supply. By this means, anionic species migrate in the same direction as the EOF. This mode of operation has been applied to the separation of phenolic species by several authors [210–213] with the main advantage that fast separations can be achieved.

Table 6.22 summarizes the most relevant studies of the CZE and MEKC separation of phenols. Several authors separated the 11 EPA priority pollutant phenols using fused-silica capillaries and phosphate or borate buffers or mixtures of them [203–205,207]. As an example, Fig. 6.12 shows the separation of these compounds by CZE using a phosphate–borate buffer. Other buffers are used such as diethylmalonic acid [209] and the ACES buffer (N-(2-acetamido)-2-aminoethanesulfonic acid) [214]. Cyclohexylaminoethanesulfonic acid (CHES) [207] and diethylmalonic acid [234] have been used in CZE–ES–MS because of its volatility. CHES has also been used in electrochemical detection because it gives a low intensity current (1–4 μA) [203].

As can be seen in Table 6.22, all the CE separations for the 11 EPA phenols were performed at basic pH. Nevertheless, for other phenolic compounds different working pHs have been proposed to take advantage of differences in pK_a values. In fact, the influence of pH on the separation of phenols has been extensively studied by several authors, who

TABLE 6.22

SUMMARY OF CE WORKS FOUND IN THE LITERATURE FOR THE ANALYSIS OF PHENOLIC COMPOUNDS

Compounds	CE mode	Buffer electrolyte	Sample type/volume	LOD	Detection	Ref.
11 EPA	CZE	sodium borate 15 mM + fluorescein 1 mM pH 10.1	standard solutions	0.01–0.75 mg l^{-1}	LIF	[201]
11 EPA	CZE	CHES 20 mM pH 10.1	standard solutions	0.03–7 mg l^{-1}	ED (+1100 mV)	[203]
11 EPA	CZE (SPE)	Na$_2$B$_4$O$_7$ 20 mM pH 9.9	tap/river (500 ml)	0.3–1.0 µg l^{-1}	UV ($\lambda = 220$ nm)	[204]
11 EPA	CZE	NaH$_2$PO$_4$ 45 mM + NaBO$_3$ 15 mM pH 8	standard solutions	0.3 mg l^{-1}	UV ($\lambda = 210$ nm)	[205]
11 EPA	CZE	CHES 20 mM pH 10	standard solutions	low mg l^{-1}	ES–MS	[207]
	CZE (stacking)		industry effluents	50 µg l^{-1}		
15 RPs	CZE	borate 1 mM, 30% ACN pH 11.2	standard solutions	–	UV ($\lambda = 254$ nm)	[208]
16 CPs	CZE	DEM 30 mM pH 7.2	standard solutions	0.3–2.0 mg l^{-1}	DAD ($\lambda = 214$ nm)	[209]
18 CPs	CZE coelectroosmotic	NaH$_2$PO$_4$ 15 mM + Na$_2$B$_4$O$_7$ 1.25 mM + 2-butanol + ethylene glycol + ACN + HDB 0.001% pH 8.25	standard solutions	0.3–0.5 mg l^{-1}	UV ($\lambda = 214$ nm)	[210]
19 CPs	CZE coelectroosmotic	30% methanol–10% ACN, 15 mmol l^{-1} phosphate, 1.25 mmol l^{-1} tetraborate + HDB 0.001% pH 7.8	standard solutions	0.5–1.0 mg l^{-1}	DAD ($\lambda = 215$ nm)	[213]
16 RPs	MEKC	NaH$_2$PO$_4$ 50 mM + Na$_2$B$_4$O$_7$ 25 mM + SDS 1 mM	standard solutions	–	UV ($\lambda = 220$ nm)	[216]
19 CPs	MEKC	NaH$_2$PO$_4$ 50 mM + Na$_2$B$_4$O$_7$ 25 mM + SDS 70 mM	standard solutions	–	UV ($\lambda = 220$ nm)	[217]
11 EPA	CZE (stacking)	Na$_2$B$_4$O$_7$ 20 mM pH 9.9	standard solutions	35–50 µg l^{-1}	DAD ($\lambda = 195$ nm)	[225]
6 NPs	MEKC	Na$_2$B$_4$O$_7$ 25 mM–H$_3$PO$_4$ 20 mM, SDS 50 mM pH 8	standard solutions	0.6–1 mg l^{-1}	UV ($\lambda = 220$ nm)	[226]
	MEKC (SPE–FAI)		river water	5.5–10 µg l^{-1}		

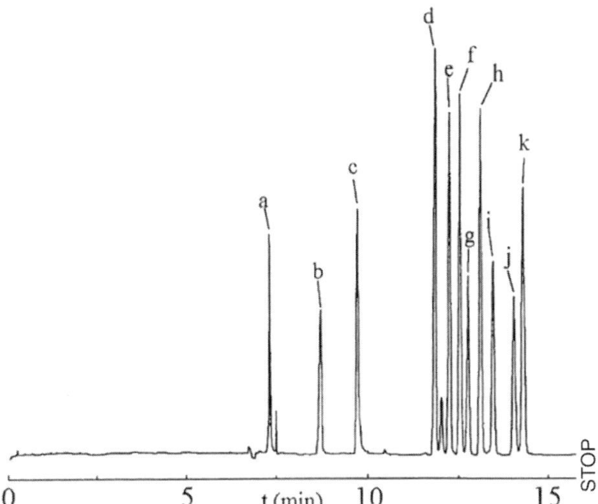

Fig. 6.12. CZE separation of eleven priority phenols. Electropherogram of a standard solution of phenols (25 mg l^{-1}). Buffer: [Na$_3$PO$_4$] = [Na$_2$B$_4$O$_7$] = 10 mM (pH 9.8); applied voltage 22.5 kV; vacuum injection time 10 s. Peaks: (a) 24DMP; (b) P; (c) 4C3MP; (d) PCP; (e) 246TCP; (f) 24DCP; (g) 2M46DNP; (h) 2CP; (i) 24DNP; (j) 4NP; (k) 2NP. (Reprinted from [205].)

showed that pH is the most important parameter in the optimization of CE separation of phenols and that the extent of their dissociation, which determines the overall electrical charge of the solute, is governed by the pH buffer [200,202,204,205,209,213].

Apart from the 11 EPA phenols, applications of CZE to separate the positional isomers of some phenols have been published. For instance, the separation of some alkylphenols [208], some dichlorophenols [206] and some trichlorophenols [215] were reported with phosphate and/or borate buffers. In addition, sixteen chlorophenols were separated using diethylmalonic acid buffer that provides total resolution among these isomers [209]. Fig. 6.13 shows the electropherogram of a mixture of 17 positional isomers of chlorophenols.

The 19 isomers of chlorophenols have not been resolved by counterelectroosmotic CZE [200], although Liu and Frank [213] claim that with the co-electroosmotic mode their separation is possible using a phosphate–borate buffer with 30% methanol, 10% acetonitrile and 0.001% hexadimetrine to reverse the EOF. Coelectroosmotic CZE can also separate six phenols in less than 1 min with a phosphate–borate buffer at pH 11 and CTAB as EOF modifier [211]. The results found in the literature confirm that the main application of coelectroosmotic CE is the separation of positional isomers which are difficult to separate by other less efficient techniques.

Micellar electrokinetic chromatography (MEKC) has also been applied to the separation of phenolic species (Table 6.22). In this case, an ionic surfactant is added to the CZE buffer at a concentration exceeding the critical micelle concentration, thereby expanding CE's enormous power to the separation of both charged and uncharged solutes. Analytes are separated according to their differential partitioning between the buffer phase and the micelles. Studies of electrokinetic separations in a micellar medium

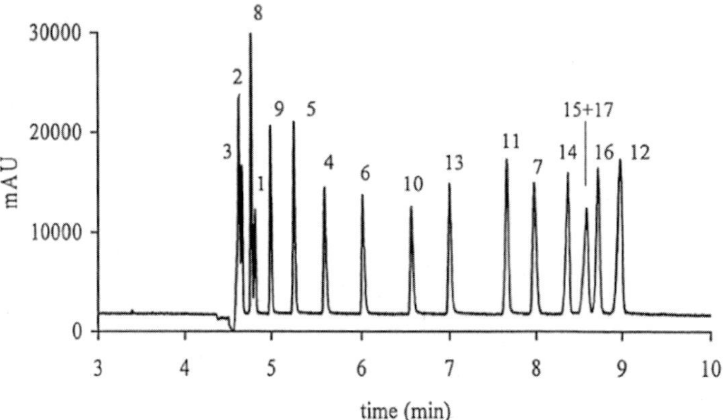

Fig. 6.13. CZE separation of chlorophenols. Electropherogram of a standard solution of chlorophenols (20 mg l^{-1}). Buffer: diethylmalonic acid 30 mM pH 7.25; applied voltage 20 kV; hydrodynamic injection, 4s; Peaks: (1) 2CP; (2) 3CP; (3) 4CP; (4) 23DCP; 5) 24DCP; (6) 25DCP; (7) 26DCP; (8) 34DCP; (9) 35DCP; (10) 234TCP; (11) 235TCP; (12) 236TCP; (13) 245TCP; (14) 246TCP; (15) 2346TeCP; (16) 2356TeCP; (17) PCP. (Reprinted from [209], with permission of Wiley VCH Verlag.)

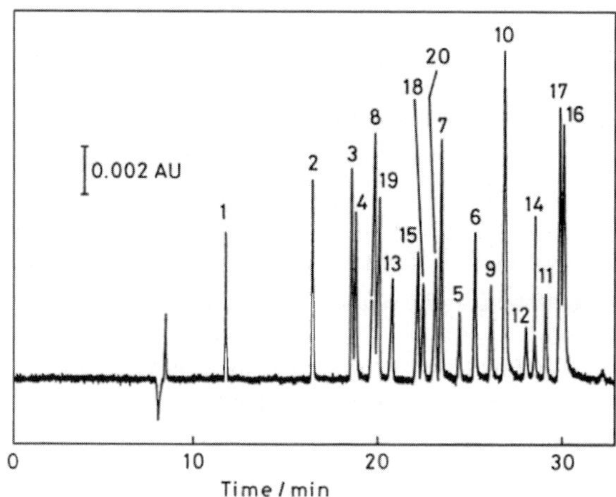

Fig. 6.14. MEKC separation of nineteen positional isomers of chlorophenols and phenol. Micellar solution, 0.07 M SDS, 0.05 M NaH_2PO_4–0.025 M $Na_2B_4O_7$ pH 7.0; applied voltage 10 kV. Peaks: (1) P; (2) 2CP; (3) 3CP; (4) 4CP; (5) 23DCP; (6) 4DCP; (7) 25DCP; (8) 26DCP; (9) 34DCP; (10) 35DCP; (11) 234TCP; (12) 235TCP; (13) 236TCP; (14) 245TCP; (15) 246TCP; (16) 345TCP; (17) 2345TeCP; (18) 2346TeCP; (19) 2356TeCP; (20) PCP. (Reprinted from [217].)

were performed by Terabe and co-workers who separated the 19 positional isomers of chlorophenols [216–218] using phosphate–borate pH 7.0 buffer and sodium dodecyl sulfate (SDS) as anionic surfactant (Fig. 6.14). This group of researchers also separated fourteen alkylphenols (*o*-, *m*- and *p*-cresol, among them) with a borate–phosphate buffer at pH 7 and 1 mM SDS [216]. Coelectroosmotic separations have also been reported for 9 cresol and xylenol isomers (including the *o*-, *m*- and *p*-cresol, which are only separated by LC with difficulty) as well as mixtures of 18 chlorophenols, 11 phenols (EPA priority pollutant) and 9 phenols (EPA method 8040) by Zemann and Volgger. For instance the nine cresol and xylenol isomers could be separated in less than 1 min with phosphate–borate buffer at pH 10.7, 25% acetonitrile and CTAB in coelectroosmotic MEKC [210].

The 11 EPA priority pollutant phenols have also been separated by MEKC with phosphate–borate buffer at pH 6.6 and potassium dodecyl sulfate (KDS) as surfactant [219]. Non-ionic surfactants (Brij 35 and Tween 40) have also been tested for the separation of phenols [220]. Some comparisons between CZE and MEKC have been published showing that the main advantage of MEKC is to allow the separation of ions with very similar electrophoretic mobilities, such as chlorophenol and alkylphenol isomers, because the partition between the aqueous and micellar phases increases selectivity, but it is less sensitive, less efficient and less stable than CZE and is also more difficult to couple to an MS system [200]. As such, CZE must be the technique of choice because it is more simple and sensitive than MEKC and can be easier coupled to an MS system. Nevertheless, complex mixtures of isomers of phenolic compounds can be better resolved by MEKC.

The detection system most commonly used is the on-column UV due to the strong absorption of phenols in the UV region (210–280 nm). The main drawback is its low sensitivity because of the short optical path length, which makes this technique inappropriate for determining phenols at low concentrations, as for instance in water at the legislated levels. Typical detection limits for direct injection of phenols are in the low mg l^{-1} range (Table 6.22). Sensitivity in terms of injected mass is extremely low because the injection volume in CE is often several nanoliters, but sensitivity in terms of concentration is, in general, 10 to 100 times higher than in LC. Increasing the injection time was not a good solution because, although the peak area increased, broadened and distorted peaks were obtained which indicated capillary overloading [221,222]. Different solutions to this problem have been proposed: the application of more sensitive detection devices such as fluorescence [201] or electrochemical devices [203,214,223]. As can be seen in Table 6.22, these detection modes lead to a slight decrease in detection limits. For instance, LODs between 5 and 11 μg l^{-1} were obtained for direct injection of chlorophenols and electrochemical detection [214]. Using laser-induced indirect fluorescence detection, LODs between 10 and 750 μg l^{-1} were obtained for the determination of phenols after the addition of fluorescein to the running borate buffer [201].

Another solution to the problem of the lack of sensitivity in CE systems is the preconcentration of the samples prior to analysis by off-line solid-phase extraction (SPE), which gave improved LODs. For instance, Martínez et al. found LODs between 0.3 and 1 μg l^{-1} for phenols in CZE after SPE (500 ml) [204]. If SPE is combined with highly sensitive detection devices such as the electrochemical device, relatively low LODs can be reached. For instance, LODs in river water between 0.07 and 0.2 μg l^{-1} were reported via SPE after preconcentrating 100 ml of sample [214].

Field-amplified injection (FAI) has been shown to be a solution to the poor sensitivity of CE–UV systems. Some papers dealing with different strategies for enrichment in CZE have been published [224,225]. Briefly, FAI involves the injection of large volumes of sample dissolved in a lower conductivity buffer matrix than those used for CE separation. FAI is suitable for concentrating analytes from relatively clean matrixes with low and reproducible ionic strength resulting in a decrease in LODs. For instance, Martínez et al. found an improvement in LODs from 0.1 to 0.25 mg l^{-1} for direct injection [204] to 35–50 μg l^{-1} after sample stacking [225], whereas Cela and co-workers reported a LOQ of 1.9 μg l^{-1} when the sample stacking procedure was applied together with an extended light-path detection window also called bubble capillary [221]. The main drawback of these preconcentration procedures is the low concentration factors and the poor precision when used for treating dirty samples with high ionic strength. To address these problems, combined use of SPE and FAI has been proposed for the analysis of some pollutants and some applications in the field of phenolic compounds analysis have been found. The general trend consists in modifying slightly the established SPE protocols in order to obtain a final solution with low salinity for applying the stacking. Among the solutions, the dilution of the extract with water [227] or with a ten-times diluted buffer [226] or the re-extraction of the organic solvent (e.g. CH_2Cl_2) with pure water [221] have been proposed. With SPE–FAI a big drop in LODs could be attained. For instance, nitrophenols showed a LOD of 0.6–1 mg l^{-1} in normal CZE and a LOD of 5.5–10 μg l^{-1} after SPE–FAI [226]. Another example: LOD for PCP for standard solutions and direct injection in a CZE system is 2 mg l^{-1}, whereas after preconcentration (100 ml) in a GCB cartridge, and application of stacking, a LOQ of 60 ng l^{-1} was achieved [221]. This mode of operation (combination of SPE with FAI) provides not only an enhancement of detection sensitivity but also an improvement in separation selectivity because enrichment in SPE is based on the difference in polarity between interferences and target analytes, while enrichment by FAI is based on the difference in charge. The procedure has been used to analyze phenols in river and tap waters [221,226,227].

Mass spectrometry has also been coupled to CE and some reviews have been published with a detailed description of the different designs [228,229]. As an on-line separation method, CE–MS distinguishes analytes by both their differences in electrophoretic mobilities and molecular masses, but several limitations associated with CE–MS have precluded the technique being widely accepted for routine analysis. The major limitation, as in the case of UV detection, is its relatively poor concentration sensitivity with LODs in the mg l^{-1} range although sample stacking can be applied with a considerable improvement in LODs, 50 μg l^{-1}. Moreover, the difficulty for performing quantitative analysis and the loss of resolution if compared with the separation obtained in CE–UV systems are other factors that make routine use of this coupling difficult. Nevertheless, some selected applications have been found in the literature. Fig. 6.15 gives an example of the selected ion electropherograms of the EPA phenols at their different m/z. As occurred in LC–ES/MS, the addition of a base is necessary in order to have the ionic species in solution. In the case of CE–ES/MS this feature was accomplished by adding ammonia [207] or isopropanol–dimethylamine [234] to the sheath liquid.

In conclusion, CE has been applied to analysis of phenolic compounds in different modes (CZE, MEKC). In our view, its main field of application is the separation of complex mixtures of isomers of phenols (chlorophenols, alkylphenols). However,

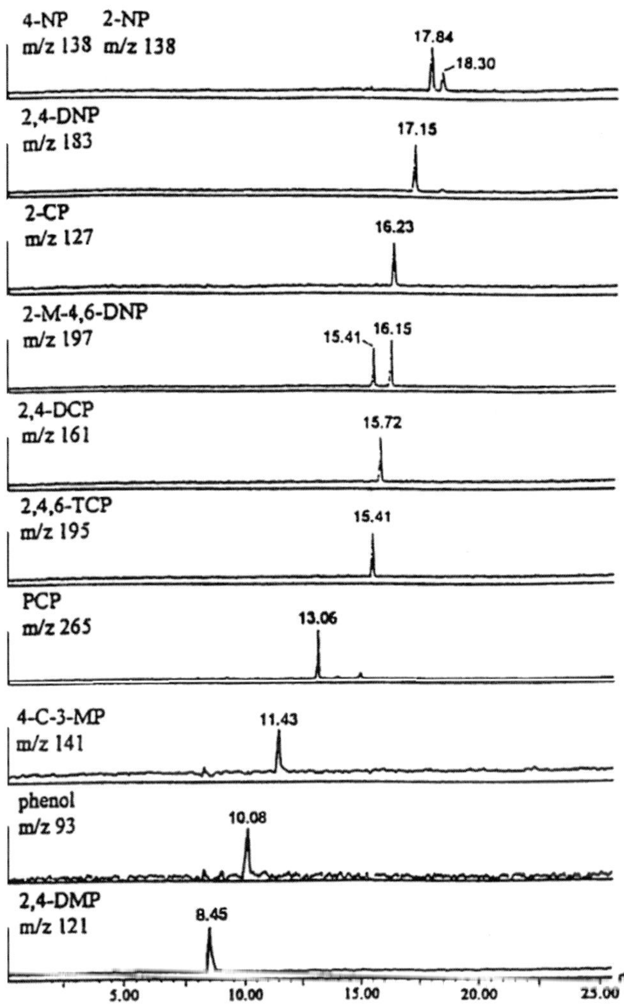

Fig. 6.15. Selected ion electropherograms of a standard solution of the eleven priority phenols (10 mg l^{-1}). CE conditions: buffer solution CHES 20 mM pH 10; applied voltage 20 kV, electrokinetic injection (10 kV for 10 s); sheath liquid: 20 : 80 0.5% ammonia-2-propanol. (Reprinted from [207].)

widespread use of CE as a routine quantitative technique in environmental analysis is still very limited because its LODs are not as good as those attained with chromatographic techniques. Some solutions to this problem (SPE, FAI) have been proposed but the problem of the much greater influence of the sample matrix than in chromatographic techniques is still not resolved. Nevertheless, the shorter analysis time with CZE than with LC is an important advantage of its use as an alternative method for determining phenols.

References pp. 231–236

6.5 IMMUNOCHEMICAL METHODS

Immunochemical methods were first developed in clinical chemistry in which a large number of samples have to be processed to reduce the time and cost of classical analytical methods. Their use in environmental chemistry has been limited due to that the generation of antibodies for small molecules is difficult because small molecules are unable to produce an immune response. Recent progress in the strategies for the rational design of haptenes, for the development of antibodies to small molecules, partly explains the strong increase observed in recent years. A recent review discussed the strengths and weaknesses of immunoassays for the analysis of pesticides in water [230]. Enzyme-linked immunosorbent assay (ELISA) is the most widely used type of immunochemical analysis for pesticide residues, and commercial ELISA kits are available for more than 25 pesticides. The EPA has recently released an official method for PCP screening by immunoassay (EPA method 4010A) [231]. This method is recommended for screening samples to determine whether PCP is likely to be present at defined concentrations, i.e. commercial kits are available which give positive results at 5 μg l^{-1} for aqueous samples and at 0.5, 10 or 100 mg kg^{-1} in solid samples. Briefly, the method is performed using a water sample or an extract of a water sample. The sample/extract and an enzyme conjugate reagent are added to immobilized antibody. The enzyme conjugate 'competes' with PCP present in the sample for binding to immobilized anti-PCP antibody. The test is interpreted by comparing the response produced by testing a sample to the response produced by testing standard(s) simultaneously. Nevertheless, compounds that are chemically similar may cause a positive test (false positive) for PCP. For instance, for the test of PCP at 5 μg l^{-1}, 2,6-dichlorophenol may cause a false positive at a concentration of 600 μg l^{-1} but 2,3,5,6-tetrachlorophenol at a concentration of 7 μg l^{-1}. Other phenols such as 2,3-dichlorophenol, 2,4-dichlorophenol, 3,5-dichlorophenol or 4-chlorophenol have been found to yield negative results even at 1000 mg l^{-1} level. So, EPA recommends that in cases where the exact concentration of PCP is required, additional chromatographic techniques should be used. Similarly, Barceló and co-workers evaluated the use of ELISA kits for the determination of PCP in waters and soils. These authors studied the cross-reactivity of other phenols such as 2,3,4,6-tetrachlorophenol and 2,3,5,6-tetrachlorophenol in the determination of pentachlorophenol and also analyzed several water and soil samples with both the ELISA kits and on-line SPE–LC–UV and LC–APCI–MS. Their results showed discrepancies between ELISA and LC when the samples contained concentrations of 2,4,6-trichlorophenol 1.5–2 times higher than that of PCP, so ELISA can 'alert' about the presence of relevant contaminants, and afterwards LC techniques will give the accurate measurement of all individual analytes present in complex samples, thus reducing time and cost of analysis mainly when monitoring programs are applied [232,233].

6.6 CONCLUSIONS

Nowadays, different approaches can be applied to the analysis of phenolic compounds in environmental samples. First of all, extraction of analytes from samples by solid-phase extraction both in off-line and on-line modes can be recommended. Styrene divinylbenzene copolymers are the most suitable sorbents for phenols extraction because the higher breakthrough volumes so, the lower detection limits were obtained. In respect to the chromatographic analysis, gas chromatography has been classically the technique of choice because its great separation efficiency and, in addition, it is the official method of analysis in EPA methods. For GC analysis it is advisable to form a less polar derivative of phenolic compounds in order to improve chromatographic performance. Obviously, this derivatization procedure is not straightforward and also is time- and chemicals-consuming. Liquid chromatography has grown in interest because of no need of derivatization and, in consequence, the important reduction in the analysis time. Among the different detection devices, electrochemical detection is very suitable for phenols because it allows to reach low detection limits and also selectivity is improved compared to UV. Nevertheless, the sole technique that allows unequivocal identification or confirmation for analytes is the coupling liquid chromatography–mass spectrometry. On-line SPE–LC–MS approaches gave the lowest detection limits for phenols in environmental samples which are very similar to those with GC–MS/MS systems but without derivatization of phenols. Capillary electrophoresis shows a high separation efficiency very useful for the analysis of positional isomers of phenols, i.e. alkylphenols, chlorophenols or cresols which are difficult to separate by LC, but has the main drawback of its lack of sensitivity (two orders of magnitude compared to LC–ED or LC–API–MS systems). Finally, the use of immunoassays is practically limited to screening purposes or as alarm devices in field samples.

6.7 REFERENCES

1   J.W. Moore and S. Ramamoorthy, Organic Chemicals in Natural Waters. Applied Monitoring and Impact Assessment, Springer-Verlag, New York, 1984, pp. 141–164.
2   K. Levsen, P. Mußmann, E. Berger-Preiß, A. Preiß, D. Volmer and G. Wünsch, Acta Hydrochim. Hydrobiol., 21 (1993) 153.
3   Y. He and H.K. Lee, J. Liq. Chromatogr. Rel. Technol., 21 (1998) 725–739.
4   M. Alber, H.B. Bohm, J. Brdesser, J. Feltes, K. Levsen and H.F. Schöler, Fresenius' Z. Anal. Chem., 334 (1989) 540–545.
5   H.B. Böhm, J. Feltes, D. Volmer and K. Levsen, J. Chromatogr., 478 (1989) 399–407.
6   M.E. Fernández, J.L. Pérez and B. Moreno, Chromatogr. A, 852 (1999) 395–406.
7   J. Jensen, Rev. Environ. Contam. Toxicol., 146 (1996) 25–51.
8   Y. Samiullah, Prediction of the Environmental Fate of Chemicals, Elsevier, New York, 1990.
9   J. Knuutinen, H. Palm, H. Hakala, J. Haimi, V. Huhta and H. Salminen, Chemosphere, 20 (1990) 609.
10   Z. Vasilic, S. Fingler and V. Drevenkar, Fresenius' J. Anal. Chem, 341 (1991) 732–737.
11   J. Angerer, B. Heinzow, K.H. Schaller, D. Weltle and G. Lehnert, Fresenius' J. Anal. Chem., 342 (1992) 433–438.
12   D.D. Perrin, B. Dempsey and E.P. Serjeant, pK_a Prediction for Organic Acids and Bases, Chapman and Hall, London, 1981.
13   E.E. McConnell, J.A. Moore, B.N. Gupta, A.H. Rakes, M.I. Luster, J.A. Goldstein, J.K. Haseman and C.E. Paiker, Toxicol. Appl. Pharmacol., 52 (1980) 468–490.
14   N.I. Kerkvliet, L. Baecher-Steppan and J.A. Schmitz, Toxicol. Appl. Pharmacol., 62 (1982) 55–64.
15   T. Kishino and K. Kobayshi, Water Res., 30 (1996) 393–399.
16   E. Eljarrat, J. Caixach and J. Rivera, Environ. Sci. Technol., 33 (1999) 2493–2498.

17   T. Kishino and K. Kobayashi, Water Res., 30 (1996) 387–392.
18   US EPA, Toxic Substance Control Act, Washington, DC, 1979.
19   Directive 76/464/CEE Commission of the European Communities, Brussels, 1982.
20   Drinking Water Directive 80/778/EEC, Commission of the European Communities, Brussels, 1980.
21   Directive 76/160/EEC, Commission of the European Communities, Brussels, 1975.
22   Directive 86/280/EEC, Commission of the European Communities, Brussels, 1986.
23   ACGIH, Threshold Limit Values and Biological Exposure Indices, American Conference of Government Industrial Hygienists, Cincinatti, OH, 1996.
24   EPA method 604, Phenols in Federal Register, Friday October 26, 1984, Environmental Protection Agency, Washington, DC, part VIII, 40 CFR, Part 136, pp. 58–66.
25   EPA method 625, Base/Neutrals and Acids in Federal Register Friday October 26, 1984, Environmental Protection Agency, Washington, DC, part VIII, 40 CFR, Part 136, pp. 153–174.
26   R.S.K. Buisson, P.W.W. Kirk and J.N. Lester, J. Chromatogr. Sci., 22 (1984) 339.
27   G. Lamprecht and J.F.K. Huber, J. Chromatogr. A, 667 (1994) 47.
28   C. Baiocchi, M.A. Roggero, D. Giacosa and E. Marengo, J. Chromatogr. Sci., 33 (1995) 338–346.
29   P.A. Realini, J. Chromatogr. Sci., 19 (1981) 124–129.
30   M.T. Galceran and F.J. Santos, in: G. Argelletti and A. Bjørseth (Eds.), Organic Micropollutants in the Aquatic Environment, Commission of the European Community, Kluwer, Dordrecht, 1987, pp. 46–51.
31   R.T. Coutts, E.E. Hargesheimer and F.M. Pasutto, J. Chromatogr., 179 (1979) 291–299.
32   S. Shang-Zhi and A.M. Duffield, J. Chromatogr., 284 (1984) 157–165.
33   S. Dupeyron, M. Astruc and M. Marbach, Analusis, 23 (1995) 470–473.
34   M.T. Galceran and O. Jáuregui, Anal. Chim. Acta, 304 (1995) 75–84.
35   P. de Voogt, E.A. Maier and A. Chollot, Fresenius' J. Anal. Chem., 361 (1998) 158–163.
36   EPA method 3650, Acid–Base Partition Clean-Up, in: Federal Register, Revison 1, December 1987, Environmental Protection Agency, Washington, DC.
37   I. Liska, J. Chromatogr. A, 655 (1993) 163–176.
38   R.E. Majors, LC–GC Int., 11 (1998) 416–426.
39   E. Rostad, W.E. Pereira and S.M. Ratcliff, Anal. Chem., 56 (1984) 2856–2860.
40   E. Chladek and R.S. Marano, J. Chromatogr. Sci., 22 (1984) 313–320.
41   M.L. Bao, F. Pantani, K. Barbieri, D. Burrini and O. Griffini, Chromatographia, 42 (1996) 227–233.
42   D. Jahr, Chromatographia, 47 (1998) 49–56.
43   S. Fingler, V. Drevenkar and Z. Vasilic, Mikrochim. Acta [Wien], II (1987) 163–175.
44   P. Hua-Tang and J.S. Ho, J. High Resolut. Chromatogr., 17 (1994) 509–518.
45   P. Mußmann, K. Levsen and W. Radeck, Fresenius' J. Anal. Chem., 348 (1994) 654.
46   I. Rodríguez, M.H. Bollaín, C.M. García and R. Cela, J. Chromatogr. A, 733 (1996) 405–416.
47   I. Rodríguez and R. Cela, J. Chromatogr. A, 786 (1997) 285–292.
48   I. Turnes, I. Rodríguez, C.M. García and R. Cela, J. Chromatogr. A, 743 (1996) 283.
49   A. Chudziak and M. Trojanowicz, Chem. Anal. (Warsaw), 40 (1995) 39.
50   J. Ruana, I. Urbe and F. Borrull, J. Chromatogr. A, 655 (1993) 217–226.
51   J. Frebortová and V. Tartaricová, Analyst, 119 (1994) 1519–1523.
52   D. Puig and D. Barceló, Chromatographia, 40 (1995) 435–444.
53   E. Pocurull, M. Calull, R.M. Marcé and F. Borrull, Chromatographia, 38 (1994) 579–584.
54   A. Di Corcia, S. Marchese and R. Samperi, JAOAC Int., 77 (1994) 446–453.
55   C. Markell, D.F. Hagen and V.A. Bunnelle, LC–GC Int., 4 (1991) 10–14.
56   L. Schmidt, J.J. Sun, J.S. Fritz, D.F. Hagen, C.G. Markell and E.E. Wisted, J. Chromatogr., 641 (1993) 57.
57   E. Pocurull, M. Calull, R.M. Marcé and F. Borrull, J. Chromatogr. A, 719 (1996) 105–112.
58   C.G. Markell and D.F. Hagen, Proceedings of the 7th Annual Waste Testing and Quality Assurance Symposium, Washington, DC, July 1991, Vol. II, EPA, p. 27.
59   D. Puig and D. Barceló, J. Chromatogr. A, 733 (1996) 371–381.
60   A. Di Corcia and M. Marchetti, Anal. Chem., 63 (1991) 580–585.
61   A. Di Corcia, A. Bellioni, M.G. Madbouly and S. Marchese, J. Chromatogr. A, 733 (1996) 383–393.
62   N. Masqué, R.M. Marcé and F. Borrull, J. Chromatogr. A, 793 (1998) 257–263.

63   J. Gawdzik, B. Gawdzik and V. Czerwinska-Bil, Chromatographia, 25 (1988) 504–506.

64   B. Gawdzik, J. Gawdzik and V. Czerwinska-Bil, J. Chromatogr., 509 (1990) 135–140.

65   L. Renberg and K. Lindström, J. Chromatogr., 214 (1981) 327–334.

66   E.R. Brouwer, H. Lingeman and U.A.Th. Brinkman, Chromatographia, 29 (1990) 415–418.

67   O. Jáuregui, E. Moyano and M.T. Galceran, J. Chromatogr. A, 787 (1997) 79–89.

68   E. Pocurull, R.M. Marcé and F. Borrull, Chromatographia, 40 (1995) 85–90.

69   D. Puig and D. Barceló, Anal. Chim. Acta, 311 (1995) 63–69.

70   E. Pocurull, R.M. Marcé and F. Borrull, J. Chromatogr. A, 738 (1996) 1–9.

71   N. Masqué, E. Pocurull, R.M. Marcé and F. Borrull, Chromatographia, 47 (1998) 176–182.

72   D. Puig and D. Barceló, J. Chromatogr. A, 778 (1997) 313–319.

73   O. Jáuregui and M.T. Galceran, Anal. Chim. Acta, 340 (1997) 191–199.

74   P. Trippel, W. Maasfeld and A. Kettrup, Int. J. Environ. Anal. Chem., 23 (1985) 97–110.

75   N. Cardellicchio, S. Cavalli, V. Piangerelli, S. Giandomenico and P. Ragone, Fresenius' J. Anal. Chem, 358 (1997) 749–754.

76   M.W.F. Nielen, J. de Jong, R.W. Frei and U.A.Th. Brinkman, Int. J. Environ. Anal. Chem., 25 (1986) 37–43.

77   D. Puig, I. Silgoner, M. Grasserbauer and D. Barceló, Anal. Chem., 69 (1997) 2756–2761.

78   C.E. Werkhoven-Goewie, U.A.Th. Brinkman and R.W. Frei, Anal. Chem., 53 (1981) 2072–2080.

79   N. Masqué, R.M. Marcé and F. Borrull, J. Chromatogr. A, 793 (1998) 257–263.

80   N. Masqué, M. Galià, R.M. Marcé and F. Borrull, J. Chromatogr. A, 803 (1998) 147–155.

81   M. Castillo, D. Puig and D. Barceló, J. Chromatogr., 778 (1997) 301–311.

82   E.R. Brouwer, I. Liska, R.B. Geerdink, P.C.M. Frintrop, W.H. Mulder, H. Lingeman and U.A.Th. Brinkman, Chromatographia, 32 (1991) 445–452.

83   E. Pocurull, R.M. Marcé and F. Borrull, Chromatographia, 41 (1995) 521–526.

84   I. Liska, E.R. Brouwer, H. Lingeman and U.A.Th. Brinkman, Chromatographia, 37 (1993) 13–19.

85   E.H.R. Van der Waal, E.R. Brouwer, H. Lingeman and U.A.Th. Brinkman, Chromatographia, 369 (1994) 239–245.

86   EPA method 555, Determination of chlorinated acids in water by high performance liquid chromatography with a photodiode array ultraviolet detector, US Environmental Protection Agency, Rev. 1.0, Aug. 1987, Cincinnati, OH.

87   N. Masqué, R.M. Marcé and F. Borrull, Chromatographia, 48 (1998) 231–236.

88   A.J.H. Louter, J.J. Vreuls, R.T. Ghijsen and U.A.Th. Brinkman, Int. J. Environ. Anal. Chem., 56 (1994) 49.

89   A.J.H. Louter, J.J. Vreuls and U.A.Th. Brinkman, J. Chromatogr. A, 842 (1999) 391–426.

90   M.A. Crespín, E. Ballesteros, M. Gallego and M. Valcárcel, Chromatographia, 43 (1996) 633–639.

91   V. Pichon, M. Bouzige, C. Miège and M.-C. Hennion, Trends Anal. Chem., 18 (1999) 219–235.

92   N. Masqué, R.M. Marcè, F. Borrull, P.A.G. Cormack and D. Sherrington, Anal. Chem., 72 (2000) 4122–4126.

93   R.P. Belardi and J. Pawliszyn, Water Pollut. Res. J. Can., 24 (1989) 179–191.

94   Z. Zhang, M.J. Yang and J. Pawliszyn, Anal. Chem., 66 (1994) 844A–853A.

95   R. Eisert and J. Pawliszyn, Anal. Chem., 69 (1997) 3140.

96   H.L. Lord and J. Pawliszyn, LC–GC Int., 11 (1998) 776–785.

97   H. Prosen and L. Zupancic-Kralij, Trends Anal. Chem., 18 (1999) 272–282.

98   A. Peñalver, E. Pocurull, F. Borrull and R.M. Marcé, Trends Anal. Chem., 18 (1999) 557–568.

99   J. Chen and J. Pawliszyn, Anal. Chem, 67 (1995) 2530–2533.

100   C. Whang and J. Pawliszyn, Anal. Commun., 35 (1998) 353–356.

101   K. Buchholz and J. Pawliszyn, Environ. Sci. Technol., 27 (1993) 2844–2846.

102   K. Buchholz and J. Pawliszyn, Anal. Chem., 66 (1994) 160–167.

103   M. Möder, S. Schrader, U. Franck and P. Popp, J. Anal. Chem., 357 (1997) 326–332.

104   M. Lee, R. Lee, Y. Lin, Ch. Chen and B. Hwang, Anal. Chem., 70 (1998) 1963–1968.

105   P. Barták and L. Cáp, J. Chromatogr. A, 767 (1997) 171–175.

106   A.A. Boyd-Boland and J. Pawliszyn, Anal. Chem., 68 (1996) 1521–1529.

107   M. Moder, P. Popp and J. Pawliszyn, J. Microcolumn. Sep., 10 (1998) 225–234.

108   Y. Wu and S. Huang, J. Chromatogr. A, 835 (1999) 127–135.

109   J.H. Phillips, M. Zabik and R. Leavitt, Int. J. Environ. Anal. Chem., 16 (1983) 81–93.

110   H.-B. Lee, Y.D. Stokker and A.S.Y. Chau, J. Assoc. Off. Anal. Chem., 70 (1987) 1003–1008.

111   S.J. Harrad, Th.A. Malloy, M. Ali Khan and Th.D. Goldfarb, Chemosphere, 23 (1991) 181.

112   I. Cruz and D.E. Wells, Int. J. Environ. Anal. Chem., 48 (1992) 101–113.

113   S.B. Hawthorne and D.J. Miller, Anal. Chem., 66 (1994) 4005.

114   T.S. Reighard and S.V. Olesik, J. Chromatogr. A, 737 (1996) 233.

115   S.R. Wild, S.J. Harrad and K.C. Jones, Water Res., 27 (1993) 1527–1534.

116   O. Jáuregui, F.J. Santos, F.J. Pinto and M.T. Galceran, Quim. Anal., 16 (1997) 247–251.

117   A.J. Wall and G.W. Stratton, Chemosphere, 22 (1991) 99.

118   EPA method 3540B, Soxhlet Extraction, in: Federal Register, Revision 2, September 1984, Environmental Protection Agency, Washington, DC.

119   V. Lopez-Avila, K. Bauer, J. Milanes and W.F. Beckert, JAOAC Int., 76 (1993) 864–880.

120   J. Frébortová, Biosci. Biotech. Biochem., 59 (1995) 1930.

121   M.A. Crespín, M. Gallego and M. Válcárcel, Anal. Chem., 71 (1999) 2687–2696.

122   M.-R. Lee, Y.-Ch. Yeh, W.-S. Hsiang and B.-H. Hwang, J. Chromatogr. A, 806 (1998) 317–324.

123   O. Jáuregui, E. Moyano and M.T. Galceran, J. Chromatogr. A, 823 (1998) 241–248.

124   T.J. Cardwell, I.C. Hamilton, M.J. McCormick and R.K. Symons, Int. J. Environ. Anal. Chem., 34 (1988) 167–178.

125   M. Richards and R.M. Campbell, LC–GC Int., 9 (1991) 358–364.

126   F.J. Santos, O. Jáuregui, F.J. Pinto and M.T. Galceran, J. Chromatogr. A, 823 (1998) 249–258.

127   M.P. Llompart, R.A. Lorenzo, R. Cela, K. Li, J.M.R. Bélanger and J.R.J. Paré, J. Chromatogr. A, 774 (1997) 243–251.

128   H.-B. Lee, T.E. Peart and R.L. Hong-You, J. Chromatogr., 605 (1992) 109–113.

129   H.-B. Lee, T.E. Peart and R.L. Hong-You, J. Chromatogr., 636 (1993) 263–270.

130   M.P. Llompart, J. Chromatogr. Sci., 34 (1996) 43–51.

131   M.H. Liu, S. Kapila, K.S. Nam and A.A. Elseewi, J. Chromatogr., 639 (1993) 151–157.

132   M.C. Alonso, D. Puig, I. Silgoner, M. Grasserbauer and D. Barceló, J. Chromatogr. A, 823 (1998) 231–239.

133   H. Yuan and S.V. Olesik, J. Chromatogr. A, 764 (1997) 265–277.

134   J.A. Fisher, M.J. Scarlett and A.D. Scott, Environ. Sci. Technol., 31 (1997) 1120–1127.

135   P.G. Jorens and P.J.C. Schepens, Human Exp. Toxicol., 12 (1993) 479–495.

136   M.-R. Lee, Y.-Ch. Yeh, W.-S. Hsiang and Ch.-Ch. Chen, J. Chromatogr. B, 707 (1998) 91–97.

137   K.D. McCurtrey, A.E. Holcomb, A.U. Ekwenchi and N.C. Fawcett, J. Liq. Chromatogr., 7 (1984) 953–960.

138   E.M. Lores, T.R. Edgerton and R.F. Moseman, J. Chromatogr. Sci., 19 (1981) 466–469.

139   M.L. Menezes and A.C.C.O. Demarchi, J. Liq. Chromatogr. Rel. Technol., 21 (1998) 2355–2363.

140   M. Guidotti and M. Vitali, J. High Resolut. Chromatogr., 21 (1998) 137–138.

141   EPA method 8041, Phenols by Gas Chromatography: Capillary Column Technique, Environmental Protection Agency, Washington, DC, 1995, pp. 1–28.

142   J. Hajslová, V. Kocourek, I. Zemanová, F. Pudil and J. Davídek, J. Chromatogr., 439 (1988) 307–316.

143   L. Wennrich, J. Efer and W. Engewald, Chromatographia, 41 (1995) 361–366.

144   P.G. Nielsen, Chromatographia, 18 (1984) 323–325.

145   I. Liska and J. Slobodník, J. Chromatogr. A, 733 (1996) 235–258.

146   H.-B. Lee, L.-D. Weng and A.S.Y. Chau, J. Assoc. Off. Anal. Chem., 67 (1984) 1086–1091.

147   H.-B. Lee and A.S.Y. Chau, J. Assoc. Off. Anal. Chem., 66 (1983) 1029–1038.

148   K. Nick and H.F. Schöler, Fresenius' J. Anal. Chem., 343 (1992) 304–307.

149   I. Rodríguez and R. Cela, Trends Anal. Chem., 16 (1997) 463–475.

150   H. Malissa, G. Szölgyényi and K. Winsaver, Fresenius' Z. Anal. Chem., 321 (1985) 17–26.

151   H.-B. Lee, R.L. Hong-You and A.S.Y. Chau, J. Assoc. Off. Anal. Chem., 68 (1985) 422–426.

152   I. Tolosa, J.M. Bayona and J. Albaigés, Mar. Pollut. Bull., 22 (1991) 603–607.

153   M.A. Crespín, S. Cárdenas, M. Gallego and M. Valcárcel, J. Chromatogr. A, 830 (1999) 165–174.

154   EPA method 8040, Phenols, US Environmental Protection Agency, Rev. 1, Dec. 1987, Washington, DC.

155   EPA method 1653, Chlorinated phenolics in wastewater by in situ acetylation and GCMS, US Environmental Protection Agency, Nov. 1996, Washington, DC.

156   EPA method 515.1, Determination of chlorinated acids in water by gas chromatography with an electron capture detector, US Environmental Protection Agency, Rev. 4.1 (1995) Cincinnati, OH.

157   EPA method 515.2, Determination of chlorinated acids in water using liquid–solid extraction and gas chromatography with an electron capture detector, US Environmental Protection Agency, Rev. 1.1 (1995), Cincinnati, OH.

158   EPA method 525.1, Determination of organic compounds in drinking water by liquid–solid extraction and capillary columns gas chromatography/mass spectrometry, US Environmental Protection Agency, Rev. 2.2-EPA EMSL-Ci, May 1991.

159   EPA method 3550, Sonication Extraction, US Environmental Protection Agency, Rev. 0, Sep. 1986, Washington, DC.

160   EPA method 3640, Gel permeation clean-up, US Environmental Protection Agency.

161   H.A. Mc Leod and G. Laver, J. Chromatogr., 244 (1982) 385–390.

162   G.A. Marko-Varga, in: D. Barceló (Ed.), Techniques and Instrumentation in Analytical Chemistry, Vol. 13, Environmental Analysis, Techniques, Applications and Quality Assurance, Elsevier, Amsterdam, 1993, pp. 225–271.

163   E. Tesarová and V. Pacáková, Chromatographia, 17 (1983) 269.

164   H.K. Lee, S.F.Y. Li and Y.H. Tay, J. Chromatogr., 438 (1988) 429–432.

165   J. Nair, K.M. Munir and S.V. Bhide, J. Liq. Chromatogr., 6 (1983) 2829–2837.

166   E.R. Brouwer and U.A.Th. Brinkman, J. Chromatogr. A, 678 (1994) 223–231.

167   B. Schultz, J. Chromatogr., 269 (1983) 208–212.

168   O. Busto, J.C. Olucha and F. Borrull, Chromatographia, 32 (1991) 423–428.

169   O. Busto, J.C. Olucha and F. Borrull, Chromatographia, 32 (1991) 566–572.

170   H. Guolan, Z. Weihua and Z. Zhiren, J. Liq. Chromatogr. Rel. Technol., 19 (1996) 899–909.

171   N. Cardellicchio, S. Cavalli, V. Piangerelli, S. Giandomenico and P. Ragone, Fresenius' J. Anal. Chem., 358 (1997) 749–754.

172   U. Lewin, J. Efer and W. Engewald, J. Chromatogr. A, 730 (1996) 161–167.

173   E. Pocurull, G. Sánchez, F. Borrull and R.M. Marcé, J. Chromatogr. A, 696 (1995) 31–39.

174   T.J. Cardwell, I.C. Hamilton, M.J. McCormick and R.K. Symons, Int. J. Environ. Anal. Chem., 24 (1986) 23–35.

175   E. Sooba, T. Tenno, O. Jáuregui and M.T. Galceran, Oil Shale, 14 (1997) 544–553.

176   A. Hagen, J. Mattusch and G. Werner, Fresenius' J. Anal. Chem., 339 (1991) 26–29.

177   R.E. Shoup and G.S. Mayer, Anal. Chem., 54 (1982) 1164–1169.

178   D.A. Baldwin and J.K. Debowski, Chromatographia, 26 (1988) 186–190.

179   E.C.V. Butler and G. Dal Pont, J. Chromatogr., 609 (1992) 113–123.

180   W.A. MacCrehan and J.M. Brown Thomas, Anal. Chem., 59 (1987) 477–479

181   P.J. Rennie and S.F. Mitchell, Chromatographia, 24 (1987) 319–323.

182   M.T. Galceran and F.J. Santos, Water Supply, 7 (1989) 69–75.

183   C.E. Werkhoven-Goewie, W.M. Boon, A.J.J. Praat, R.W. Frei, U.A.Th. Brinkman and C.J. Little, Chromatographia, 16 (1982) 53–59.

184   C. Ruiter, J.F. Bohle, G.J. de Jong, U.A.Th. Brinkman and R.W. Frei, Anal. Chem., 60 (1988) 666.

185   C. De Ruiter, J.-H.W. Brinkman, R.W. Frei, H. Lingeman, U.A.Th. Brinkman and P. Van Zoonen, Analyst, 115 (1990) 1033–1036.

186   D. Barceló, Chromatographia, 25 (1988) 295–299.

187   D. Barceló, G. Durand, R.J. Vreeken, G.J. de Jong, H. Lingeman and U.A.Th. Brinkman, J. Chromatogr., 553 (1991) 311–328.

188   A. Farran, J.L. Cortina, J. De Pablo and D. Barceló, Anal. Chim. Acta, 234 (1990) 119–126.

189   D. Puig, D. Barceló, I. Silgoner and M. Grasserbauer, J. Mass Spectrom., 31 (1996) 1297–1307.

190   A. Cappiello, G. Famiglini, P. Palma, A. Berloni and F. Bruner, Environ. Sci. Technol., 29 (1995) 2295.

191   M. Carini, G. Aldini, S. Furlanetto, R. Stefani and R. Maffei Facino, J. Pharm. Biomed. Anal., 24 (2001) 517–526.

192   C. Crescenzi, A. Di Corcia, S. Marchese and R. Samperi, Anal. Chem., 67 (1995) 1968–1975.

193   M.A. Aramendía, V. Boráu, Y. García, C. Jiménez, F. Lafont, J.M. Marinas and F.J. Urbano, Rapid Commun. Mass Spectrom., 10 (1996) 1585–1590.

194   J.M.E. Quirke, C.L. Adams and G.J. Van Berkel, Anal. Chem., 66 (1994) 1302–1315.

195   B.M. Hughes, D.E. McKenzie and K.L. Duffin, J. Am. Soc. Mass Spectrom., 4 (1993) 604–610.

196   D. Temesi and B. Law, LC–GC Int., 12 (1999) 175–180.

197   J.W. Jorgenson and K.D. Lukacs, Anal. Chem., 53 (1981) 1298–1302.

198   J.W. Jorgenson and K.D. Lukacs, Science, 222 (1983) 266–272.

199   E. Dabek-Zlotorzynska, Electrophoresis, 18 (1997) 2453–2464.

200   A.L. Crego and M.L. Marina, J. Liq. Chromatogr. Rel. Technol., 20 (1997) 1–20.

201   Y.-Ch. Chao and Ch.-W. Whang, J. Chromatogr. A, 663 (1994) 229–237.

202   M.F. Gonnord and J. Collet, J. Chromatogr., 645 (1993) 327–336.

203   I.-Ch. Chen and Ch.-W. Whang, J. Chin. Chem. Soc., 41 (1994) 419.

204   D. Martínez, E. Pocurull, R.M. Marcé, F. Borrull and M. Calull, J. Chromatogr. A, 734 (1996) 367–373.

205   G. Li and D.C. Locke, J. Chromatogr. B, 669 (1995) 93–102.

206   Ch. Lin, W.-Ch. Lin and W. Chiou, J. Chromatogr. A, 705 (1995) 325–333.

207   Ch.-Y. Tsai and G.-R. Her, J. Chromatogr. A, 743 (1996) 315–321.

208   N.J. Benz and J.S. Fritz, J. High Resolut. Chromatogr., 18 (1995) 175–178.

209   O. Jáuregui, L. Puignou and M.T. Galceran, Electrophoresis, 21 (2000) 611–618.

210   A. Zemann and D. Volgger, Anal. Chem., 69 (1997) 3243–3250.

211   S.M. Masselter and A.J. Zemann, J. Chromatogr. A, 693 (1995) 359–365.

212   S.M. Masselter and A.J. Zemann, Anal. Chem., 67 (1995) 1047–1053.

213   X. Liu and H. Frank, J. High Resolut. Chromatogr., 21 (1998) 309–314.

214   M. Van Bruijnsvoort, S.K. Sanghi, H. Poppe and W.Th. Kok, J. Chromatogr. A, 757 (1997) 203–213.

215   Ch.-E. Lin, Ch.-Ch. Hsueh, W.-Ch. Lin and Ch.-Ch. Chang, J. Chromatogr. A, 746 (1996) 295–299.

216   S. Terabe, K. Otsuka, K. Ichikawa, A. Tsuchiya and T. Ando, Anal. Chem., 56 (1984) 111–113.

217   K. Otsuka, S. Terabe and T. Ando, J. Chromatogr., 348 (1985) 39–47.

218   S. Terabe, K. Otsuka and T. Ando, Anal. Chem., 57 (1985) 834–841.

219   C.P. Ong, C.L. Ng, N.C. Chong, H.K. Lee and S.F.Y. Li, J. Chromatogr., 516 (1990) 263–270.

220   G. Li and D.C. Locke, J. Chromatogr. A, 734 (1996) 357–365.

221   I. Turnes, M.C. Mejuto and R. Cela, J. Chromatogr. A, 733 (1996) 395–404.

222   Ch.A. Lucy, K.K.-C. Yeung, X. Peng and D.D.Y. Chen, LC–GC Int., 11 (1998) 148–156.

223   C.D. Gaitonde and P.V. Pathak, J. Chromatogr., 514 (1990) 389.

224   M.W.F. Nielen, Trends Anal. Chem., 12 (1993) 345–356.

225   D. Martínez, F. Borrull and M. Calull, Trends Anal. Chem., 18 (1999) 282–291.

226   Y. He and H.K. Lee, J. Liq. Chromatogr. Rel. Technol., 21 (1998) 725–739.

227   I. Rodríguez, M.I. Turnes, M.H. Bollaín, M.C. Mejuto and R. Cela, J. Chromatogr. A, 778 (1997) 279–288.

228   J. Cai and J. Henion, J. Chromatogr. A, 703 (1995) 667–692.

229   J.F. Banks, Electrophoresis, 18 (1997) 2255–2266.

230   M.-C. Hennion and D. Barceló, Anal. Chim. Acta, 362 (1998) 3–34.

231   EPA Method 4010A, Screening for Pentachlorophenol by Immunoassay, US EPA, Washington, DC, January 1995, pp. 1–17.

232   M. Castillo, A. Oubiña and D. Barceló, Environ. Sci. Technol., 32 (1998) 2180–2184.

233   A. Oubiña, D. Puig, J. Gascón and D. Barceló, Anal. Chim. Acta, 346 (1997) 49–59.

234   O. Jáuregui, E. Moyano and M.T. Galceran, J. Chromatogr. A, 896 (2000) 125–133.

CHAPTER 7

Polychlorinated biphenyls

Jacob de Boer

*Netherlands Institute for Fisheries Research, P.O. Box 68,
1970 AB IJmuiden, The Netherlands*

7.1 INTRODUCTION

Polychlorinated biphenyls (PCBs) are compounds derived from biphenyl by substitution of one to ten hydrogen atoms by chlorine atoms. Each homologue group has a particular number of isomers: mono-chlorobiphenyl 3, di- 12, tri- 24, tetra- 42, penta- 46, hexa- 42, hepta- 24, octa- 12, nona- 3 and decachlorobiphenyl 1. In total there are 209 possible PCB congeners [1,2]. The coding system normally used for the PCB congeners (chlorobiphenyls (CBs) is that of Ballschmiter et al. [3], although Guitart et al. [4] have criticised the logics of the numbering of the CBs 107, 108 and 109.

PCBs have been produced and used world-wide in large quantities for many years as transformer oils, cutting oils, hydraulic oils, heat transfer fluids, additives in plastics, dyes and carbonless copying paper, and metal-casting release oils [2,4–7]. The production figures of PCBs are estimated at 1.3 million metric ton [8]. The production of PCBs was terminated world-wide around the late 1970s to early 1980s after authorities became aware of the adverse effects of PCBs on the environment, due to their persistency, bioaccumulative properties and toxicity.

Commercial PCB products are produced under the names of Aroclor (USA), Chlophen (Germany), Kanechlor (Japan), Fenclor (Italy), and others [1,2,8]. The trademark is usually followed by a four-digit number, the first two (12) indicate the type of compound (biphenyl) and the other two indicate the average percentage of chlorine. An exception is Aroclor 1016, which contains about 40% of chlorine and is similar to Aroclor 1242 [2].

During the last decade, much attention has been paid to the toxicity of PCBs, particularly to the congeners which show the same type of toxicity as polychlorinated dibenzo-*p*-dioxines (PCDDs) and dibenzofurans (PCDFs). Certain PCBs, which lack chlorine substituents in the *ortho*-position, show a particularly high '*dioxin-like*' toxicity, viz. PCB-77, PCB-126 and PCB-169 [1,5,6,9–11]. Also, mono-*ortho* substituted CBs show such a type of dioxin-like toxicity, although to a minor extent [12]. The toxicity of dioxin like PCB congeners is compared to that of the most toxic dioxin 2,3,7,8-TCDD

References pp. 260–262

TABLE 7.1

TOXIC EQUIVALENCY FACTORS FOR DIOXIN-LIKE COMPOUNDS FOR HUMANS AND WILD-LIFE DERIVED AT A WHO MEETING IN STOCKHOLM, SWEDEN, 15–17 JUNE 1997 [13]

Congener	Toxic equivalency factor (TEF)		
	Humans/mammals [a]	Fish [b]	Birds [b]
2,3,7,8-TCDD	1	1	1
1,2,3,7,8-PeCDD	1	1	1 [g]
1,2,3,4,7,8-HxCDD	0.1 [b]	0.5	0.05 [g]
1,2,3,6,7,8-HxCDD	0.1 [b]	0.01	0.01 [g]
1,2,3,7,8,9-HxCDD	0.1 [b]	0.01 [f]	0.1 [g]
1,2,3,4,6,7,8-HpCDD	0.01	0.001	<0.001 [g]
OCDD	**0.0001** [b]	–	–
2,3,7,8-TCDF	0.1	0.05	1 [g]
1,2,3,7,8-PeCDF	0.05	0.05	0.1 [g]
2,3,4,7,8-PeCDF	0.5	0.5	1 [g]
1,2,3,4,7,8-HxCDF	0.1	0.1	0.1 [d,g]
1,2,3,6,7,8-HxCDF	0.1	0.1 [d]	0.1 [d,g]
1,2,3,7,8,9-HxCDF	0.1 [b]	0.1 [d,f]	0.1 [d]
2,3,4,6,7,8-HxCDF	0.1 [b]	0.1 [d]	0.1 [d]
1,2,3,4,6,7,8-HpCDF	0.01 [b]	0.01 [c]	0.01 [c]
1,2,3,4,7,8,9-HpCDF	0.01 [b]	0.01 [c,f]	0.01 [c]
OCDF	**0.0001** [b]	0.0001 [c,f]	0.0001 [c]
3,3',4,4'-TCB (77)	**0.0001**	0.0001	0.05
3,4,4',5-TCB (81)	**0.0001** [b,c,d,f]	0.0005	0.1 [f]
3,3',4,4',5-PeCB (126)	0.1	0.005	0.1
3,3',4,4',5,5'-HxCB (169)	0.01	0.0005	0.001
2,3,3',4,4'-PeCB (105)	0.0001	<0.000005	0.0001
2,3,4,4',5-PeCB (114)	0.0005 [b,c,d,e]	<0.000005 [c]	0.0001 [h]
2,3',4,4',5-PeCB (118)	0.0001	<0.000005	0.00001
2',3,4,4',5-PeCB (123)	0.0001 [b,d,e]	<0.000005 [c]	0.00001 [h]
2,3,4',4,4',5-HxCB (156)	0.0005 [c,d]	<0.000005	0.0001
2,3,4',4,4',5'-HxCB (157)	0.0005 [c,d,e]	<0.000005 [c,d]	0.0001
2,3',4,4',5,5'-HxCB(167)	0.00001 [b,e]	<0.000005 [c]	0.00001 [h]
2,3,3',4,4',5,5'-HpCB (189)	0.0001 [b,d]	<0.000005	0.00001 [h]

–, No TEF because of lack of data.

[a] TEF values differing from the list presented in Table 2.1 are given in bold.

[b] Limited data set.

[c] Structural similarity.

[d] QSAR modelling prediction from CYP1A induction (monkey, pig, chicken, or fish).

[e] No new data from 1993 review.

[f] In vitro CYP1A induction.

[g] In vivo CYP1A induction after in ovo exposure.

[h] QSAR modelling prediction from class specific TEFs.

(tetrachlorodibenzo-*p*-dioxin). Those congeners have been given a toxic equivalency factor (TEF). The TEF value of 2,3,7,8-TCDD is defined as 1.0. These TEF values have been regularly updated according to the most recent information from toxicological studies. The most recent TEF values derived by the World Health Organisation (WHO) [13] are given in Table 7.1.

7.2 SAMPLING AND SAMPLE-PRETREATMENT

PCBs are analysed in all sorts of matrices: air, seawater, freshwater, sediments, sewage sludge, benthic organisms, fish, marine mammals, human blood and adipose tissue and others. All samples require a specific approach. A few examples will be discussed, focussing on the matrices sediment and biota, because these belong to the most frequently analysed matrices in environmental analysis. The analysis of PCBs in water is known to be extremely difficult because PCBs tend to migrate to a fatty environment. Their solubility in water is low. Consequently, PCB concentrations in water are extremely low and the analysis is hindered by background contamination. PCB concentrations in air can be determined after sampling by filters or passive sampling systems [14].

Surface sediments can be collected intertidally or by means of a variety of grab samplers from a vessel. Because PCB concentrations in sediments can show a patchy distribution, several grabs from one location are normally combined to one poled sediment sample. The buckets in which the material sampled by the grab samplers is collected should be made from stainless steel to avoid contamination from e.g. plastic materials which contain plasticisers such as phthalates. PCBs are often analysed by gas chromatography (GC) with electron capture detection (ECD). This detector is very sensitive for oxygen containing compounds such as many plasticisers. The grab samplers such as van Veen sampler normally collect the upper few centimetres of a sediment layer, but may disperse the unconsolidated floc at the sediment–water surface during descent. If this floc is to be sampled, then multiple damped-array shallow coring devices are to be preferred [15]. Passive sediment traps can be used to collect samples of material sedimenting through the water column or resuspended from the seabed [15]. In depositional areas, the changing fluxes of PCBs to the seabed over time can be reconstructed by analysis of dated sediment cores [15–17]. For this purpose cores of more than a meter in length may be required.

PCB concentrations in biota can be very variable, particularly at higher levels of the food chain. It is therefore difficult to obtain a representative sample from a certain population. Benthic organisms, shellfish and fishes are normally pooled to obtain a representative sample and to save expensive analyses of individual samples. PCBs are metabolised relatively slowly, which means that biota samples can be stored at −25°C for relatively long periods (years) prior to analysis. However, changes in the fat composition may take place at those temperatures, but the influence of those processes on the final result of the PCB analysis is negligible compared to the analytical error. Biota samples can be taken in various ways: benthic organisms can be taken intertidally from rocks or beachers, or sampled by means of a grab dredge or trawl in deeper water. Fish can be taken by rod and line, or in a variety of trawls and other nets depending of the habitat and habits of the species under investigation. Electric fishery techniques are particularly useful for collecting eel samples [18]. Guidelines describing the collection of biota samples for contaminant analysis were issued by the Oslo and Paris Commissions [15].

Sediment samples are transferred to wide-mouthed, solvent-cleaned glass jars for storage in a freezer at −25°C prior to analysis. The jars should not be overfilled because breakage can easily occur for samples with a high water content. The plastic lids of the jars should be covered inside with solvent-rinsed aluminium foil to avoid the introduction of plasticisers in the sample. Sediment samples can also be dried before analysis, but several procedures carry a risk for the integrity of the samples due to evaporation losses or introduction of interferences or cross-contamination [15,19].

Animals can be stored in a freezer at −25°C. The tissues needed for analysis should be removed under appropriate conditions of cleanliness in the laboratory. The glass jars need to be treated in the same way as for sediment samples (see above). The samples are normally extracted as wet samples, after drying with e.g. sodium sulphate. Freeze-drying is not recommended because losses of lower chlorinated CBs can occur. In addition, freeze-drying carries a risk of cross-contamination.

7.3 EXTRACTION

The extraction method used depends on the type of sample to be analysed and on the expected PCB levels. In this paragraph, Soxhlet extraction, solid phase extraction (SPE), supercritical fluid extraction (SFE), solvent extraction and several other extraction methods are discussed.

7.3.1 Soxhlet extraction

Soxhlet extraction has been used for a wide variety of samples like soils, sediments, animal and plant tissues (Table 7.3) [10]. Smedes and de Boer [19] described a Soxhlet extraction method for PCB analysis in sediment. A wide variety of solvents like dichloromethane (DCM), pure or mixed with acetone or hexane, and acetone–hexane mixtures can be used. The use of non-polar solvents only is not recommended [19]. The minimum time needed for a regular Soxhlet extraction is normally ca. 8 h. Sulphur present in the sediment- and soil-samples is also extracted, and must be removed by a later clean-up step [10]. Hess et al. [10] recommend Soxhlet extraction for fish samples because of its high recovery.

7.3.2 Solid phase extraction

Solid phase extraction has been used for PCB containing air- and water-samples. Cleghorn et al. [20] used 10 g of XAD-2 resin in a glass cartridge to collect PCB samples in air. The XAD-2 resin samples were air dried and Soxhlet extracted with dichloromethane for 16 h, then evaporated to near dryness. Ten (10) ml hexane (used as a keeper) was added and evaporated to near dryness again. The sample was concentrated to 1 ml in a centrifuge tube and subjected to a multi absorbent column clean-up. Tanabe et al. [21], Lang [2] and Hess et al. [10] described a method for the solid phase extraction of PCBs in (sea) water using a XAD-2 column. After collecting the PCBs on the XAD-2 sorbent (in some cases Tenax [2] or polyurethane foam [2,10] is used instead of XAD-2) the samples were subjected to solvent extraction. The amounts of water sampled are usually between a litre and hundreds of litres. According to Hess et al., this is an useful technique, but it puts a heavy demand on the cleaning procedure to ensure a low blank from the resin matrix [10]. Huetstis et al. [9] used a column extraction technique for fish. After homogenisation, 5 g sample was ground with anhydrous sodium sulphate (Na_2SO_4), packed in a chromatography column and eluted with DCM. The next step, after concentration, was gel permeation chromatography (GPC) to remove the bulk of the lipids. Schmidt and Hesselberg not only added Na_2SO_4, but also sea sand to improve the grinding of the material [22].

7.3.3 Supercritical fluid extraction (SFE)

The use of SFE as an extraction technique is related to the unique properties of the supercritical fluid. These fluids have a low viscosity, high diffusion coefficients, low toxicity and low flammability, all clearly superior to the organic solvents used in SPE-extraction. The most common fluid used is carbon dioxide [10]. Langenfeld et al. [23] used SFE for the extraction of PCBs from sediment. The PCBs were extracted at a temperature of 20°C at any pressure in the range of 150–650 atm. Each of the PCBs studied by Langenfeld et al. [23] were spiked onto sand, then extracted and then collected similarly to the sediment samples. Recoveries of the spiked analytes were found to be quantitative (>95%) for all the individual PCBs. Alexandrou et al. [24] also used SFE for the determination of PCBs. They concluded that the separation of PCDDs from PCBs and chlorinated benzenes is quite difficult because of their similar solubility. Van der Velde et al. [25] used an SFE technique for the extraction of PCBs (at ng/kg level) from fish, fish oil and total diet. Samples, mixed in different weight ratios with silica/silver nitrate 10%, were placed in 7-ml extraction cells. Spikes with several PCBs were added on top of the extraction cell to check the retention behaviour of fat/adsorbent combinations. The analytes were recovered by elution with 1.5–1.8 ml of hexane. With the correct fat : silica ratio and SFE conditions, no additional clean-up procedure is necessary for GC–ECD analyses. In a later review, van der Velde et al. used, depending on the type of fat, a basic alumina mixture, instead of silica/silver nitrate (1 : 10, w/w), to extract PCBs from fatty samples [26]. On the other hand SFE is a technique which has still to be developed further to obtain better results. A general drawback of SFE may be that the methods developed are valid for a specific matrix, but as soon as e.g. the fat content of a biota sample or the type of lipids changes, the method has to be adapted.

7.3.4 Other extraction methods

Other extraction methods like blending-, microwave-, ultra sonic- and solvent-extraction are mentioned in the literature [2,10,27–31]. Storr-Hansen et al. [27] used a blending method for the extraction of PCBs from seals. The main features were as follows: 10 g of homogenised sample were blended three times with DCM/methanol (2 : 1, v/v) in an Ultra-Turrax blender. The combined extracts were shaken with acidified water in a separating funnel. The organic phase was filtered through anhydrous sodium sulphate and the solvent evaporated. The residue was dissolved in hexane and treated with sulphuric acid absorbed on silica gel. The final clean-up of the extract was by chromatography on basic aluminium oxide, deactivated with 1% (w/w) water. Draper et al. [28] used an almost identical method of extraction as Storr-Hanssen [27]. Five grams of thawed fish tissue was combined with 20 g of sodium sulphate and 60 ml of petroleum ether in a blender jar. The solvent was homogenised for 2 min with a blender. The solvent was dried over sodium sulphate. Tissue extraction was repeated three times with petroleum ether. After the extraction silica column clean-up was carried out. Larsson et al. [31] used an ultrasonic bath for 10 min to extract the PCBs from pike. A segment of the muscle (15–20 g) was weighed and homogenised in 30 ml of acetone, and 30 ml of *n*-hexane was added as a keeper. The homogenate was then treated in an ultrasonic bath for 10 min and thereafter washed 60 ml of NaCl (2%). The hexane phase, containing fat and lipophilic pollutants, was separated and evaporated to dryness in a water bath (70°C). For milk and blood absorption at a solid sorbent has been used. Milk, for example, was first mixed with fibrous cellulose and Florisil and, after evaporation of water, the mixture was extracted with hexane. In some cases, the milk was mixed with Lipidex 5000 gel and the chemicals of interest were successfully eluted [2]. Duarte-Davidson et al. [29] and Dewailly et al. [30] used a solvent extraction method for the purification of PCBs in milk. Approximately 2 g of freeze dried milk is mixed with dry sodium sulphate. The samples are then extracted with about 170 ml *n*-hexane for a minimum period of 8 h. The recoveries found for the mono-, di- and tri-PCBs were between 22 and 87%.

Accelerated solvent extraction (ASE) has been applied more frequently over the last years [5,32,33]. ASE was performed by adding 10 ml toluene or hexane/acetone (1 : 1) solvent to the cell containing 4 g sample, pre-heating the cell for 5 min to 150°C, holding at 150°C for 5 min, removing the solvent, and rinsing with 5 ml of solvent [5]. The extracts were evaporated to near dryness and brought to 4 ml with toluene or methanol.

Another technique mentioned by Donnelly et al. [5] is microwave assisted extraction (MAE). Samples to which 100 ml of purified water had been added were extracted for 15 min in the microwave oven. When 150 ml of saline solution was substituted for purified water, a recovery of 97.1% was obtained. In Table 7.2, an evaluation of the most frequently used extraction techniques is given.

7.4 CLEAN-UP

Because the amount of lipophilic material in the extract can affect the active surface of the stationary phase and degrade the resolving power of the GC columns, an effective clean-up procedure is essential. Clean-up of extracts can be performed in either a non-destructive or destructive way [10]. In the following paragraphs several methods are discussed.

TABLE 7.2

EVALUATION OF EXTRACTION TECHNIQUES [10]

| | Ease of automation | Maximum sample intake | Extraction time | Cost | | Ease of optimisation |
				Initial set-up	Running expenses	
Soxhlet	Possible	Moderate/large	Moderate	Moderate	Low	Easy
Blending/ultrasonic	Difficult	Large	Rapid	Moderate	Low	Difficult
Chemical modification	Difficult	Large	Slow	Very low	Very low	Moderate
Column extraction	Possible	Large	Slow	Low	Low	Difficult
ASE	Easy	Moderate/large	Rapid	High	Low	Moderate
SFE	Easy	Small	Rapid	Very high	High	Complex

7.4.1 Non-destructive lipid removal

7.4.1.1 Gel permeation chromatography

GPC is a method used for the clean-up of fatty samples. Smaller molecules such as PCBs are retained while larger molecules such as fat are eluted earlier. The PCBs in the sample can be isolated in this way. Glausch et al. [34] used a GPC method for the clean-up procedure of PCBs in milk. The fat in the milk was separated by centrifugation. It was afterwards mixed with sodium sulphate, treated with petroleum ether and than filtered. The solvent was evaporated and the obtained fat was purified by GPC, followed by silica gel adsorption chromatography. According to Lang [2], the most widely applied gel for GPC is BioBeads S-X3, but other gels, such as Sephadex LH-20, PLRP-S, BioBeads S-X2 [35], BioBeads S-X4 [2], PL-gels (Polymer Laboratories) and other polystyrene–divinylbenzene copolymer columns [2,19,36,37] are used. Eluents used for the GPC clean-up are usually mixtures such as DCM–hexane [19], ethyl acetate–toluene [2,19], cyclohexane–DCM, cyclohexane–ethyl acetate, toluene–ethyl acetate and 2-propanol–heptane [2]. One main disadvantage of the GPC system is that it is difficult to remove all the lipids. The remaining traces of lipids have to be removed in a second clean-up procedure, e.g. on an additional silica column or by a second GPC step [2,10]. PL-gels may offer the best result of the current choice of gels. Another disadvantage is that GPC does not separate the PCBs from the other compounds in the same molecular range, such as organochlorine pesticides [19]. Therefore, an additional fractionation is often required.

7.4.1.2 Column chromatography

Column clean-up procedures have been designed to cope with two types of interference; the co-extracted bulk components comprising predominantly lipids and, secondly, co-extracted organochlorine pesticides (which is described in Section 7.5). Alumina columns normally provide an excellent clean-up of PCBs from bulk components such as lipids [37–39].

7.4.2 Destructive lipid removal

7.4.2.1 Sulphuric acid treatment

Because PCBs and most pesticides are generally resistant to sulphuric acid (H_2SO_4), sulphuric acid treatment is used for degradation of most aliphatic and many aromatic compounds in environmental samples [29]. 20 ml of concentrated H_2SO_4 are added to 10 ml extract and the mixture is shaken vigorously for 30 s. The phases are allowed to separate, and the cleaning of the hexane extract is repeated twice with 10 ml H_2SO_4. The collected hexane-sample is subsequently evaporated to 1 ml for further clean-up [29]. H_2SO_4 treatment has been used by Larson et al. [31]. According to Lang [2], problems may be caused by the fact that sulphuric acid affects compounds (some halogenated pesticides such as dieldrin) that are determined together with PCBs in one extract.

7.4.2.2 Saponification

According to Hess et al. [10], lipids can be saponified by heating the extract in a small volume of solvent with 20% ethanolic potassium hydroxide at ca. 70°C for 30 min. Saponification is not only used for lipids but is also used for the removal of sulphur from sediment extracts [19]. The conditions of saponification are critical. Too high temperatures and too long saponification times can cause decomposition of higher chlorinated compounds, such as hexa–deca PCBs, in particular when trace metals are present, e.g. in sediment samples, that can act as a catalyst [19]. Perdih and Jan [41] used a saponification method for the determination of PCBs in silicone rubber. Van der Valk and Dao [40] used a saponification method to separate PCBs from environmental samples. Because very low recoveries for higher chlorinated biphenyls were sometimes observed in inter-laboratory studies of sewage sludge, they started an investigation on the degradation of highly chlorinated biphenyls. In this investigation they concluded that on aromatic compounds, with four or more chlorine atoms at one ring, the chlorine can be substituted by an ethoxy group under hot saponification conditions. Therefore, care must be taken with respect to these congeners during saponification and alternative clean-up procedures without saponification must be considered. Although the percentage of degradation is normally low, this process plays an important role in the analysis of non-*ortho* chlorinated PCBs. The small percentage of degradation of higher chlorinated PCBs can lead to a relatively high bias in the concentration of non-*ortho* chlorinated PCBs which can be formed after degradation of the higher chlorinated mono-*ortho* substituted PCBs [42].

7.5 PRE-FRACTIONATION

Pre-fractionation is an important method in the analysis of PCBs because in single-column GC only 100–150 different compounds can be separated [2,6,11]. Because after the clean-up PCBs are normally present in the extract together with other halogenated compounds, such organochlorine pesticides, brominated biphenyls and diphenyl ethers [18,43], it is important to apply a pre-fractionation. Co-elution problems are avoided in this way. In addition, a pre-fractionation may be important in the analysis of non-*ortho*

PCBs. In that case, non-*ortho* PCBs are separated from the other PCBs. The reason for the latter pre-separation is the difference in concentration of a factor 100–1000, which leads to problems in separation and detection.

7.5.1 Adsorption column chromatography

For the clean-up of PCBs in several matrices, columns like alumina, silica and Florisil [2,10,19], in different mesh sizes, levels of activity and column sizes, can be used. Florisil is one of the oldest materials used in the clean-up procedure for PCBs [19]. Florisil is a mixture of several inorganic oxides with SiO_2 and MgO as the main substances. Because the Florisil is a mixture, the composition varies from batch to batch. Silica and alumina have been applied to clean-up extracts from air, water, sediments, rice bran oil and animal or human tissue [2]. Smedes and de Boer [19] and de Boer [39] reported the use of a silica and alumina clean-up for PCBs and other pesticides in fish, while Dewailly et al. [30] and Duarte-Davidson et al. [29] used such columns for the clean-up of PCBs in milk. Rebbert et al. [44] used a disposable silica SPE column (SepPak, Waters Associates, Milford, MA, USA), which had been pre-cleaned with hexane for the clean-up of PCB containing samples. Athanasiadou et al. [45] used an alumina column for the separation of polychlorinated dibenzofurans (PCDFs) and naphthalenes (PCNs) in the PCB containing samples. After the alumina clean-up, a carbon column was used to separate the planar PCBs from the non-planar PCBs. Duinker et al. [46] also used an alumina column to obtain different fractions of organic compounds like aliphatics, polyaromatic hydrocarbons (PAHs), toxaphene and PCBs in environmental samples. Tanabe et al. [21] used a Florisil column to separate the PCBs and *p,p'*-DDE which come in the first fraction from the DDT and HCH isomers which come in the second fraction of sediment samples. According to Smedes and de Boer [19], silica is an ideal material for the isolation of PCBs from organochlorine pesticides. Pre-heating and deactivation of the silica with a few percent of water is necessary. De Boer [39] used for the first fraction (chlorobenzenes, chlorostyrenes, PCBs and *p,p'*-DDE) an 1.8 g SiO_2–3% H_2O column eluted with 11 ml iso-octane, while for the second fraction (remaining chlorinated pesticides and chlordanes) 10 ml of 15% diethylether/iso-octane was used. In this way, all the PCBs are eluted in the first fraction. In Fig. 7.1, a scheme of methods used for the isolation clean-up and fractionation of PCBs in wet and dry sediment is shown [19].

7.5.2 Carbon column chromatography

Carbon column chromatography (CCC) exhibits excellent selectivity for those CBs of which a planar conformation is assumed. The majority of reported methods uses different forms of carbon to isolate these congeners [47]. The sorbents used for CCC are, e.g. activated carbon, activated carbon/polyurethane foam, activated carbon/glass fibres, activated carbon/silica gel, activated carbon/celite, activated carbon/Chromosorb W-HP and porous graphitic carbon. Falandysz et al. [48] used an activated carbon

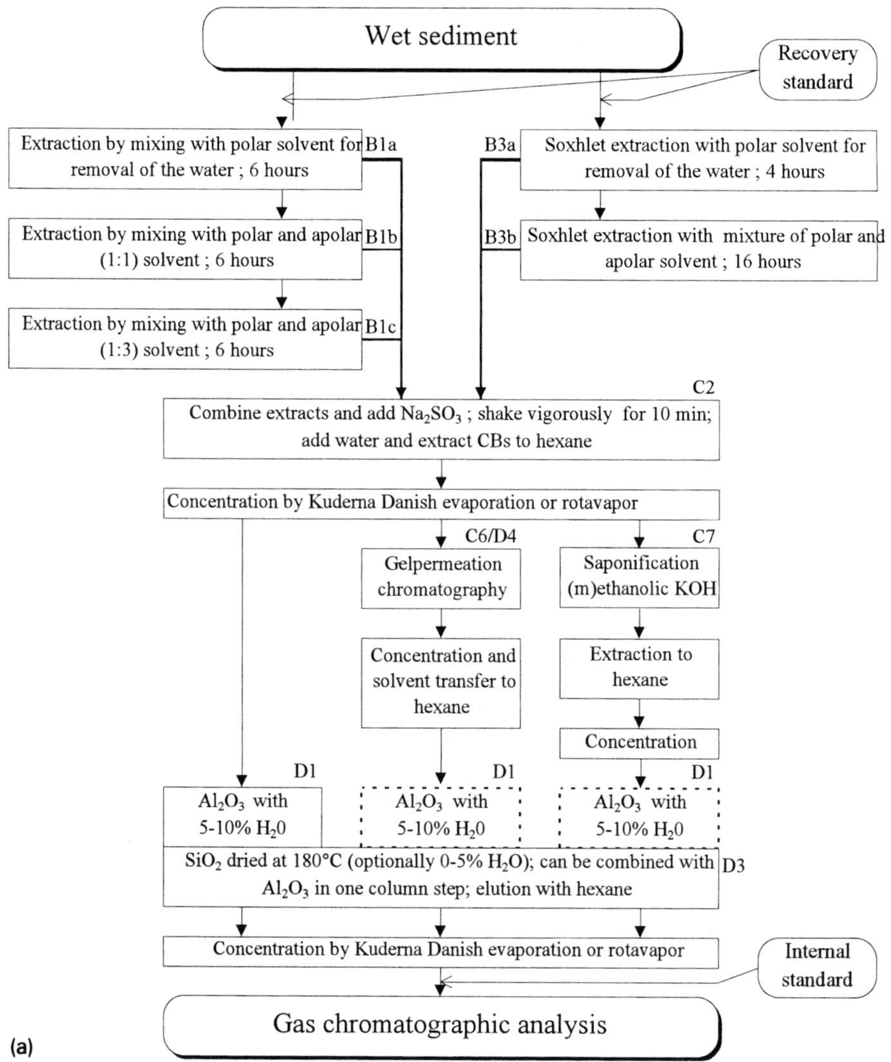

Fig. 7.1. Methods used for the isolation, clean-up and fractionation of PCBs in (a) wet sediment and (b) dry sediment [19].

column from Wako Pure Chemicals (Japan), to separate the CBs 77, 126 and 169 in a technical PCB mixture named Chlorofen from Poland. Those three CBs were quantified at a concentration of 0.52, 0.25 and 0.43 mg/g, respectively. According to Larsen et al. [6], activated charcoal is the most popular choice. These methods are mostly used for the enrichment of planar PCB congeners. But activated charcoal has some serious drawbacks such as low efficiency, severe tailing of elution profiles, irreversible adsorption and large batch-to-batch variations. The use of HPLC with a porous graphitic carbon column may help to overcome these drawbacks. An overview of

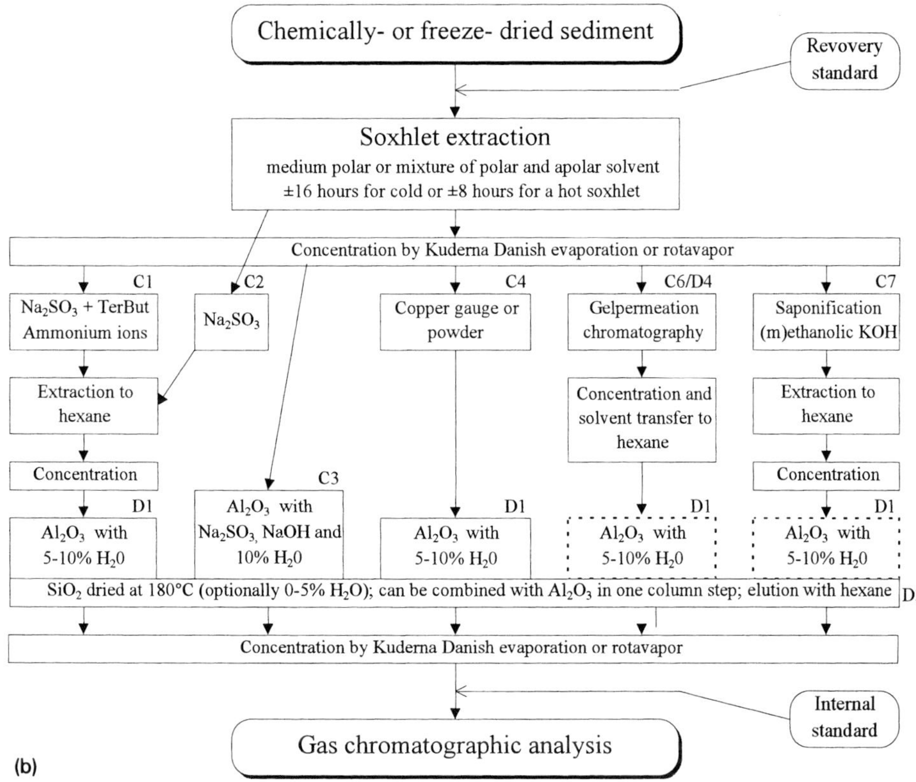

Fig. 7.1 (continued).

methods used for fractionation and determination of non-*ortho* CBs is given by Creaser et al. [47]. Duinker et al. [46] reported separation of *ortho-* and non-*ortho* CBs with an activated charcoal column. The non-*ortho*'s were recovered subsequently with aromatic solvents like toluene, benzene and dichloromethane. Differences in efficiency between brands of charcoal were observed. Schwartz et al. [49] used a carbon/glass fibre PX-21 carbon column, which was firstly introduced by Smith [47,49]. Interference of *p,p'*-DDE (2,2-bis(4-chlorophenyl)-1,1-dichloroethylene) with CB 77, on a GC with an electron capture detector (ECD) using a DB-1 column, is almost entirely removed due to the use of carbon/glass fibre enrichment. Recoveries found of CB 77 were 65 ± 4, 63 ± 15, 86 ± 10 and 82 ± 10 for, respectively, 100, 10, 2 and 0.05 ng/g spiked fish lipids [49]. An AX-21–Celite column was used for the separation of PCBs and other pesticides by Kocan et al. [50]. In the second fraction (fraction 1b) the multi-*ortho* CBs were eluted with 7 ml cyclohexane–dichloromethane–methanol (2:2:1). The mono-*ortho* and non-*ortho* CBs eluted in the fifth fraction (fraction 2b) with 10 ml of toluene. Kannan et al. [51] made an extensive and thorough comparison of six activated charcoals (Wako, Baker analytical, Serva-SK-4, Anderson, AX-21, Alltech, SK4-type and Merk) using multidimensional (MD) GC–ECD. They concluded that none of the charcoals tested was able to completely separate the non-*ortho* CBs from the *ortho* CBs and, consequently, significant interferences in the quantification of CBs 77 and 126 could be expected [47].

7.5.3 Other HPLC columns

Haglund et al. [52] reported the separation of non-*ortho* CBs by using an electron donor–acceptor HPLC method using a 2-(1-pyrenyl)ethyldimethylsilylated silica (PYE) HPLC column instead of a carbon column. The PYE column separates PCBs according to the number of chlorine atoms in the *ortho* position [47]. The selectivity of the PYE-phase may be explained by a charge transfer mechanism, in which electron-density acceptor (EDA) and donor regions of the CBs induce a change in the localisation of the p-electron cloud of the pyrene moieties of the phase so that an EDA complex is formed. This type of mechanism could account for the observed retention behaviour in the following three ways: (1) highly chlorinated biphenyls would be expected to form strong EDA complexes with the PYE phase because chlorinated compounds are very good electron-density acceptors; (2) CBs with many *ortho* chlorines should be less retained owing to steric interaction between *ortho* chlorines, leading to twisting of the biphenyls-bond, an increase in the distance between the biphenyl and the pyrene moieties, and thus to weaker EDA complex; and (3) CBs with half-ring structures with the chlorines close together offer naturally better acceptor pockets for EDA complexing than those have half-ring structures with the chlorines spread over the rings, and are therefore more retained [53]. According to Haglund et al. [53] separation of the non-*ortho* CBs on a PYE column takes less than 15 min. This is much quicker than some carbon column separations. A disadvantage may be that the PYE-column is considerably higher in price than a carbon column, and the PYE-phase is more sensitive to lipids [54]. Grimvall et al. [55] reported the use of a 2,4-dinitroanilinopropyl silica (DNAP) phase for the isolation of CB 77, 126 and 169 in human blood plasma. They studied the retention characteristics of individual PCB congeners on a DNAP, porous graphitic carbon and PYE stationary phases, and concluded that the best results were obtained on a DNAP column.

7.6 GC ANALYSIS

PCBs are almost without exception determined by GC. Apart from a choice for injection or detection technique, there are also different possibilities for the type of columns used, either in single-column GC or in multi-column approaches.

7.6.1 Injection techniques

The most frequently used injection techniques in PCB analysis are splitless and on-column injection. Split injection is not recommended because strong discrimination effects can occur [19]. The presence of high boiling compounds can also increase the effect of discrimination. On-column injection is a technique in which the sample is deposited directly into the column with a syringe. On-column injection is well suited for high boiling compounds. According to Lang [2], on-column injection yields better results, but is much more sensitive to contamination than splitless injection. De Boer et al. [56] reported that splitless injection was the most frequently used injection technique in a large multi-step international interlaboratory study on CBs (±80%) followed by on-column injection (±15%). In splitless injection, a sample is injected in a hot liner while the splitter is closed. The sample evaporates and enters the GC-column. During the time the splitter is closed, the oven is relatively cold 80–100°C, dependent of the solvent; the ideal temperature is ca. 10°C below the boiling point of the solvent. After the splitter is opened, the oven is rapidly heated and the chromatographic analysis begins. If the liner is not big enough, memory effects can occur due to contamination of the gas tubing attached to the injector [19]. Large volume injection is becoming more popular. The increase of the injection volume is directly related to a decrease of sample intake. Staniewski et al. [57] used a programmed temperature injector (PTV) equipped with a deactivated liner containing a porous bed of sintered glass beads for the determination of PCBs. Sample volumes smaller than 25 µl were introduced manually. Large volumes were introduced into the liner of the injector by means of a microprocessor-controlled syringe pump. With this system, sample volumes up to 1 ml could be introduced at rates between 1 and 2000 µl/min. Disadvantages of large volume injection are the stronger influence of solvent-impurities in the chromatogram and contamination of the column and detector. The use of mass spectrometric detection with negative chemical ionisation (NCI–MS) is in particular sensitive for contamination due to large volume injection. A regular cleaning of the ion source may be necessary.

7.6.2 Detection techniques

The most frequently used detection system for the analysis of PCBs in various matrices is the ECD. This extremely sensitive detector is particularly selective for halogenated compounds [2,19,37]. According to Smedes and de Boer [19] the very high sensitivity of ECD for PCBs theoretically allows detection limits below 1 ng/kg. Its extreme sensitivity, on the other hand, makes ECD vulnerable to dirt and overloading. The ECD response is variable and it varies from one detector to another [2]. For the seven mono-*ortho* PCBs studied by de Boer et al. [58], absolute detection limits between 5 and 10 pg were found for MDGC–ECD.

The second most frequently used detection system is mass spectrometry (MS). There are several modes in which the MS is employed to detect PCBs: electron impact (EI) ionisation, chemical ionisation with negative-ion detection (NCI), or chemical ionisation with positive-ion detection (PCI) [2,10]. According to Kannan et al. [51] detection difficulties may rise for the determination of non-*ortho* CBs when co-eluting peaks, which had to be determined by single column with an ECD or an MS detection system, have a large difference in concentration. Due to these concentrations differences (sometimes more than 1000-fold) a pre-separation technique was developed (pre-separation on carbon columns or LC), in which the non-*ortho* PCBs are separated from the other PCBs prior to GC-analyses.

EI mass spectra are fairly reproducible. They show relatively intense molecular ions and the natural isotopic distribution of chlorine gives rise to typical clusters, which are easily recognisable [2]. The response factors in EI–MS differ between isomers by no more than about two-fold, which is much less than in ECD. It is therefore possible to use one surrogate standard within each isomer group [2]. Larsen et al. [59] used an EI mode from 50 to 550 a.m.u. every second for the detection of PCBs in Aroclor mixtures. Full scan spectra were obtained. Detection limits varied from 25 to 50 pg depending on the GC column bleeding.

According to Bøwadt et al. [60], who compared NCI with PCI, the sensitivity of CBs in NCI increases dramatically with an increase in number of chlorine atoms. The opposite occurs for the CBs determined on PCI [10]. Detection limits found for CB 28 in full scan mode are for PCI and NCI 200 and 700 pg/ml, respectively, while the detection limits for PCB 180 for PCI and NCI were, 253 and 2.2 pg/ml, respectively [60]. The choice, NCI or PCI, depends on the PCBs to be analysed, but as most PCBs found in the environment are relatively highly chlorinated, NCI is more often used. Actually, PCI applications for PCB analysis are rather rare. De Boer et al. [54] calculated the following detection limits for the GC–MS method with NCI: 1.0, 0.1 and 0.05 ng/kg for, respectively, CB 77, 126 and 169. Because of the blank values, presumably caused by the use of a porous graphitised carbon column, these detection limits could not be lowered. Absolute detection limits of the GC–MS of 100, 100 and 50 fg were calculated for the CBs 77, 126 and 169, respectively. GC/NCI–MS has a more than 10-fold higher sensitivity for PCBs with more than four chlorine atoms. EI–MS has a ca. 10-fold lower sensitivity than ECD. The low concentrations of non- and mono-*ortho* PCBs will in general not allow a proper determination by. Fig. 7.2 shows relative response factors (RRF) of six CBs with a different number of chlorine atoms for GC/ECD, GC/EI–MS, GC/NCI–MS (total ion chromatogram (TIC)) and GC/NCI–MS (single ion chromatograms (SIC)) [61]. The highly different NCI signals of CBs with less than five chlorine atoms and higher chlorinated CBs are obvious.

7.6.3 GC separation

7.6.3.1 Single-column GC

For the analysis of CBs, several GC-columns can be used. The most frequently used GC column is the SE-54 column (5% diphenyl–1% vinyl dimethylsiloxane) [62,63]. A complete set of retention time data on an SE-54 column is available for all 209 CBs on this column [62]. Other 5% phenyl phases comparable to the SE-54 phase are: CP-Sil-8, DB-5, HP-5, Ultra-2, SE-52 and OV-73 [64]. One disadvantage of the SE-54 phase is the number of coeluting PCB congeners: CBs 15/17/18, 28/31, 37/42/59, 47/48/75, 52/73, 66/95, 67/100, 77/110, 82/151, 84/90/101, 87/115, 105/132/153, 114/122/131, 118/123/148/149, 126/129/178, 137/176, 138/158/160/163, 156/171/202, 157/173/201, 170/190, 195/208 and 196/203 [11,19]. A longer column and particularly a smaller diameter may help to resolve some of these congener pairs [65]. A good alternative for the SE-54 column is the CP-Sil-13 column (15% diphenyl dimethyl-siloxane) which is the only published thermo-stable column reported till 1993 which

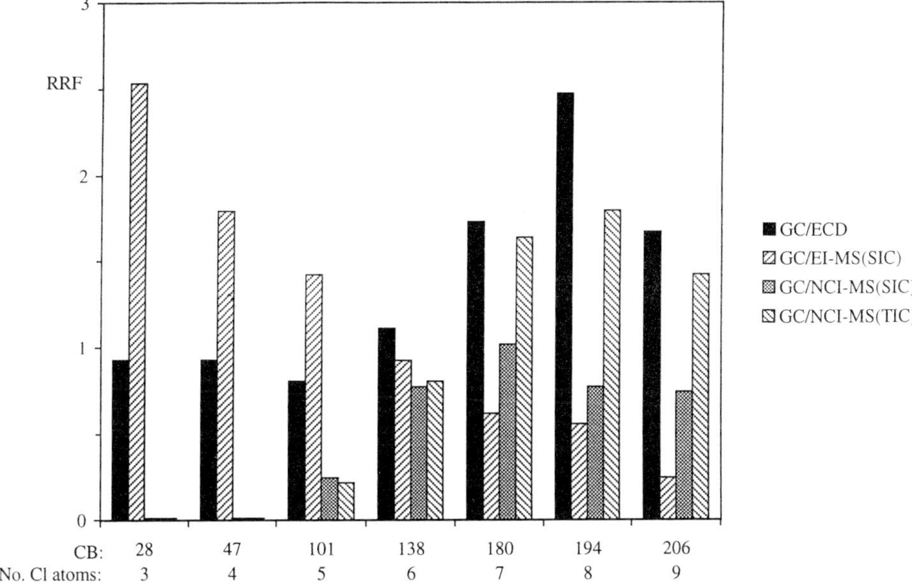

Fig. 7.2. Relative response factors (RRF) of six CBs with a different number of chlorine atoms for GC/ECD, GC/EI–MS and GC/NCI–MS [61].

allows interference-free analyses of the CBs 43, 132 and 196 [66]. From the work of de Boer et al. [65] and Larsen et al. [6], it was concluded that a CP-Sil 19 (14% cyanopropyl phenyl 1% vinyl dimethylsiloxane) column offers better possibilities for the separation of the most important PCBs than a SE-54 type column. A disadvantage is that the production of stable batches of CP-Sil 19 is difficult. Other phases used, comparable to CP-Sil-19, are: OV-1701, DB-1701, BP-10 and SFB-7 [64]. According to Larsen et al. [59], more than two columns are needed for true confirmation of the major part of the PCBs. A single CP Sil 19 column offers accurate analyses of 84 congeners by GC–ECD (Fig. 7.3) and 100 congeners by GC–MS. Of the priority PCBs (36 congeners which have been indicated as most environmentally threatening, based on their frequency of reported occurrence in environmental samples) 20 can be accurately analysed by GC–ECD and 25 by GC—MS [59]. Co-elution on a CP-Sil-19 column also occurs: CBs 7/9, 16/29, 17/18, 20/33, 24/27, 32/34, 37/41/64/71, 44/59, 46/49, 55/92, 60/99, 66/95, 81/87/97/115, 82/124/149, 84/119, 107/123, 114/146, 129/183, 132/179, 138/158/160/163/178, 141/176, 144/147, 157/180/197, 169/196/203, 170/190 and 174/200 [59]. Other stationary phases used for the analysis of PCB congeners are OV-210, 50% trifluoropropyl methylsiloxane) or comparable phases like QF-1, SP-2401 and RSL-400. Also 25% phenyl 25% cyanopropyl (OV-225) and comparable phases such as CP-Sil-43, DB-225, BP-15, XE-60 and CS-5 have been used [64]. Larsen et al. [59] and de Boer et al. [65] used, in addition to the columns mentioned before, also a highly polar *bis*-cyanopropyl (CP-Sil-88) column, commonly used in dioxin analysis, for the analysis of PCBs (Fig. 7.4). A drawback of this strongly

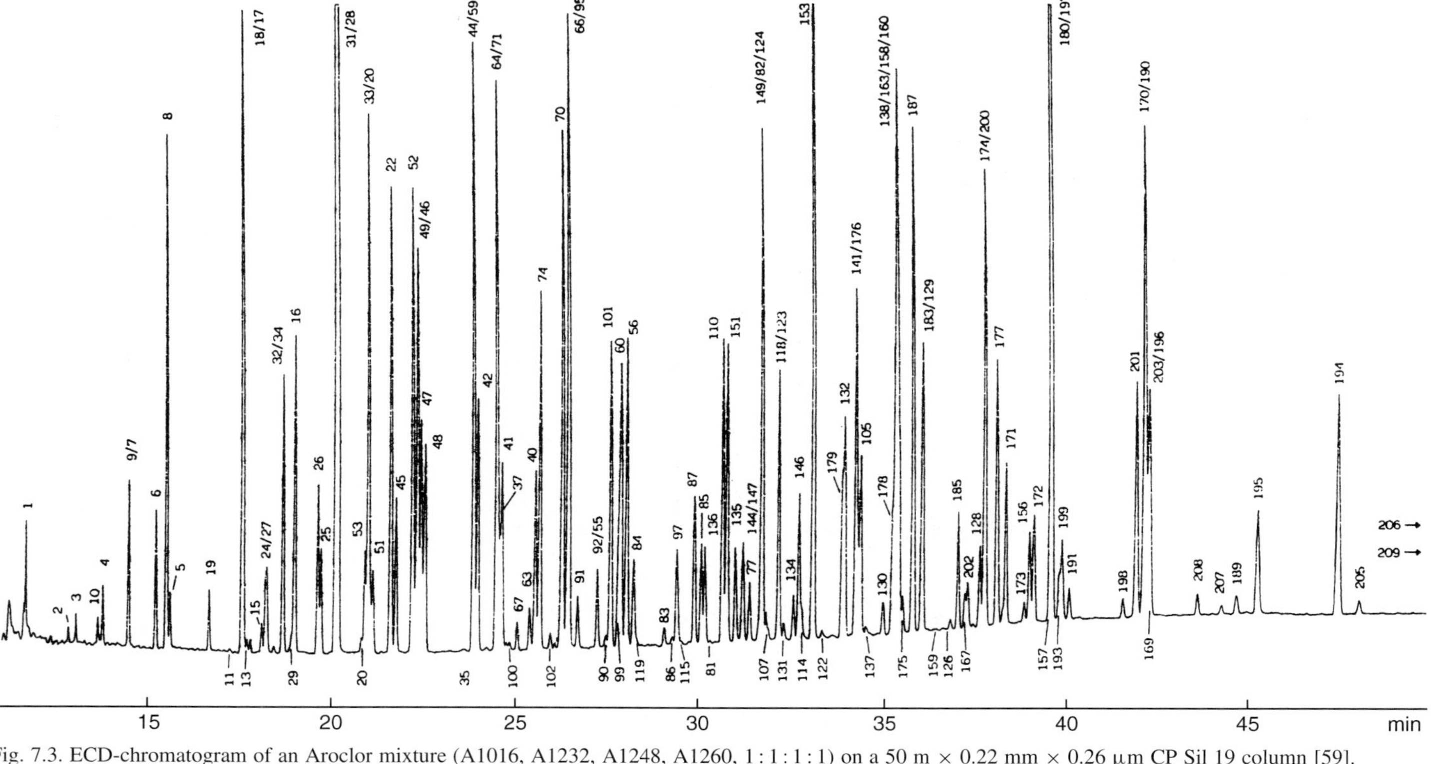

Fig. 7.3. ECD-chromatogram of an Aroclor mixture (A1016, A1232, A1248, A1260, 1:1:1:1) on a 50 m × 0.22 mm × 0.26 μm CP Sil 19 column [59].

polar phase is its low maximum temperature of 240°C which increases the analysis time. This phase offers good separation for the most toxic PCBs (77, 126 and 169) [59]. Co-elution also occurs at this phase (Fig. 7.4): CBs 2/7, 11/25, 16/28, 20/45, 31/51, 35/70, 42/59, 47/49, 55/101, 60/144, 63/64, 66/119, 67/71, 81/82/183/187, 83/84/115/135/136/151, 85/87/149, 92/95/99, 97/147, 105/129, 107/124, 110/153, 114/137, 118/123/200, 128/193/201, 138/158/160/185, 146/176, 170/198, 172/207, 180/197 [19,59]. In a later, study Larsen [67] compared eleven stationary phases: 50 m × 0.25 mm, 0.20 μm 50% *n*-octyl dimethylsiloxane (OCTYL), 50 m × 0.25 mm, 0.15 μm dimethylsiloxane (Sil-5), 50 m × 0.25 mm, 0.25 μm 5% diphenyl 1% divinyl dimethylsiloxane (SE-54), 50 m × 0.25 mm, 0.25 μm 5% diphenyl dimethylsiloxane (Sil-8), 50 m × 0.25 mm, 0.20 μm 15% diphenyl dimethylsiloxane (Sil-13), 50 m × 0.25 mm, 0.25 μm 50% diphenyl dimethylsiloxane (DB-17), 25 m × 0.25 mm, 0.10 μm 1,7-dicarba-closo-dodecarborane dimethylpolysiloxane (HT-5), 60 m 0.25 mm, 0.15 μm or 0.25 μm 1,7-dicarba-closo-dodecarborane phenylmethyl polysiloxane (HT-8), 50 m × 0.25 mm, 0.20 μm 14% cyanopropyl 1% vinyl methylsiloxane (Sil-19), 50 m × 0.25 mm, 0.20 μm 70% cyanopropyl dimethylsiloxane (BPX70) and 50 m × 0.22 mm, 0.20 μm 100% cyanopropylsiloxane (Sil-88). None of the 11 phases was able to resolve all 209 congeners, neither the approximately 150 congeners present in technical PCB mixtures, nor the 36 priority congeners and the seven indicator PCBs: 28, 52, 101, 118, 138, 153 and 180 (these congeners have been selected on the basis of their presence at high concentration in technical PCB mixtures and in the environment). The number of separated PCBs on the 11 phases are shown in Table 7.3. This table shows that the HT-8 column offers the best resolution (Fig. 7.5). In combination with MS detection, interference-free analysis is offered for all seven indicator CBs and all 36 priority CBs.

TABLE 7.3

NUMBER OF CBS WHICH CAN BE ANALYSED WITH LESS THAN 10% INTERFERENCE FROM ANY COELUTING CONGENER ON 11 THOROUGHLY CHARACTERISED STATIONARY PHASES [67][a]

Name	Indicator CBs		Priority CBs		Aroclor CBs	
	MS	ECD	MS	ECD	MS	ECD
OCTYL[c]	6	6	33	31	100	ND
SIL-5	6	5	29	27	102	84
SIL-8	5	5	24	22	81	65
SE-54[b]	5	4	18	15	75	69
SIL-13	5	4	25	23	104	93
HT-5	6	5	24	22	91	78
HT-8	7	6	36	28	157	106
DB-17	5	4	22	19	93	86
SIL-19	5	4	25	20	100	84
BPX70	4	2	19	13	89	72
SIL-88	5	2	24	18	99	74

[a] Unless stated otherwise, the temperature program was optimised for a 1-h total run time.

[b] 2 h total run time.

[c] 1.5 h total run time.

References pp. 260–262

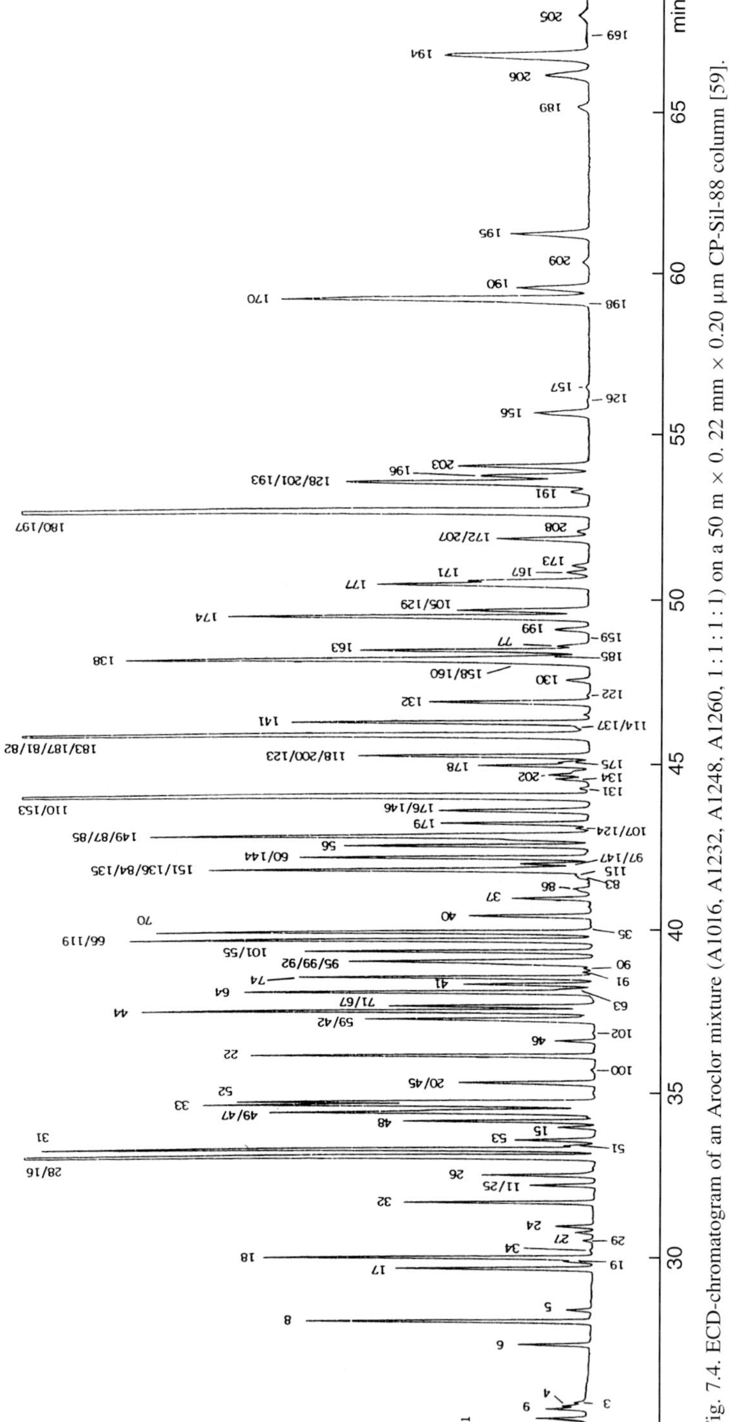

Fig. 7.4. ECD-chromatogram of an Aroclor mixture (A1016, A1232, A1242, A1248, A1260, 1 : 1 : 1 : 1 : 1) on a 50 m × 0. 22 mm × 0.20 μm CP-Sil-88 column [59].

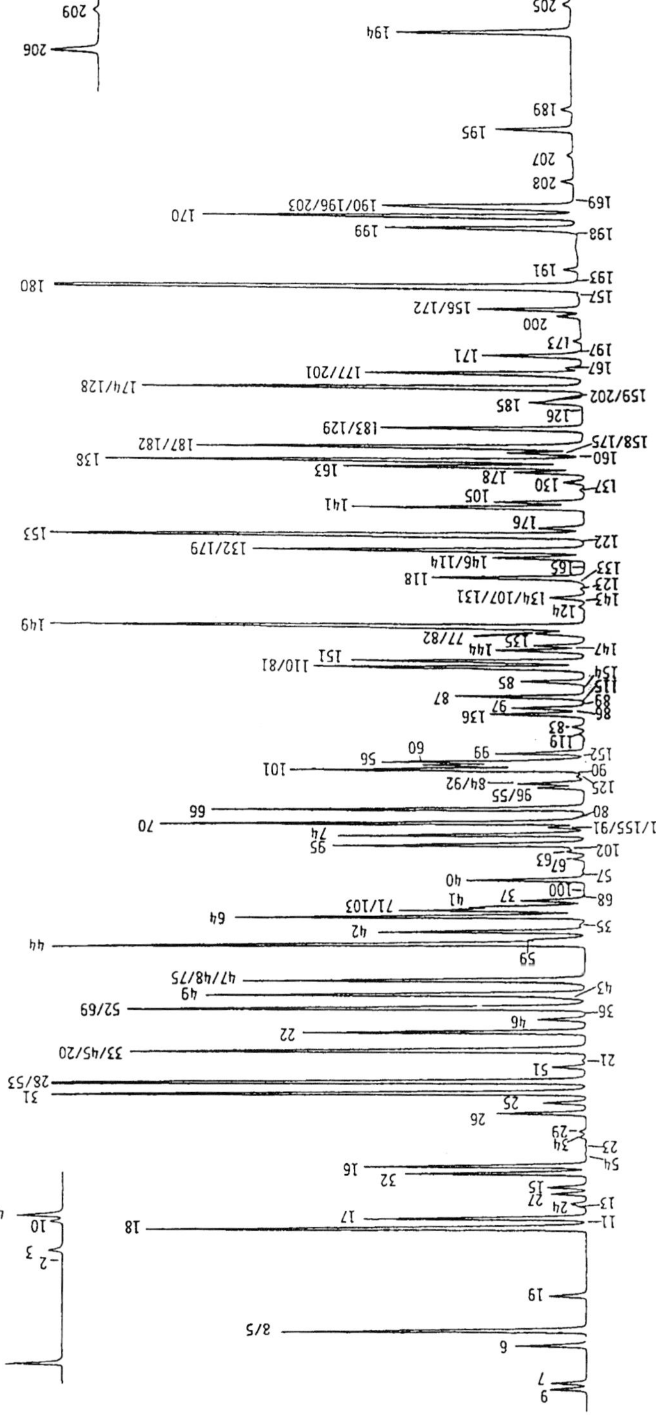

Fig. 7.5. ECD-chromatogram of an Aroclor mixture (A1016, A1232, A1248, A1260, 1:1:1:1:1) at an 60 m × 0.25 mm × 0.15 μm HT-8 (1,7-dicarba-closo-dodecacarborane phenyl methyl siloxane) column [67].

Its high temperature stability and its low bleeding open new possibilities for routine analyses of PCBs [67].

Haglund et al. [68] studied the elution sequence of the (+)- and (−)-enantiomers of nine stable atropisomeric CBs (84, 91, 95, 132, 135, 136, 149, 174 and 176) on a Chirasil-Dex (per-methylated β-cyclodextrin) column. In this study, the CBs 95, 132 and 149 were completely resolved on the Chirasil-Dex column (resolution (R) = 1.25; 1.50; and 1.25, respectively) and the CBs 84, 91, 135, 136, 174 and 176 were partially separated (R = 0.7–0.9). They concluded that this column was capable of separating nine of the 19 stable atropisomeric PCBs.

7.6.3.2 Multi-column GC

7.6.3.2.1 Parallel column GC. Parallel column GC is a method in which two columns are used in parallel and are connected via a glass T-split to a retention gap which is led into the injector. At the other end the columns are each connected to a detector. In this way the PCB analyses can be carried out on two different stationary phases at the same time [27]. An advantage of dual-GC is that only one GC-oven, one auto-sampler and one injector are necessary. Storr-Hansen [27] used a 60 m × 0.25 mm DB-5 (5% phenylmethylsilicone) and a 60 m × 0.25 mm DB-1701 (14% cyanopropylphenyl) for the parallel column GC determination of PCBs. They concluded that the parallel GC analysis on a CP-Sil-19 column gives a more satisfactory result for all toxic CBs than serial GC [6].

7.6.3.2.2 Serial GC. Bøwadt et al. [6,69] and Larsen et al. [70] used an in serial connected gas chromatographic column system. Two columns (CP-Sil-8 and HT-5) were coupled in series via a Chrompack quick-seal glass tube. Later, a combination of HT-5 with Sil-5 or with Sil-19 resulted in a separation of 129 and 127 congeners, respectively, by GC–MS [59]. Bøwadt and Johansson [69] used a method to analyse CBs in sulphur containing sediment by SFE and GC–ECD with a dual column system. They used a DB-17 (50% diphenyl dimethylsiloxane) (60 m × 0.25 mm) in parallel with a CP-Sil-8 (5% diphenyl dimethylsiloxane) (25 m × 0.25 mm) column and a HT-5 (1,7-dicarba-closo-dodecarborane dimethylsiloxane) (25 m × 0.22 mm) column in serial to obtain the best chromatographic separation [6,10,69]. According to Larsen et al. [70], the highest number of separated PCB congeners in dual column GC (104 with GC–ECD, 127 with GC–MS) can be obtained with this column combination. One hundred and fifty seven PCB congeners were separated in Aroclor with GC–MS. Detection limits obtained with a quadrupole mass spectrometer were around 15 pg. Frame et al. [71] used an HP-5 column instead of a CP-Sil-8 serial coupled with a HT-5 column. This column (HP-5) consists of a 5% diphenyl dimethylpolysiloxane phase. The two columns used by Frame et al. [71] were connected with a Restek press-tight fused-silica column connector. 47 CBs eluted uniquely on those in serial connected columns.

7.6.3.2.3 Multi-dimensional GC. The resolving power of capillary gas chromatography can be enormously increased by using MDGC. MDGC can be distinguished in two techniques, heart-cut MDGC and comprehensive MDGC. According to Hess et al. [10] MDGC is one of the most effective techniques to separate PCBs and to avoid interference from other congeners or other compounds. In MDGC normally a non-polar phase, such as SE-54 or CP-Sil-8 is used in the first column to make the initial, well characterised separation. The column flow is then switched for the duration of the elution of the peak(s) of interest only to the second column of a different, usually more polar character, e.g. CP-Sil-19 or CP-Sil-88. This technique is called heart-cut technique and has a precision of tenths of seconds [58]. During one run multiple heart-cuts can be made [8]. According to de Boer et al. [65] the use of longer columns and smaller internal diameters will improve the separation of the PCB congeners. The character of the two stationary phases will also strongly influence the final separation. De Boer et al. [58] reported a determination of seven mono-*ortho* CBs (60, 74, 114, 123, 157 and 167) using MDGC. A Siemens Sichromat 2–8 GC with two independently controlled ovens was used. Heart-cuts from the first column, an HP-Ultra 2 (5% phenyl–95% methylpolysiloxane), were transferred to the second column, an HP-FFAP (polyethylene glycol terephthalic acid ester), by use of a valveless switching technique, in which instead of the gas flow, the pressure drop along the coupling piece is used for column switching. These columns were selected on the basis of the relatively long retention times of mono-*ortho* CBs on the FFAP column, which should enhance resolution in combination with a SE-54 type stationary phase [58]. Kannan et al. [51,72] used a multidimensional GC system with a 25 m fused silica SE-54 (0.25 mm, 0.32 mm i.d.) as a first column and a 25 m fused silica OV-210 (0.25 mm, 0.32 mm i.d.) as second column. The columns were placed in a separate temperature-controlled oven and each connected to an ECD. A small fraction of the first column is cut from the eluate and transferred to the second column. This can be achieved efficiently, quantitatively and reproducibly. One hundred and thirty-two PCB congeners were detected an quantified from the technical mixtures by MDGC-ECD [72]. Blanch et al. [73] also used MDGC for the determination and quantification of the chiral CBs 95, 132 and 149 in shark liver. Duinker et al. [46,74] used MDGC–ECD for the separation of PCB congeners in the commercial mixtures Clophen and Aroclor. It appeared that congeners that were separated on an SE-54 column by only 0.01–0.1 min difference elute from the second column with peak separations of 0.5–1 min difference. Thus, each of the poorly separated CBs on the SE-54 column, can be completely separated on the second column. With MDGC all of the congeners in Aroclor and Clophen were determined [74]. Fig. 7.6 shows the chromatograms of the heart-cuts of the seven mono-*ortho*-PCBs of interest present in Aroclor. De Boer et al. [58] concluded that MDGC–ECD is recommended for the determination of the seven mono-*ortho* CBs in biological samples. The combination of an HP-Ultra and an FFAP column can be used to determine all these PCBs except CB 60. For CB 60 MDGC using a DB-5 and an SB-Smectic column is recommended [58]. The use of a heart-cut technique is relatively laborious when more peaks per run need to be checked for co-elution. When the speed of the secondary separation is high enough to separate a cut from the first separation while the next cut is being collected, it is possible to record a complete set of secondary chromatograms. From the secondary chromatograms the complete two-dimensional chromatogram can be constructed. A method capable of doing this is called comprehensive MDGC [75–77]. This technique is rather new and still in development. Recently, applications for PCBs have been shown [78,79]. The promising potential of comprehensive MDGC, also

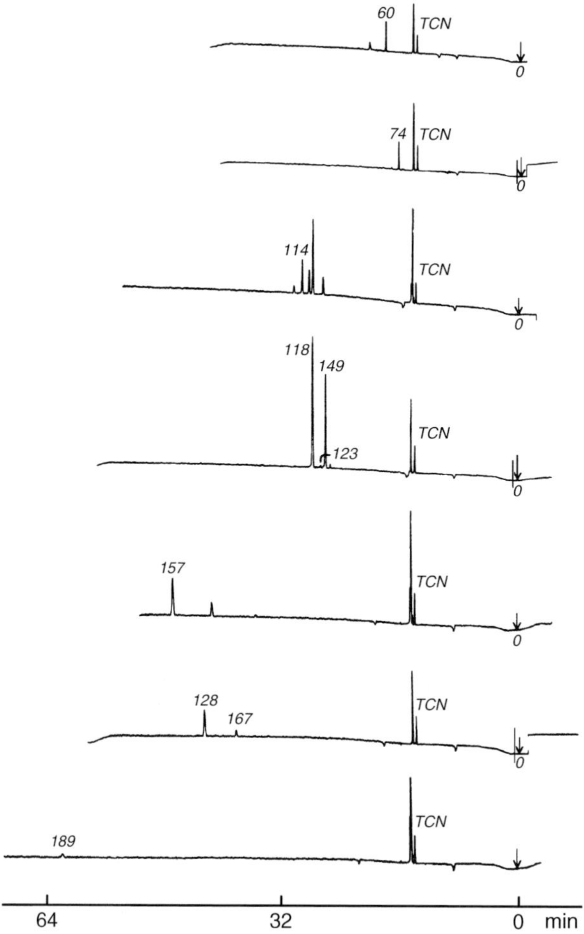

Fig. 7.6. Separate heart-cuts of PCBs 60, 74, 114, 123, 157, 167 and 189 from Aroclor 1254. TCN = tetra-chloronaphthalene (internal standard) [58].

called GCxGC, has now initiated a rapid development. The heart of the comprehensive MDGC is the modulator of which two types are now available: one based on a slotted heater (sweeper), the other one based on a cryogenic modulator [80–84]. The latter is meanwhile available in different versions, using either carbon dioxide or nitrogen as cooling agent. The peaks eluting from the comprehensive MDGC are extremely narrow, which improves the sensitivity, but which requires a detector which is able to respond very quickly to the analyte. FID is suitable, but lacks, of course, the sensitivity required for PCB analysis. A micro-ECD developed by Hewlett Packard is able to handle the GCxGC peaks and has enough sensitivity for PCBs. A time-of-flight (ToF)–MS may be the ideal detector as it couples a high selectivity to a rapid response, and, with the GCxGC, with a high sensitivity.

7.7 QUALITY ASSURANCE

A number of measurements should be taken to ensure a sufficient quality of the analysis. Smedes and de Boer [19] identified four main areas: calibrants and calibration, system performance, control of extraction and clean-up and control of long-term stability.

7.7.1 Calibrants and calibration

A CB determination should always be carried out using calibration solutions prepared from crystalline CBs. Preferably certified CBs should be used. A number of certified crystalline CBs can e.g. be obtained from the EU Measurements and Testing Programme (BCR), Brussels. Utmost attention should be paid to the preparation of calibration solutions. Always two independent solutions should be prepared simultaneously. There are a number of precautions that should be paid attention to in weighing and diluting calibration solutions of CBs. Detailed information can be obtained from Wells et al. [85]. Calibration solutions should preferably be stored in ampoules. It is recommended to use one or more internal standards, added in a fixed volume or weight to all standards and samples, to control the final concentrations step and the GC analysis. Preferably a CB, which does not occur in the samples and does not coelute with other CBs, should be used, for example CB 29, CB 112, CB 155, CB 198. Alternatively 1,2,3,4-tetrachloronaphthalene (TCN) or homologs of dichloroalkylbenzylether [86] can be used. For an extra check on the recovery of the extraction an extra internal standard should be added before extraction.

7.7.2 System performance

By regularly measuring the resolution of two closely eluting CBs, the performance of the system can be monitored. A decreasing resolution points to deteriorating conditions. The signal-to-noise ratio gives information on the condition of the detector. A dirty ECD detector can be recognised from a higher signal together with a lower signal-to-noise ratio [19].

7.7.3 Control of extraction and clean-up

For the control of extraction and clean-up it is recommended to let a standard solution pass the whole procedure, from the extraction to the final determination. This standard is used to determine the recovery of the sample series. Additionally, to all samples and to the standard solution mentioned above, an internal standard should be added to check whether in individual samples losses have occurred. It is recommended that this recovery-standard is not used to correct any data, but only for a control on the whole procedure. If major losses have occurred, results should not be reported. CB 29 is suggested as recovery standard. Because of its high volality, losses due to evaporation are easily detected. CB 29 elutes relatively late from alumina and silica columns. Therefore, also losses due to clean-up may easily be detected with this CB. Small peaks that may be present in the chromatogram at the retention time of CB 29 do not hinder the use of this CB because the recovery standard only controls major errors in extraction or clean-up [19].

7.7.4 Long-term stability

One internal reference sample should be included in each series of samples. This sample should be taken from a large, homogeneous batch of sediment that can serve as an internal reference material over a long period. A quality chart should be recorded [54]. In case the warning limits are exceeded, the method used should be checked on possible errors. When the alarm limits are exceeded, the results obtained should not be reported. It is recommended that at least annually a certified reference material is being analysed. On a regular basis, a laboratory analysing PCBs should take part in interlaboratory studies on the determination of CBs [19].

ACKNOWLEDGEMENTS

The assistance of Mr. Aschwin van der Horst in the preparation of this manuscript was much appreciated.

7.8 REFERENCES

1   M. Bolgar, J. Cunningham, R. Cooper, R. Kozloski and J. Hubball, Chemosphere, 31 (1995) 2687.
2   V. Lang, J. Chromatogr., 595 (1992) 1.
3   K. Ballschmitter, R. Bacher, A. Mennel, R. Fischer, U. Riehle and M. Swerev, J. High Resolut. Chromatogr., 15 (1992) 260.
4   R. Guitart, P. Puig and J. Gómez-Catalán, Chemosphere, 27 (1993) 1451.
5   J.R. Donnelly, A.H. Grange, N.R. Herron, G.R. Nichol, J.L. Jeter, R.J. White, W.C. Brumley and J.M. van Emon, J. Assoc. Off. Anal. Chem., 79 (1996) 953.
6   B. Larsen, S. Bøwadt and S. Facchetti, Int. J. Environ. Anal. Chem., 47 (1992) 147.
7   V. Ivanov and E. Sandell, Environ. Sci. Technol., 26 (1992) 2012.
8   C.R. Pearson, Halogenated aromatics, in: O. Hutzinger (Ed.), The Handbook of Environmental Chemistry, Vol. 3, Part B: Anthropogenic Compounds, Springer Verlag, Heidelberg, 1982, p. 90.
9   S.Y. Huetstis, M.R. Servos, D.M. Whittle and D.G. Dixon, J. Great Lakes Res., 22 (1996) 310.
10   P. Hess, J. de Boer, W.P. Cofino, P.E.G. Leonards and D.E. Wells, J. Chromatogr., 703 (1995) 417.
11   B. Larsen, S. Bøwadt, R. Tilio and S. Facchetti, Chemosphere, 25 (1992) 1343.
12   J. de Boer, Q.T. Dao, P.G. Wester, S. Bøwadt and U.A.Th. Brinkman, Anal. Chim. Acta, 300 (1995) 155.
13   A.K.D. Liem and R.M.C. Theelen, Dioxins, chemical analysis, exposure and risk assessment, Ph. D. Thesis, University of Utrecht, Utrecht, The Netherlands, 1997.
14   N. Wågman, B. Strandberg and M. Tysklind, Organohalogen Compounds, 35 (1998) 209.
15   R. Law and J. de Boer, Quality assurance of analysis of organic compounds in marine matrices: application to analysis of chlorobiphenyls and polycyclic aromatic hydrocarbons, in: Ph. Quevauviller

(Ed.), Quality Assurance for Environmental Monitoring, VCH Verlag, Weinheim, 1995, pp. 129–156.

16 J.C. Duinker, D.E. Schulz and G. Petrick, Chemosphere, 23 (1991) 1009.

17 T.R. Schwartz, D.E. Tillitt, K.P. Feltz and P.H. Peterman, Chemosphere, 26 (1993) 1443.

18 J. de Boer and P. Hagel, Sci. Total Environ., 141 (1994) 155.

19 F. Smedes and J. de Boer, Trends Anal. Chem., 16 (1997) 503.

20 H.P. Cleghorn, R.B. Caton, N.W. Groskopf and C.W. Pilger, Chemosphere, 20 (1990) 1517.

21 S. Tanabe, A. Nishimura, S. Hanaoka, T. Yanagi, H. Takeoka and R. Tatsukawa, Marine Poll. Bull., 22 (1991) 344.

22 L.J. Schmidt and R.J. Hesselberg, Arch. Environ. Contam. Toxicol., 23 (1992) 37.

23 J.J. Langenfeld, S.B. Hawthorne, D.J. Miller and J. Pawliszyn, Anal. Chem., 65 (1993) 338.

24 N. Alexandrou, M.J. Lawrence and J. Pawliszyn, Anal. Chem., 64 (1992) 301.

25 E.G. van der Velde, W.C. Hijman, S.H.M.A. Linders and A.K.D. Liem, Organohalogen Comp., 27 (1996) 247.

26 E.G. van der Velde, S.M.H.A. Linders, W.C. Hijman, J.A. Marsman, J.R.S. den Hartog den and A.K.D. Liem, Organohalogen Comp., 35 (1998) 1.

27 E. Storr-Hansen, J. Chromatogr., 558 (1991) 375.

28 W.M. Draper and S. Koszdin, J. Agric. Food Chem., 39 (1991) 1457.

29 R. Duarte-Davidson, V. Burnett, K.S. Waterhouse and K.C. Jones, Chemosphere, 23 (1991) 119.

30 E. Dewailly, H. Tremblay-Rousseau, G. Carrier, S. Groulx, S. Gingras, K. Boggess, J. Stanley and J.P. Weber, Chemosphere, 23 (1991) 1831.

31 P. Larsson, L. Collvin, L. Okla and G. Meyer, Environ. Sci. Technol., 26 (1992) 346.

32 D.E. Knowles, B.E. Richter, J. Ezzel, F. Höfler, D.S. Waddell, T. Gobran and V. Khurana, Organohalogen Comp., 23 (1995) 13.

33 C. Bandh, E. Björklund, L. Matiasson, C. Näf and Y. Zebühr, Organohalogen Comp., 35 (1998) 17.

34 A. Glausch, J. Hahn and V. Schurig, Chemosphere, 30 (1995) 2079.

35 C.R. Macdonald and C.D. Metcalfe, Can. J. Fish. Aquat. Sci., 48 (1991) 371.

36 P.H. Peterman and J.J. Delfino, Biomed. Environ. Mass Spectr., 19 (1990) 755.

37 K. Robards, Food Addit. Contam., 7 (1990) 143.

38 J. de Boer, Chemosphere, 17 (1988) 1803.

39 J. de Boer, Chemosphere, 17 (1988) 1811.

40 F. van der Valk and Q.T. Dao, Chemosphere, 17 (1988) 1735.

41 A. Perdih and J. Jan, Chemosphere, 28 (1994) 2197.

42 Ph. Hess, The determination and environmental significance of planar aromatic compounds in the marine environment, Ph.D Thesis, Robert Gordon University, Aberdeen, UK.

43 J. de Boer, P.G. Wester, H.J.C. Klamer, W.E. Lewis and J.P. Boon, Nature, 394 (1998) 28.

44 R.E. Rebbert, S.N. Chesler, F.R. Guenther, B.J. Koster, R.M. Parris, M.M. Schantz and S.A. Wise, Fresenius J. Anal. Chem., 342 (1992) 30.

45 M. Athanasiadou, S. Jensen and E. Klasson Wehler, Chemosphere, 23 (1991) 957.

46 J.C. Duinker, D.E. Schulz and G. Petrick, Chemosphere, 23 (1991) 957.

47 C.S. Creaser, F. Krokos and J.R. Startin, Chemosphere, 25 (1992) 1981.

48 J. Falandysz, N. Yamashita, S. Tanabe and R. Tatsukawa, Int. J. Environ. Anal. Chem., 47 (1992) 129.

49 T.R. Schwartz, D.E. Tillitt, K.P. Feltz and P.H. Peterman, Chemosphere, 26 (1993) 1443.

50 A. Kocan, J. Petrík, J. Chovancová and B. Drobná, J. Chromatogr., 665 (1994) 139.

51 N. Kannan, G. Petrick, D. Schulz, J. Duinker, J. Boon, E. van Arnhem and S. Jansen, Chemosphere, 23 (1991) 1055.

52 P. Haglund, L. Asplund, U. Järnberg and B. Jansson, Chemosphere, 20 (1990) 887.

53 P. Haglund, L. Asplund, U. Järnberg and B. Jansson, J. Chromatogr., 507 (1990) 389.

54 J. de Boer, C.J.N. Stronck, F. van der Valk, P.G. Wester and M.J.M. Daudt, Chemosphere, 25 (1992) 1277.

55 E. Grimvall, A. Colmsjö, K. Wrangskog, C. Östman and M. Eriksson, J. Chromatogr. Sci., 35 (1997) 63.

56 J. de Boer, J.C. Duinker, J.A. Calder and J. Meer van der, J. Assoc. Off. Anal. Chem., 75 (1992) 1054.

57 J. Staniewski, H.-G. Janssen, J.A. Rijks and C.A. Cramers, J. Microcol. Sep., 5 (1993) 429.

58   J. de Boer, Q.T. Dao, P.G. Wester, S. Bøwadt and U.A. Th Brinkman, Anal. Chim. Acta, 300 (1995) 155.
59   B. Larsen, S. Bøwadt and R. Tilio, Int. J. Environ. Anal. Chem., 47 (1992) 47.
60   S. Bøwadt, E. Frandsen, A. Weimann, B.R. Larsen and J. Møller, Proceedings of the Fifteenth Intern. Symposium Capillary Chromatogr., Riva del Garda, Italy, Vol. 1, 1993, p. 280.
61   P.G. Wester, J. de Boer and U.A.Th. Brinkman, Environ. Sci. Technol., 30 (1995) 473.
62   M.D. Mullin, C. Pochini, S. McGrindle, M. Romkes, S. Safe and L. Safe, Environ. Sci. Technol., 18 (1984) 468.
63   J. de Boer and Q.T. Dao, J. High Resolut. Chromatogr., 14 (1991) 593.
64   W. Vetter, B. Luckas and W. Haubold, Chemosphere, 23 (1991) 193.
65   J. de Boer, Q.T. Dao and R. van Dortmond, J. High Resol. Chromatogr., 15 (1992) 249.
66   S. Bøwadt, H. Skejø-Andresen, L. Montanarella and B. Larsen, Int. J. Environ. Anal. Chem., 56 (1994) 87.
67   B.R. Larsen, J. High Resol. Chromatogr., 18 (1995) 141.
68   P. Haglund and K. Wiberg, J. High Resol. Chromatogr., 19 (1996) 373.
69   S. Bøwadt and B. Johansson, Anal. Chem., 66 (1994) 667.
70   B. Larsen, M. Cont, L. Montanarella and N. Platzner, J. Chromatogr., 708 (1995) 115.
71   G.M. Frame, J.W. Cochran and S. Bøwadt, J. High Resol. Chromatogr., 19 (1996) 657.
72   N. Kannan, D.E. Schulz-Bull, G. Petrick and J.C. Duinker, Int. J. Environ. Anal. Chem., 47 (1992) 201.
73   G.P. Blanch, A. Glausch, V. Schurig, R. Serrano and M.J. Gonzalez, J. High Resol. Chromatogr., 19 (1996) 392.
74   J.C. Duinker, D.E. Schulz and G. Petrick, Anal. Chem., 60 (1988) 478.
75   H.-J. de Geus, J. de Boer and U.A.Th. Brinkman, Encyclopedia Environmental Analysis Remediation, 1998, John Wiley & Sons, New York, NY, p. 4909.
76   Z. Liu, S.R. Sirimanne, D.G. Patterson, L. Needham and J.B. Phillips, Anal. Chem., 66 (1994) 3086.
77   J. Blomberg, P.J. Schoenmakers, J. Beens and R. Tijssen, J. High Resolut. Chromatogr., 20 (1997) 539.
78   H.-J. de Geus, A. Schelvis, J. de Boer and U.A.Th. Brinkman, J. High Resolut. Chromatogr., 23 (2000) 189.
79   J. de Boer, H.-J. de Geus and U.A.Th. Brinkman, Organohalogen Comp., 45 (2000) 1.
80   Ph.J. Marriott and R.M. Kinghorn, J. Chromatogr. A, 866 (2000) 203.
81   Ph.J. Marriott, R.M. Kinghorn, R. Ong, P. Morison, P. Haglund and M. Harju, J. High Resolut. Chromatogr., 23 (2000) 253.
82   R.M. Kinghorn, Ph.J. Marriott and P.A. Dawes, J. High Resolut. Chromatogr., 23 (2000) 245.
83   E.B. Ledford and C. Billesbach, J. High Resolut. Chromatogr., 23 (2000) 202.
84   W. Bertsch, J. High Resolut. Chromatogr., 23 (2000) 167.
85   D.E. Wells, E.A. Maier and B. Griepink, Int. J. Environ. Anal. Chem., 46 (1992) 255.
86   D.E. Wells, M.J. Gillespie and A.E.A. Porter, J. High Resolut. Chromatogr., 8 (1985) 443.

W. Kleiböhmer (Ed.), *Environmental Analysis*
Handbook of Analytical Separations, Vol. 3

CHAPTER 8

Metal Species

Rolf-Dieter Wilken

ESWE-Institute for Water Research and Water Technology, D-65201 Wiesbaden, Germany

8.1 INTRODUCTION TO METAL SPECIES

Up to now, most chemical analysis aims at the determination of the total contents of metals, whereas the determination of organic molecules nowadays is carried out by both structure and behaviour in separation. Differentiation of the chemical forms of metals and elements in total is necessary in order to predict transport behaviour, to affect patterns of toxicity and to develop remediation in the case of contamination.

For species analysis, therefore, the whole classical analytical process must be revised. Speciation requires new analytical strategies for the determination of the bonding and a new philosophy of quality management in determination. Samples cannot be stabilized by preservatives: these change the speciation of the elements in a given matrix. Care must be taken to ensure that the bioactivity in the samples does not affect the speciation in a natural sample. The analytical tools for speciation analysis are a derivatization utilizing 'hyphenated methods', normally the combination of a chromatographic process with an atomic specific detector or a molecular identification after separation.

8.2 THE MEANING OF 'SPECIATION'

It is a long lasting question: what means speciation? The answer could be, as the IUPAC defines: "Speciation is the process yielding evidence of the atomic or molecular form of an analyte." In this chapter the term 'speciation' is used in its extended meaning: binding forms of elements, exactly definable or only operationally defined. As an example of this definition Table 8.1 is given.

8.3 SPECIATION OF ELEMENTS

Most of the elements known are able to form species, even the noble gases under special circumstances. The speciation normally depends on the oxidation state and the chemical surroundings of the central atom, which may be ionic or covalent.

References pp. 274–275

TABLE 8.1

PARENT SPECIES, SELECTED MATRIX SPECIES AND ONE OF THE MOST FREQUENTLY OCCURRING PARENT SPECIES OF THE ANALYTICAL SPECIES OF THE METHYL-MERCURY(II) (AFTER BERNHARD ET AL., 1986 [1])

Parent species	Matrix	Matrix species	Analytical species
CH_3Hg^+	Air	CH_3HgL (L = weak ligand)	
	Water	$CH_3Hg(OH)$	$\Rightarrow CH_3HgCl$
	Biol. tissue	CH_3Hg-S-protein	
	Sediment	CH_3Hg-humics	

There are many elements, which form species in the environment. Much research has been done on the Cr^{3+} and Cr^{6+} species, because it is well known that Cr^{6+} is carcinogenic. Organic compounds of Hg, Sn or Pb are more toxic than the inorganic ones; this is in contrast to the arsenic compounds. Much work on speciation is done in the field of bonding to humic substances. The difficulty with these agglomerates or compounds is, that they are poorly defined, but have a major influence on the transport behaviour in a natural ecosystem.

In speciation analysis two different definitions of speciation can be found. From the chemist point of view it is clear, that these are well-defined molecules normally with a metal atom to be regarded. This is not the view of an ecologist, where transport behaviour and toxicity are the most important points. These are often not depending on certain identifiable molecules, but to a group of different molecules with similar behaviour. So two different approaches to species identification are known and supported, which are outlined in the following.

8.3.1 Classically defined species

It is understood that covalently bound elements are species in its classical meaning. Elements, which are to be considered mainly, are arsenic, cadmium, chromium, copper, nickel, antimony, selenium, tin and lead. Mercury is an element with special properties, which forms not only insoluble (HgS) but also soluble (Hg^{2+}) and volatile compounds (Hg°, dimethylmercury). The organic species of mercury are mostly more toxic than the inorganic ones.

The determination of such well defined species is done generally by the scheme shown in Fig. 8.1, and consists of a separation procedure followed by the element specific detection. It is still a major problem in the analysis of metal species to get the undisturbed species into the detection system, especially, when the species are in a special balance with the matrix and the other compounds.

8.3.2 Operationally defined species

Species can also be described by their behaviour during extraction or chemical processes. This could be demonstrated with operationally defined mercury species in river

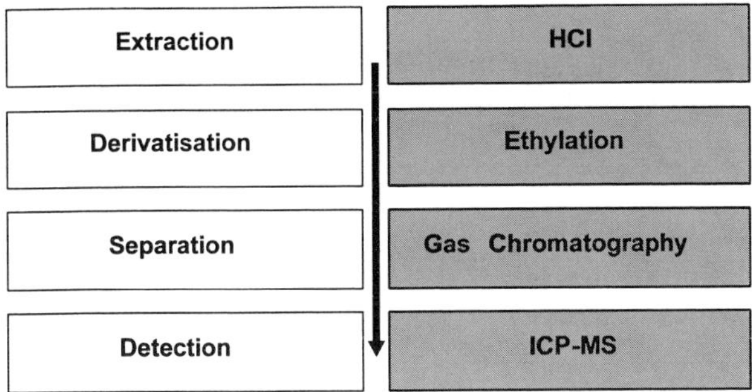

Fig. 8.1. Principle of the species analysis, explained with Hg species analysis.

systems. They can be described as 'volatile' when they can be purged by air or gases. For the chemist these operationally defined species are the sum of Hg° and dimethylmercury mainly. Another category are 'reactive' mercury species, when these species can be reduced by $SnCl_2$ to Hg°. These species are the sum of Hg°, Hg^{2+} and mercury, which could easily be detached from surfaces.

Looking at the transport behaviour of mercury through a soil passage, the importance of operationally defined species becomes clearer: in a first approach it is necessary to know the water-soluble amounts of mercury. A first result is given in Fig. 8.2.

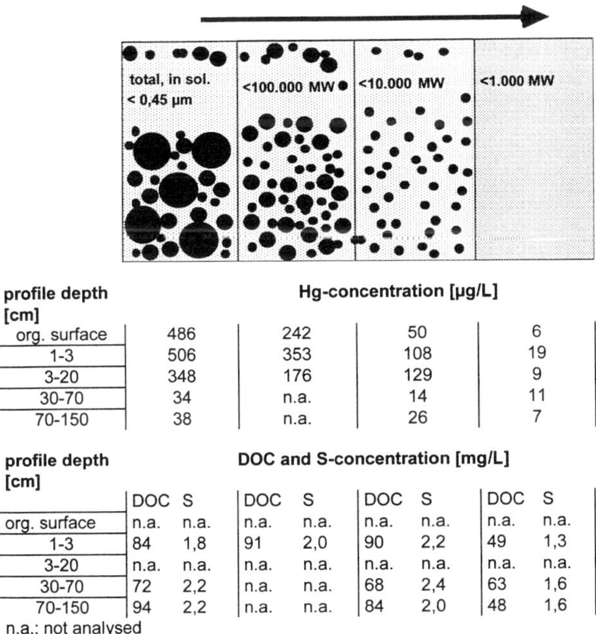

profile depth [cm]	Hg-concentration [µg/L]			
org. surface	486	242	50	6
1-3	506	353	108	19
3-20	348	176	129	9
30-70	34	n.a.	14	11
70-150	38	n.a.	26	7

profile depth [cm]	DOC and S-concentration [mg/L]							
	DOC	S	DOC	S	DOC	S	DOC	S
org. surface	n.a.	n.a.	n.a.	n.a.	n.a.	n.a.	n.a.	n.a.
1-3	84	1,8	91	2,0	90	2,2	49	1,3
3-20	n.a.	n.a.	n.a.	n.a.	n.a.	n.a.	n.a.	n.a.
30-70	72	2,2	n.a.	n.a.	68	2,4	63	1,6
70-150	94	2,2	n.a.	n.a.	84	2,0	48	1,6

n.a.: not analysed

Fig. 8.2. Operationally defined species with mercury compounds as an example: Hg penetration through a soil column depends on Hg-DOC (dissolved organic carbon) and sulphur bonds in bigger complexes.

References pp. 274–275

In this example the mercury species were extracted by mercury-free water (10:1, 1 h) in 5 separated slices of the soil column. The extract was filtered through a 0.45 mm glass fibre filter first and afterwards by ultrafiltration steps. In the whole profile the main transport of mercury in depth occurs by mercury complexes with a molecular weight < 1000 MW. Regarding also sulphur and DOC it can be concluded that there is no direct correlation between all these parameters, but between the molecular weight and the mercury, sulphur and DOC content.

In this example the water-soluble part is about 4% (14 mg from 320 mg Hg_{total}). For an evaluation, it is important to characterize the mercury compounds which could be eluted. Fig. 8.2 shows that most of the species are soluble in water; transportable in the column are to the lower extent only ionic and solvated compounds. To the higher extent they are bound to an organic matrix, with particle diameters lower than 1.5 μm and larger than 500 MW. These species are for some organisms probably more toxic than the solvated form.

Another determination of operationally defined compound classes is possible by electrochemical methods: so in some cases solvated or complexed species are separately determined before or after an oxidizing digestion.

Distinctions between species groups can be made by membrane separation, for which the experiment shown in Fig. 8.2 is an example. Another example is the penetration of methylmercury, and not ionic mercury, through certain porefree membranes [2].

8.4 DYNAMICS OF SPECIES

Some species are very stable, others are more labile. This depends on the strength of the bonding to other surroundings. Fig. 8.3 shows how rapidly the species of mercury can change from ionic mercury to volatile species under environmental conditions with the influence of bacteria [3]. Methylmercury, in the analytical process very stable, is

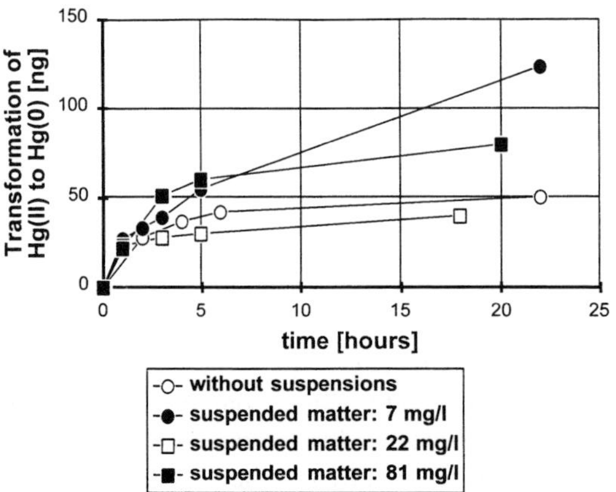

Fig. 8.3. Dynamic processes of species: change of ionic mercury to elemental mercury in Elbe river water; 750 ng $HgCl_2$ added, anaerobic conditions.

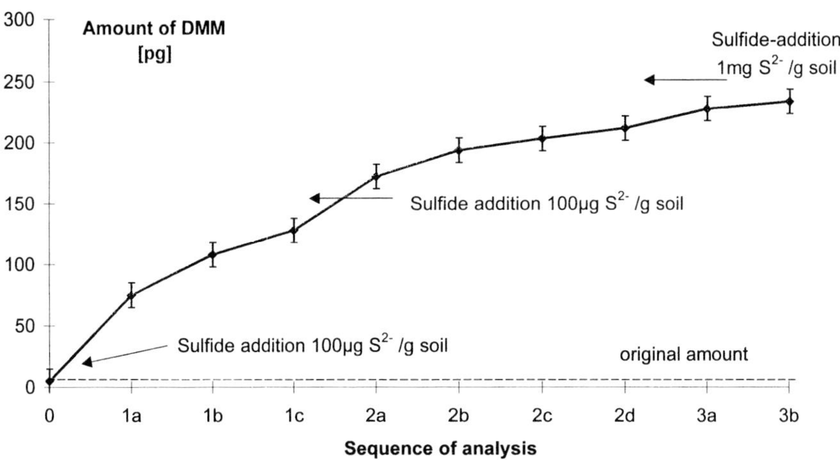

Fig. 8.4. Development of dimethylmercury from a contaminated floodplain soil.

degraded and build depending on chemical conditions of oxygen concentration and bacteria activity. The other example is the evaporation of mercury as dimethylmercury from contaminated floodplain soils after addition of water as shown in Fig. 8.4. This reaction is explained by Craig and Moreton through the equation: $2\ \text{MeHg}^+ + \text{S}^{2-} \Rightarrow (\text{MeHg})_2\text{S} \Rightarrow \text{HgS} + \text{Me}_2\text{Hg}$ [4], so that water under anoxic conditions is necessary for this reaction. Other changes are slower, but can also influence the species concentrations. So the matrix can change mercury species added to Elbe floodplain soil to other mercury species. This is shown in Fig. 8.5 [5]. In this experiment an extraction procedure was used, which extracts the soil or chemicals in the consecutive steps: water, acid (HNO_3, pH 2), alkaline water (KOH, 1 M), sulphidic water (Na_2S + KOH) and residue (HNO_3-digestion). This is also an example of the influence of the matrix on the species present in a given environment. Another example is the behaviour of atmospheric mercury, especially reactive gaseous mercury (RGM) [6] which is only <5% of the total mercury concentration in air, for modelling the fate of the element. There are convincing results from Stratton, Lindberg and Perry [6].

The dynamics of species change must be taken into consideration during sampling and pre-treatment of samples. In the next section an important example is given.

8.5 ERRORS DURING PRE-TREATMENT OF SAMPLES FOR SPECIES DETERMINATION

A special task of speciation analysis is quality control and the problem with reference materials. Michalke [7] and Adams [8] outlined it in recent publications. But there is not only a problem with QC, but also with dynamics of species during analysis.

A special example for dynamics of species during handling is the artificial build-up of methylmercury during sample preparation [9,10].

We added to a freshly taken sample of soil or sediment just before pre-treatment

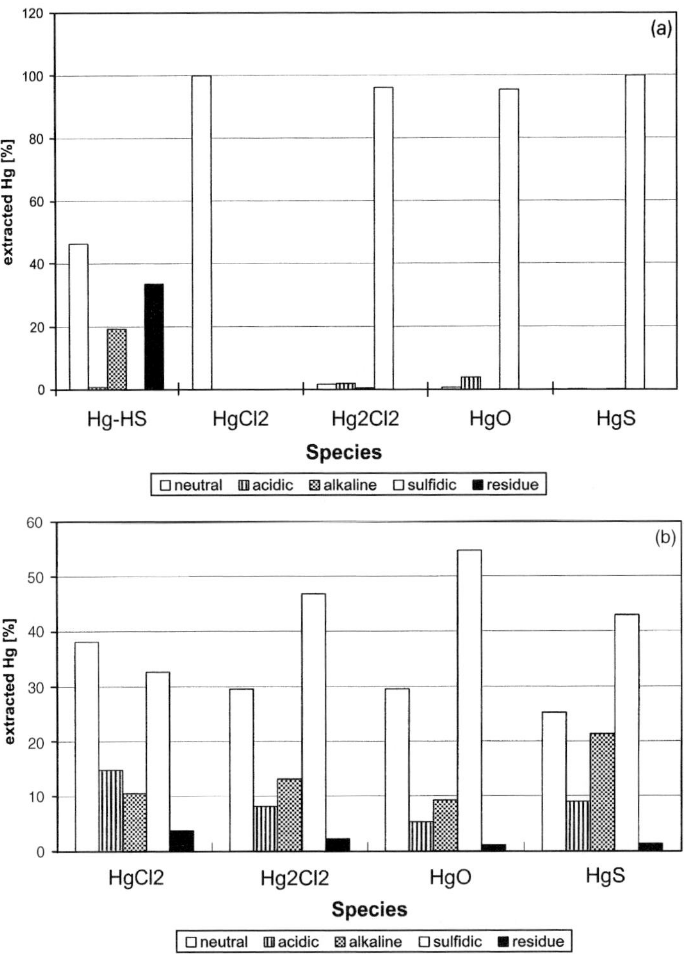

Fig. 8.5. Change of mercury species added to an Elbe river floodplain soil in about 50 h demonstrated with their solubility. Extraction steps: water, acid (HNO_3, pH 2), alkaline water (KOH, 1 M), sulphidic water (Na_2S + KOH) and residue (HNO_3-digestion). (A) The species themselves. (B) The samples after 50 h.

with HCl a spike of $^{203}Hg^{2+}$. As method the distillative extraction of methylmercury(+) was used. Unexpectedly we could measure CH_3-$^{203}Hg^+$ by ICP–MS. It is obvious that methyldonors in the sample are able to methylate Hg^{2+} instantaneously. The formation of methylmercury was up to 50% and more in comparison to the ambient methylmercury value. Therefore we postulated that under this determination protocol it is not easily possible to analyse methylmercury concentrations in the original sample correctly. This was cause for a dispute [11–13] in the scientific community, especially because CRM problems are involved.

However, this observation was not limited to sediment distillates. Other techniques and matrices prone to artifactual formation include:

- the methylmercury determination in flue gases resulted in methylation of Hg caused by a reaction of SO_2 with acetic acid, while traces of Fe(II) serve as a catalyst
- hot alkaline digestion
- use of sodium tetraethylborate as ethylation reagent lead to extra methylmercury formation owing to impurities of this compound; however, usually less than 0.01% of the Hg(II) present was converted into methylmercury
- supercritical fluid extraction gave rise to artificial methylmercury formation due to an inorganic mercury contamination
- upon distillation of a polar extract of an Everglades sediment artificial formation of ethylmercury was observed pointing to further natural alkylation processes.

We proposed a certain scheme called species-specific isotope addition correction method (SSIA) to re-calculate the correct amount of species in the sample. The correction is based on the assumption that the behaviour of the spiked isotope is the same as the ambient inorganic mercury in the sample. If the methylation behaviour is different, the real value could be higher when the spiked isotope was better methylated, or lower when the spiked isotope was methylated with poorer yields. This all is considered under the assumption that Hg^{2+} is available for methylation in the sample (Fig. 8.6) [10].

The artificial methylation of mercury ions was investigated by addition of stable mercury isotopes. The results in the ICP–MS are shown in Fig. 8.7. It could clearly be

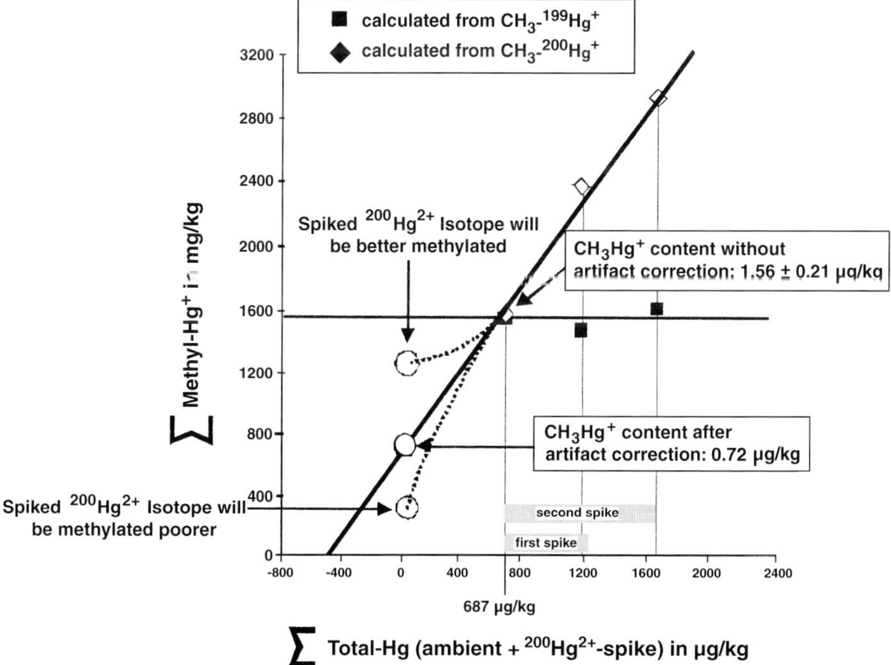

Fig. 8.6. Graphic determination of methylmercury in a river sample. The back extrapolation to zero content of total mercury gives the corrected methylmercury value.

References pp. 274–275

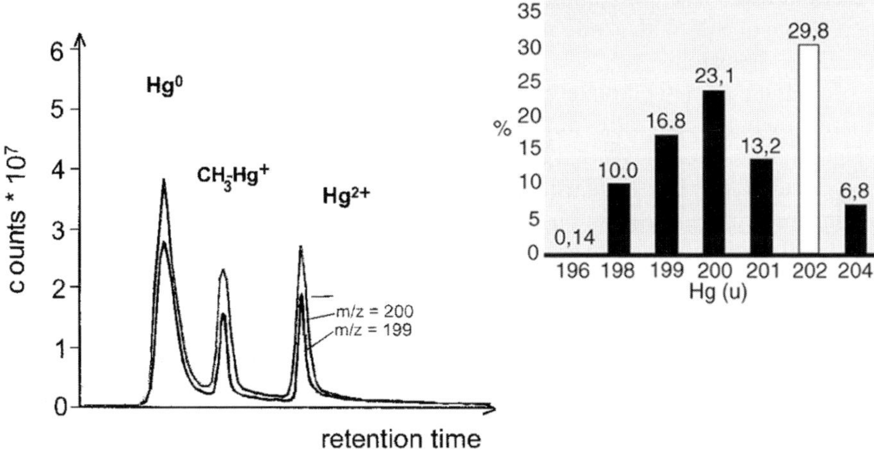

Fig. 8.7. Isotopes of mercury in the artificial mercury process.

seen, that stable mercury isotopes just added during the sample preparation process are artificially methylated.

8.6 ANALYTICAL TOOLS FOR SPECIES DETERMINATION

In most cases of species determination a separation of the different species can be achieved by chromatographic separation techniques. A derivatization of the different compounds should lead to thermal stability and an improved separation in gaschromatography and a better detectability in the detection system. An overview of the principal steps of species analysis was given in Fig. 8.1.

Often, a derivatization should be performed before separation. This is done for the improvement of volatilization in the case of a gaschromatography separation or to achieve better conditions for the HPLC separation. For many metal organyles often the alkylation for GC–AAS separation/detection [14], Fig. 8.8, is performed. In the case of difficult mercury species, derivatization by thioethanol, liquid separation and detection by atomic fluorescence is suggested [15]. The methods of species determination have been described in recently published books [16–19]. A promising new method is the coupling of GC with ICP/MS, by which the best determination limits for metal species are reported [20]. This method is shown in Fig. 8.9.

For the mercury separation in a gaschromatography system an ethylation is performed [21]; the detection is preferably the atomic fluorescence method afterwards. A new combination for the determination of many metallic species is the GC–ICP/MS combination. The ethylation is done by ethylmagnesiumchloride; the detection limits are in the range of 100 fg for tetraethyllead, 120 fg for diethylmercury and 50 fg for tetrabutyltin. The special development is the interface for the coupling of the gaschromatography system to the ICP torch, which must permit transport of the compounds without irreversible

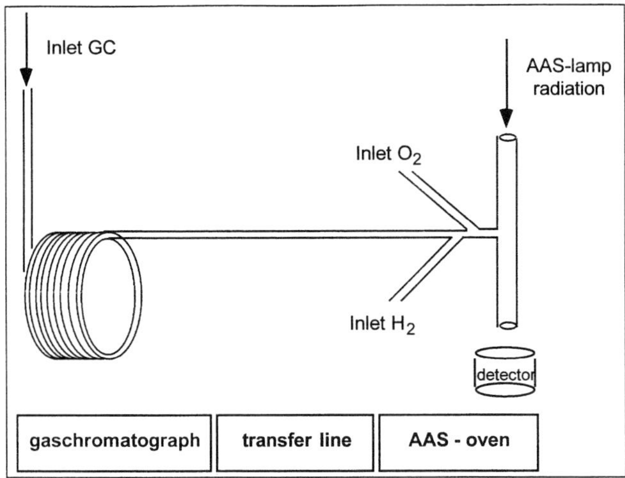

Fig. 8.8. Gaschromatography and atomic absorption coupled for element-specific determination of species.

adsorption or destruction in the transfer line to the detection system. The inner surface and the heating of the transfer line is the speciality in this combination [22]. The results obtained for the defined species of tin, mercury and lead are given in Fig. 8.10.

8.7 FUTURE ASPECTS

One of the main challenges for future species research is the determination of matrix species as defined in Table 8.1. Whereas the classically defined species and their analysis is state of the art in a concentration range sufficient for the evaluation of their impact in the ecosystem, the determination of matrix species, responsible for the behaviour of the species, is up to now not well understood. This includes also the dynamics of species changing in the environment. How is a certain compound bound in a natural surface water system? How rapidly does this change? What are the influencing parameters? What is the transport vehicle of a parent species through a river system, or, on another scale, through a membrane? This leads to the question of remediation of contaminated sites, where man can influence the degradation of toxic species to prevent further hazards to the ecosystem.

8.7.1 Measurement needed

The analysis of 'species of species', the operationally defined group of species, or as named in Table 8.1, the matrix species, is the new challenge for the analytical chemist. It is a challenge of its own, because the chemist wants clear conditions, whereas operationally defined compounds cannot be so well defined in this sense. The humic substances for example, are playing a substantial role in this field. They are not well

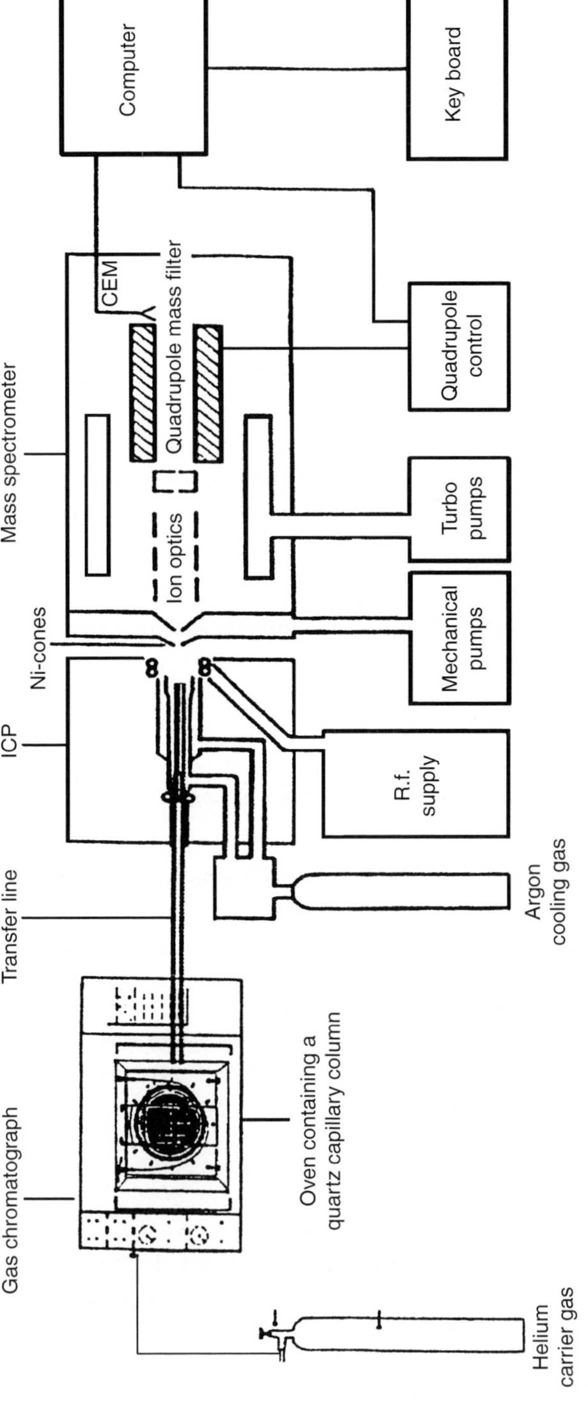

Fig. 8.9. Coupling of gaschromatography with inductively coupled plasma and mass spectroscopy.

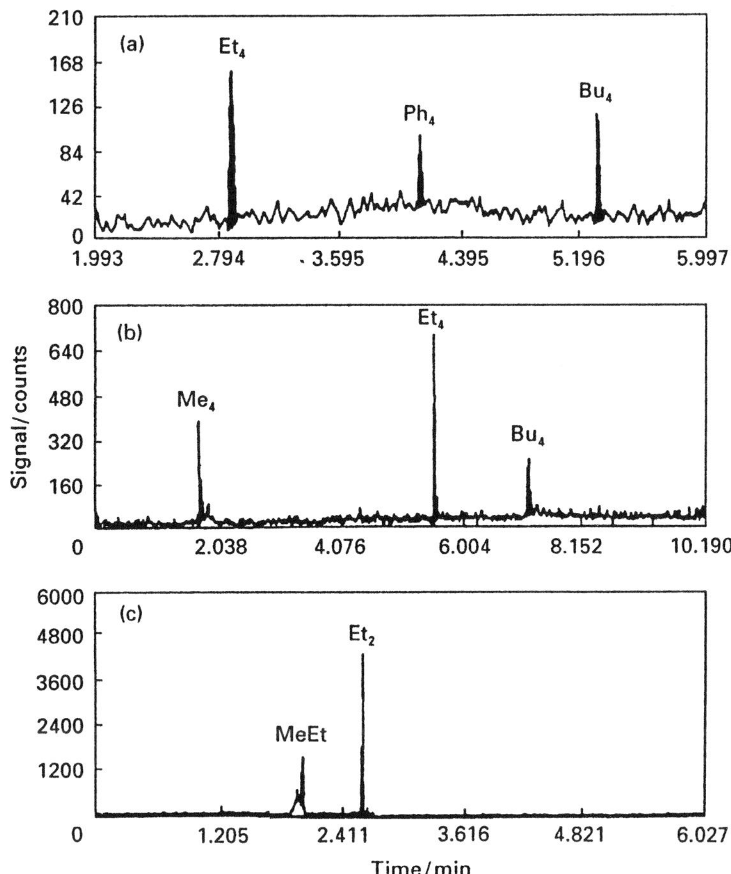

Fig. 8.10. Estimation of detection limits (LOD) of the GC–ICP/MS coupling for the determination of Sn (a), Pb (b) and Hg (c) compounds [17]. Added amounts: 150 fg Sn as Bu_4Sn, LOD (3 s) 50 fg Sn; 1000 fg Pb as Et_4Pb, LOD (3 s) 100 fg Pb; 10^4 fg Hg as Et_2Hg, LOD (3 s) 120 fg Hg.

defined because of their complex polymer structure but are important for so many transport processes as described for example in Fig. 8.2. A new approach would therefore be to develop an exact and reliable definition of separation of such operationally defined compounds. This is not easy because most of the factors influencing the species in its concentration and composition are up to now not well known.

A promising approach is therefore the elaboration of separation steps and the determination of relevant parameters on the composition and structure of these species.

The next step must be the development from one or two dimensional chemical analysis of species to the third dimension. This means to develop methods for the spatial structure of species. The problems with this kind of analysis are obvious: the concentrations are normally very low and the species in a different matrix, especially in the analytical procedure, very labile.

References pp. 274–275

8.7.2 Remediation approaches

The identification of harmful species in the ecosystem is one task; to change the threat is another. It is simple but often too expensive to clean up a contaminated site by excavation and deposition of the material in a closed system, e.g. underground storage. Under certain conditions, methods of extraction or biological remediation can be used. In some cases, and mercury decontamination is an example, thermal methods are possible [23]. The conditions under which these decontamination procedures are optimized under energy consumption aspects depend also on the species involved. In the case of mercury species, Hg° evaporates around 100°C, whereas HgS needs temperatures above 350°C [24,25]. It is therefore one of the aims of species research to reduce all mercury species to Hg° to get optimal conditions for the thermal process. This philosophy can also be developed for extraction methods, changing the species composition in such a way, that extractable species are dominant or in equilibrium with other species in that way, that the extraction is successful.

8.8 OUTLOOK

Species analysis is an important tool to predict behaviour of chemicals in the ecosystem and to develop remediation measures in the case of a contamination. It is therefore a new and promising approach to close the link between chemical analysis and environmental behaviour of compounds in the environment.

8.9 REFERENCES

1   M. Bernhard, F.E. Brinckman and K.J. Irgolic, Why 'speciation'? in: M. Bernhard, F.E. Brinckman and P.J. Sadler (Eds.), The Importance of Chemical 'Speciation' in Environmental Processes, Dahlem Konferenzen, Springer, Berlin, 1986, pp. 7–14.

2   R.-D. Wilken and H. Hintelmann, Analysis of mercury-species in sediments, in: J.A.C. Broekaert, S. Güçer and F. Adams (Eds.), Metal Speciation in the Environment, NATO ASI Series, Vol. G 23. Springer, Berlin, 1990, pp. 339–359.

3   R. Ebinghaus, H. Hintelmann and R.-D. Wilken, Mercury cycling in surface waters and in the atmosphere — species analysis for the investigation of transformation- and transport properties of mercury, Fresenius' J. Anal. Chem., 350 (1994) 21–29.

4   P.J. Craig and P.A. Moreton, Total mercury, methyl mercury and sulphide in River Carron sediments, Mar. Pollut. Bull., 14 (1983) 408–411.

5   D. Wallschläger, H. Hintelmann, R.D. Evans and R.-D. Wilken, Volatilization of dimethylmercury and elemental mercury from River Elbe floodplain soils, Water, Air, Soil Pollut., 80 (1995) 1325–1329.

6   W.J. Stratton, S.E. Lindberg and C.J. Perry, Atmospheric mercury speciation: laboratory and field evaluation of a mist chamber method for measuring reactive gaseous mercury, Environ. Sci. Technol., 35 (2001) 170–177.

7   B. Michalke, Quality control and reference material in speciation analysis, Fresenius' J. Anal. Chem., 363 (1999) 439–445.

8   F. Adams and S. Slaets, Improving the reliability of speciation analysis of organometallic compounds, Trends Anal. Chem., 19 (2000) 80–85.

9   R. Falter, Experimental study on the unintentional abiotic methylation of inorganic mercury during

analysis, Part 1. Localisation of the compounds effecting the abiotic mercury methylation, Chemosphere, 39 (1999) 1051–1073.

10 R. Falter, Experimental study on the unintentional abiotic methylation of inorganic mercury during analysis, Part 2. Controlled laboratory experiments to elucidate the mechanism and critical discussion of the species specific isotope addition correction method, Chemosphere, 39 (1999) 1075–1091.

11 N.S. Bloom, D. Evans, H. Hintelmann and R.-D. Wilken, Methylmercury revisited, Anal. Chem., 575 A (1999).

12 P. Quevauviller, personal communication, 1999.

13 R. Falter (Ed.), Sources of Error in Methylmercury Determination during Sample Preparation, Derivatisation and Detection. Selected Papers from the workshop in Wiesbaden–Schierstein, 27–29 May 1998, Chemosphere, 39 (1999).

14 R.-D. Wilken, J. Kuballa and E. Jantzen, Organotins: their analysis and assessment in the Elbe river system, Northern Germany, Fresenius' J. Anal. Chem., 350 (1994) 77–84.

15 R.-D. Wilken, Mercury analysis, a special example of species analysis, Fresenius' J. Anal. Chem., 342 (1992) 795–801.

16 A.M. Ure and C.M. Davidson, Chemical Speciation in the Environment, Blackie, London, 1995.

17 H.P. van Leeuwen and J. Buffle, Environmental Particles, Vol. II, Louis Publishers, Jersey, 1993.

18 J.R. Kramer and H.E. Allen (Eds.), Metal Speciation: Theory, Analysis and Application, Lewis Publishers, Chelsea, 1991.

19 J.A.C. Broekaert, S. Gücer and F. Adams, Metal Speciation in the Environment, NATO ASI Series, Ecological Sciences, Vol. 23, Springer, Berlin, 1990.

20 A. Prange and E. Jantzen, Determination of organometallic species by GC–ICP–MS, J. Anal. At. Spectrom., 10 (1995) 105–109.

21 N. Bloom, Determination of picogram levels of methylmercury by aqueous phase ethylation, followed by cryogenic gas chromatography with cold vapour atomic fluorescence detection, Can. J. Fish. Aquat. Sci., 46 (1989) 1131–1140.

22 A. Prange and E. Jantzen, Determination of organometallic species by gas chromatography inductively coupled plasma mass spectrometry, J. Anal. At. Spectrom., 10 (1995) 105.

23 R.-D. Wilken, M. Hempel and I. Richter-Politz, Mercury contamination and decontamination, Int. Conf. Heavy Metals in the Environment, Proc., CEP Consultants Edinburgh, Hamburg 1995, Vol. 2, pp. 42–51.

24 C.C. Windmöller, R.-D. Wilken and W. de F. Jardim, Mercury speciation in contaminated soils by thermal release analysis, Water, Air, Soil Pollut., 89 (1996) 399–416.

25 G. Bombach, K. Bombach and W. Klemm, Fresenius' J. Anal. Chem., 350 (1994) 18–20.

W. Kleiböhmer (Ed.), *Environmental Analysis*
Handbook of Analytical Separations, Vol. 3
© 2001 Elsevier Science B.V. All rights reserved

CHAPTER 9

Water quality

C. Zwiener and F.H. Frimmel

Engler-Bunte-Institut, Chair of Water Chemistry, Universität Karlsruhe (TH), Engler-Bunte-Ring 1, D-76131 Karlsruhe, Germany

9.1 INTRODUCTION

Water is the base of the existence of life on earth. This statement reveals the importance of the presence of water on our planet. Water is a major component of all organisms and can reach up to 98% of weight (Table 9.1). It is the matrix of all processes in living organism concerning transport, transformation, and excretion of inorganic and organic molecules. As a consequence, living organisms are dependent on the consumption of a certain amount of water and its ingredients (e.g. salts).

On a global view the total amount of water is the same today as it was one billion years ago and is estimated to about 1.41 billion km^3. This is an amount through which 70% of the earth's surface is covered. However, 97.3% of the water are salt water, additional 2% are trapped in the northern and southern polar regions and in glacier areas (Table 9.2). Only 0.7% or 9.71 million km^3 of the total amount of water is freshwater in the form of groundwater, rivers, and lakes as well as the humidity of atmosphere, biosphere, and pedosphere. Freshwater is in principle useful for the demands of living

TABLE 9.1

EXAMPLES FOR THE WATER CONTENT OF DIFFERENT ORGANISMS AND TISSUES

Organism	Water content (%)
Jelly fish	98
Snail	95
Human	60
muscles	70
bone	28
fat tissue	23
Insects	>50
Desert plants	2–40

References pp. 314–318

TABLE 9.2

WATER BALANCE AT THE SURFACE OF THE EARTH (ACCORDING TO [1])

	Portion	Volume (10^6 km^3)
Oceans	97.3	1370
Polar ice and glaciers	2	29
Freshwater	0.7	9.71
Total	100	1408.71

beings. In practice only a part of this amount is accessible in the form of shallow groundwater (4.2×10^6 km^3), lakes (0.125×10^6 km^3), and rivers (0.0017×10^6 km^3). The water of the different environmental compartments is connected by the hydrologic cycle, which has maintained the global amount of freshwater at our disposal at about the same level throughout the millennia. The global freshwater circulation can be considered as a huge water distillation plant with the oceans as sink, the atmosphere, the land masses as the distillation apparatus and the sun as the driving force and energy supply (Fig. 9.1). Over the oceans freshwater is evaporated (423,000 km^3/a), transported into the atmosphere and precipitated over the oceans (about 90%) and over the land (about 10% of the evaporated water). The total amount of precipitation over the land is much more than that and in the order of 110,000 km^3/a. The amount of 68,000 km^3/a is calculated for water permanently retained on and over the land masses as moisture in the atmosphere, soil and vegetation in the permanent cycle of evaporation from ground and water surfaces, transpiration from vegetation, and precipitation from the atmosphere. Balancing the global cycle the excess amount of water carried over from the evaporation of the oceans returns back to the oceans as rivers and groundwater runoff. The annual runoff from the rivers is about 20 times the volume of their water content.

In contrast to a limited and over a geological time period constant amount of freshwater there is the dramatically increasing population on earth over the last 150 years. From the first appearance of the intelligent human *Homo sapiens* about 50,000 years ago until the roman Caesar Julius (100 to 44 BC) the population reached 250 million. At about 1500 AD the population had increased to 500 million people, whereas

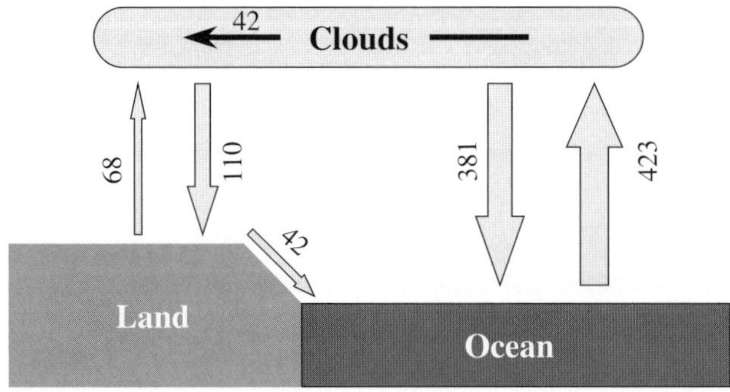

Fig. 9.1. Global water cycle. Indicated numbers represent the annual flows in 10^3 km^3/a.

TABLE 9.3

RENEWABLE AMOUNT OF WATER PER CAPITA AND YEAR FOR COUNTRIES WITH WATER SHORTAGE [3]

Country	Renewable amount of water per capita and year (m^3/a)
Botswana	797
Syria	615
Israel	372
Jordan	222
United Arab Emirates	188
Libya	154
Egypt	35

the first billion of population was reached in 1820. It took only 110 years to double this figure by the year 1930, and 30 years to add an additional billion by about 1960. Four billion people inhabited the earth 15 years later by about 1975, and five billion only 12 years later by about 1987. In 1999 we already reached a population of 6 billion people and by 2025 an estimated population in the order of 7.5 to 9.6 billion is expected. From this about 85% will live on the Southern Hemisphere. Therefore the major population growth will happen in the so-called developing countries.

However, the population growth is only one aspect of the increased demand for water, which grows many times faster than the increase in population. There has been an increased per-capita consumption of water since the beginning of the industrial age accompanied with a spectacular increase in gross national product.

The World Health Organization (WHO) states that an adult person weighing 60 kg should consume at least 2 l of water per day as drinking water or in the form of food. The daily consumption of drinking water per-capita in modern industrialized countries is in the order of 120 l to 350 l (e.g. 132 l in Germany in 1996 [2]). The whole water intake of public water supply, industry and power plants is in the order of 1500 l per-capita and day (e.g. 1650 l per-capita and day in Germany). For Germany the total amount of water intake can be calculated to about 600 m^3 per year and capita. Compared to the whole renewable amount of water in Germany of about 2000 m^3 per year and capita 28% are used for the different industrial and domestic purposes. When the renewable annual amount of water is below a level of 1000 m^3 per year the Food and Agricultural Organization (FAO) speaks of water shortage. In countries with a lower amount of water available there is a water shortage despite that the demand of drinking water is below 0.2% of the required 1000 m^3 per year and capita (Table 9.3). Mainly countries in the desert regions between 15° and 30° latitude of the Northern and Southern Hemispheres belong to that category.

In addition to the multiplied increase of water demand we have to deal with problems of industrial as well as domestic waste disposal. Due to the high percentage of water usage in industrialized as well as in developing countries the aquatic environment is heavily affected by human activities. The effects on rivers include for example an increase in temperature, the loading with nutrients (phosphate, nitrate), inorganic salts, heavy metals, and persistent organic chemicals.

References pp. 314–318

In conclusion an increasing earth population with an ever increasing demand for water will compete for the same water resources of ever poorer quality. The limited availability of drinking water is anticipated as a major reason for war in the future. However, a lot of examples reveal the positive impact of new technologies for water protection, like advanced waste water treatment, recycling of process waters, water-saving techniques in industry and households.

The task for the future for a sustainable water management is to maintain and restore the chemical, physical and biological integrity of waters with regard to different usage and under consideration of water as a natural habitat for animals and plants.

The sustainable way of water usage includes to comply with water quality standards concerning pollution emission, and to use only renewable water resources. The bases are legislative regulations and voluntary agreements in terms of threshold values for emissions and quality targets for the protection of water and its particular usage against hazardous substances. Furthermore, the distribution of water has to be improved in its original form or in the form of food.

9.2 DEFINITIONS

The term quality is well known in all areas of work and life, like quality of life, analytical quality assurance, quality management. Quality is derived from the Latin word 'qualis', which means how things are conditioned. The term quality is not independent from a certain scale or certain requirements, which define whether the required demands are fulfilled or not.

From the viewpoint of quality management quality can be defined according to the norm DIN EN ISO 8402 as "totality of attributes of a unit with regard to its suitability to fulfill defined and presupposed requirements". A unit can be a tangible or intangible product, an activity or a process, an organization, a system or a person.

That means in terms of water quality that the assessment has to be done according to certain defined and presupposed requirements, which were laid down concerning certain requirements of usage or of nature. Requirements on water quality can be part of legislative regulations or voluntary agreements.

Quality management also includes the permanent improvement of quality as an important aim of quality management systems. For water quality the adaptation of requirements to the state of the art of science and technology is mandatory. It has also to be mentioned that the quality can only be assessed by data of a certain quality. Therefore, analytical quality assurance is needed. This subject is for instance included in the drinking water ordinance. For the measurement of the single chemical parameters certain requirements are given for the precision, accuracy, and for detection limits of the analytical method.

9.3 POLICY AND LEGISLATIVE REGULATIONS

9.3.1 European Community policy

The environment of Europe is threatened by a growing extent of pollution. For instance some 2 billion tons of waste are produced in the member states every year. The quality of life especially for people living in European urban areas has declined considerably. Therefore, the protection of the environment is one of the major challenges facing Europe.

The environmental action of the Community began in 1972 with four successive action programs. The programs were based on a vertical and sectoral approach to ecological problems. About 200 pieces of legislation have been adopted concerning the limitation of pollution by introducing minimum standards for waste management, water pollution, and air pollution.

This legislative framework could not itself prevent pollution. It has become clear that concerted action at European and international levels is essential.

Community action developed over the time until the Treaty on European Community conferred to it the status of a policy. A further step was taken with the Treaty of Amsterdam, which includes the principle of sustainable development as aim of the European Community.

The principles of a European strategy of voluntary action for the period 1992–2000 are established by the Fifth Community Action Program on the Environment 'Towards Sustainability'. The beginning of the 'horizontal' Community approach was marked, and takes account of all causes of pollution (industry, energy, tourism, transport, agriculture, etc.).

This approach was confirmed by the 1998 Communication on integrating the environment into European Union policies and by the European Council in Vienna (11 and 12 December, 1998). The Community institutions are now obliged to take account of environmental considerations in all their other actions of policy [4].

Accompanied with the adopted framework legislation the Community has introduced environmental instruments. These are a financial instrument (the Life program) and technical instruments (eco-labelling, the Community system of environmental management and auditing, system for the assessment of the effects of public and private projects on the environment).

The European Environment Agency (EEA) was set up to compile and disseminate comparable environmental data. Its advisory role is important concerning adoption of new measures and assessing the impact of decisions already adopted [5].

Effective implementation of environmental legislation involves introducing incentives for economic operators. Now the introduction of environmental taxes (the 'polluter pays' principle), environmental accounting and voluntary agreements are being stressed.

Community policy on waste management involves three complementary strategies: (1) eliminating waste at the source; (2) encouraging recycling and reuse; (3) reducing pollution caused by waste incineration

At the international level this approach was adopted at the first Conference of the Parties to the *OSPAR Convention for the Protection of the Marine Environment of the*

North-East Atlantic. The parties adopted the position that dumping of offshore oil rigs and natural gas platforms should be banned and that the costs of dismantling and disposing of such installations should be borne by their owners. Furthermore, the community has already ratified the amendment of the *Convention on the Control of Transboundary Movements of Hazardous Wastes (Basle Convention)*, banning the exports of hazardous wastes from the OECD countries, the Community and Liechtenstein to non-OECD countries.

On water quality a number of directives have been adopted to introduce water quality standards (drinking water, bathing waters) and to monitor the emission of pollutants. Various international conventions aimed to protect the marine environment and watercourses are:

- the above-mentioned *OSPAR*,
- the *Barcelona Convention for the Protection of the Mediterranean Sea against Pollution*,
- the *Helsinki Convention on the Protection and Use of Transboundary Watercourses and International Lakes*, and
- the *Convention on Cooperation for the Protection and Sustainable Use of the River Danube*.

Further improving of the *ecological quality of surface waters*, introducing *Community action on fresh waters and surface waters* and *protecting Community estuaries, coastal waters and groundwater* are the aims of current proposals for directives.

Other environmental actions concerning noise pollution, air pollution, nature conservation, natural and technological hazards should not be considered here in detail (cf. [4]).

The most important regulations on water protection can be subdivided into three categories: emission-related regulations, quality-related regulations, and miscellaneous regulations. A more detailed description can be found elsewhere [6,7].

9.3.1.1 Emission-related regulations

Council Directive 76/464/EEC of 4 May 1976 (Off. J. L 129, 18.05.1976) on *pollution caused by certain dangerous substances discharged into the aquatic environment* of the Community. The objective of this directive is to harmonize the legislation of the Member States on discharges of certain hazardous substances into the aquatic environment. The aim is to terminate pollution with substances of list I and to reduce pollution with substances of list II. Amended by CD 90/656/EEC of 4 Dec. 1990 (Off. J. L 353, 17.12.1990) and CD 91/692/EEC of 23 Dec. 1991 (Off. J. L 31.12.1991).

Council Directive 80/68/EEC of 17 Dec. 1979 (Off. J. L 20, 26.01.1980) on the *protection of groundwater against pollution caused by certain dangerous substances*. The objective is to combat pollution by harmonizing the laws of the Member States on the discharge of certain dangerous substances into groundwater and by establishing systematic monitoring of the quality of such waters. Direct discharge of substances from list I are prohibited and discharge from substances of list II are limited. Amended by CD 90/565/EEC of 4 Dec. 1990 (Off. J. L 353, 17.12.1990) and CD 91/692/EEC of 23 Dec. 1991 (Off. J. L 377, 31.12.1991).

Council Directive 91/271/EEC of 21 May 1991 (Off. J. L 135, 30.05.1991) concerning *urban waste water treatment*. The aim is to protect the environment from any adverse effects due to discharge of urban and industrial waste waters. Member States are obliged to install collecting systems and sewage treatment plants for municipal waste waters and to implement tertiary treatment steps to reduce nitrogen and phosphorus in sensitive areas. Amended by 98/15/EC of 27 Feb. 1998 (Off. J. L 67, 07.03.1998).

Council Directive 96/61/EC of 24 Sept. 1996 (Off. J. L257, 10.10.96) concerning *integrated pollution prevention and control (IPPC)*. Measures are assigned for prevention and reduction of emissions in air, water and soil by certain industrial activities. Integrated environmental protection is the aim of this process-oriented approach.

9.3.1.2 Quality-related regulations

Council Directive 75/440/EEC of 16 Jun. 1975 (Off. J. L 194, 25.07.1975) concerning the *quality required of surface water intended for the abstraction of drinking water* in the Member States. Minimum quality requirements in the form of parameters (physical, chemical, microbiological) and limit values are set to be met by surface freshwater. Amended by CD 79/869/EEC of 9 Oct. 1979 (Off. J. L 271, 29.10.1979), CD 90/656/EEC of 4 Dec. 1990 (Off. J. L 353, 17.12.1990), CD 91/692/EEC of 23 Dec. 1991 (Off. J. L 377, 31.12.1991).

Council Directive 76/160/EEC of 8 Dec. 1975 (Off. J. L 31, 05.02.1976) concerning the *quality of bathing water*. Minimum quality criteria are set in the form of parameters (physical, chemical, microbiological) and mandatory limit values/indicative values to be met by bathing water. Amended by CD 90/656/EEC of 4 Dec. 1990 (Off. J. L 353, 17.12.1990) and CD 91/692/EEC of 23 Dec. 1991 (Off. J. L 377, 31.12.1991).

Council directive 78/659/EEC of 18 Jul. 1978 (Off. J. L222, 14.08.1978) on the *quality of fresh waters needing protection or improvement in order to support fish life*. Minimum quality criteria are set in the form of parameters (physical, chemical, microbiological) and their binding limit values and indicative values to designate freshwater suitable for fish-breeding. Amended by CD 90/656/EEC of 4 Dec. 1990 (Off. J. L 353, 17.12.1990) and CD 91/692/EEC of 23 Dec. 1991 (Off. J. L 377, 31.12.1991).

Council Directive 79/923/EEC of 30 Oct. 1979 (Off. J. L 281, 10.11.1979) on the *quality required of shellfish waters*. Members shall designate coastal and brackish waters to be considered as shellfish waters. Amended by CD 91/692/EEC of 23 Dec. 1991 (Off. J. L 377, 31.12.1991).

9.3.1.3 Miscellaneous regulations

Council Directive 80/778/EEC of 15 Jul. 1980 (Off. J. L 229, 30.08.1980) relating to the *quality of water intended for human consumption*. Imperative standards are set for parameters (organoleptic, physical, and undesirable chemical substances, toxic and microbiological) to be met by the quality of drinking water. Amended by CD 81/858/EEC of 19 Oct. 1981 (Off. J. L 319, 07.11.1981), CD 90/656/EEC of 4 Dec. 1990 (Off. J. L 353, 17.12.1990), CD 91/692/EEC by 23 Dec. 1991 (Off. J. L 377, 31.12.1991), CD 98/83/EC of 3 Nov. 1998 (Off. J. L 330, 05.12.1998).

Council Directive 91/676/EEC of 12 Dec. 1991 (Off. J. L 375, 31.12.1991) concerning the *protection of waters against pollution caused by nitrates from agricultural sources*. The aim is to reduce the amount of nitrate by implementing good agricultural practice, by limiting spreading fertilizers and manure.

Council Directive 91/414/EEC of 15 Jul. 1991 (Off. J. L 230, 19.08.1991) concerning the *placing of plant protection products on the market*. Uniform rules are laid down concerning the conditions and procedures for authorizing plant protection products and a positive list of active substances. Amended by CD 93/71/EEC of 27 Jul. 1993 (Off. J. L 221, 31.08.1993), CD 94/37/EC of 22 Jul. 1994 (Off. J. L 194, 29.07.1994), CD 94/43/EC of 27 Jul. 1994 (Off. J. L 227, 01.09.1994), CD 94/79/EC of 21 Dec. 1994 (Off. J. L 354, 31.12.1994), CD 95/35/EC of 14 Jul. 1995 (Off. J. L 172, 22.07.1995), CD 95/36/EC of 14 Jul. 1995 (Off. J. L 179, 29.07.1995), CD 96/12/EC of 8 Mar. 1996 (Off. J. L 65, 15.03.1996), CD 96/46/EC of 16 Jul. 1996 (Off. J. L 214, 23.08.1996), CD 96/68/EC of 21 Oct. 1996 (Off. J. L 277, 30.10.1996).

9.3.2 German policy

During the end of the sixties and the beginning of the seventies water pollution reached an alarming degree in Germany. Due to its climatic favored location Germany has sufficient water resources. It is only a matter of distribution especially in certain regions and densely populated urban areas. However, the high population density and the high degree of industrialization require the improvement of water quality and the protection of water resources.

Water management policy is based on the following principles: (1) the priority of the precautionary principle; (2) the cooperation of all participants; and (3) the polluter pays principle.

The responsibility for the environment does not end at the national border lines. Therefore, international cooperation for the protection of inland waters as well as coastal and marine waters is one of the focal points of environmental policy. The federal government is liable for reports to the European Community [8].

The objective of water management concerns the Federal Government which enacts directives of water framework. Important regulations are given by the single Federal States (Länder) amending the framework directives and making them more precise. For instance, subjects like the property of water, the supervision of water, the conditions and procedures for authorizing of water usage and discharge into sewage treatment are governed by the Federal states [7].

Water conservation actions are part of the environmental policy, set by the Federal Ministry of Environment, Nature Protection and Reactor Safety (Bundesministerium für Umwelt, Naturschutz und Reaktorsicherheit). This ministry is in charge of the Framework Directive of Water Conservation, the Act of Waste Water Taxes, the Act of Detergents and Cleansing Agents. It has the national responsibility for the Directive for Water Protection of the EC, the Marine Preservation and for the Comity of River Basins of Transboundary Waters.

The main partners of the Federal Environmental Ministry concerning water management and water preservation are the following.

- The Federal Ministry of Food, Agriculture and Forestry (Bundesministerium für Ernährung, Landwirtschaft und Forsten), e.g. responsible for water conservation tasks concerning flood and coastal protection, legislation for pesticide and fertilizer control.

- The Federal Ministry of Health (Bundesministerium für Gesundheit), responsible in the field of drinking water supply and the quality of drinking and bathing water.
- The Federal Ministry of Traffic (Bundesministerium für Verkehr), responsible for the interests of shipping on inland and marine waters, pollution protection of coastal waters.
- The Federal Ministry of Education, Science, Research and Technology (Bundesministerium für Bildung, Wissenschaft, Forschung und Technologie), which coordinates the science promotion of the Federal Government, controls fundamental and applied research, and technological development in water research and technology.
- The Federal Ministry of Economy (Bundesministerium für Wirtschaft), which represents the economic affairs.
- The Federal Ministry of Economical Cooperation and Development (Bundesministerium für wirtschaftliche Zusammenarbeit und Entwicklung), engaged in fundamental affairs and the coordination of the German cooperation of development.

9.3.2.1 Legislative instruments of water conservation

In the following, important legislative instruments with regard to German acts on water protection and management are reported.

- The Water Protection Act (Wasserhaushaltsgesetz WHG from 1957, last amendment 12. Nov. 1996; BGBl. I p. 1695) is a Framework Directive and includes the basic regulations concerning surface waters and coastal waters. The aim is to protect water as part of nature and as natural environment for animals and plants. All kinds of usage of waters underlay an authorization. Authorizations, e.g. for the discharge of waste water, can be given by the according water authorities of the Federal States, if a minimum of requirements (state of the art of technology) are fulfilled.
- Waste water Taxing Act (Abwasserabgabengesetz AbwAG of 1976; last amended 3. Nov. 1994, BGBl. I p. 3370, 1996 p. 1690, 1997 p. 582, 1998 p. 2455). It appoints taxes to be paid for the direct discharge of pollutants into surface waters. The tax is charged due to amount and harmfulness of discharged waste constituents and should set up economic incentives.
- The Groundwater Ordinance (Grundwasserverordnung of 1979; last amended 18. Mar. 1997, BGBl. I 1997 p. 542) is the implementation of the legislation for the Directive 80/68/EEC and is aimed to protect the groundwater from pollution with certain list I and list II substances.

The major acts of the Federal government are only general outlines. Therefore, important orders are also contents of the Federal States regulations. We have to speak about the following acts concerning water protection and water quality.

- The Act for Detergents and Cleansing Agents (Wasch- und Reinigungsmittelgesetz of 1975, last amended in 1994, BGBl. I p. 875, ... p. 1440 1994) states requirements on the environmental digestibility of detergents. The use of environmental hazardous substances can be banned or limited.
- On the base of the Fertilizer Act (Düngemittelgesetz of 15.11.1977, BGBl. I p. 2134 1977; last amended 27.9.1994, BGBl. I p. 2705) the Act on Good Fertilizing Practice (Verordnung über die Grundsätze der guten fachlichen Praxis beim Düngen (Düngeverordnung of 26.01.1996, BGBl. I p. 118) was issued. An improved protec-

tion of waters from diffuse pollution, especially from nitrate of agricultural sources, should be achieved.

The major requirements on drinking water quality can be found in the Epidemic Control Act.

- The Federal Act of Epidemic Control (Bundes-Seuchengesetz of 18.12.1997, BGBl. I p. 2263 1979) is an act to prevent and control human epidemics. It includes major requirements for the quality of drinking water and hygienic requirements on the removal of waste water.
- The Drinking Water Ordinance (Trinkwasserverordnung TrinkwV of 05.12.1990, BGBl. I p. 2612 1990) is based on the Federal Act of Epidemic Control and the Food and Commodity Act (Lebensmittel- und Bedarfsständegesetz of 08.07.1993, BGBl. I p. 1169 1993) and includes requirements on drinking water and on water for food industry.

9.3.2.2 Further tasks for water management

The technical requirements and long-term objectives of water management and all its aspects in an integral approach were set by the German Länder Working Party on Water (Länderarbeitsgemeinschaft LAWA) in a paper with the title Requirements for a Progressive Water Pollution Control Policy — LAWA 2000 — [9]. The central tasks of water conservation concerning the current medium-term objectives were presented in the National Water Conservation Plan [10]. In this paper the features of Germany's natural environment, the hazards threatening water, and the current standard of water conservation were taken into account.

The tasks for the future are set by the Federal Environmental Agency (Bundesumweltministerium [8]). The assignment of different tasks concerning water management can be subdivided into the following topics: (1) water supply and groundwater protection; (2) sewage treatment; (3) target for water quality; (4) water pollution from agricultural activities; (5) flood prevention and coastal conservation; (6) shipping; (7) transport of hazardous substances; (8) leisure usage of water.

9.3.2.2.1 Water supply and groundwater protection. The activities aim to preserve the water supply concerning quantity and quality of available water resources. This means a sparing usage of water and further improvement of water quality. From this point the following needs are resulting:

- development of water-saving production technologies;
- application of closed water cycles in industry;
- reduction of water losses in transport pipelines;
- reduction of water consumption during agricultural irrigation;
- remediation of waste water sewers;
- information on water-saving technologies in the households;
- removal of pathogens from communal and other waste waters before discharge.

During the next years special attention will be paid to the protection of groundwater as the most important resource for drinking water supply. Groundwater extraction is only authorized if the amount is balanced by natural renewable groundwater. Land use should be done appropriate to good farming practice.

TABLE 9.4

EXAMPLES FOR QUALITY TARGETS FOR THE PROTECTED GOODS DRINKING WATER SUPPLY (DWS) AND AQUATIC ECOSYSTEMS (AES) [11]

Substance	Quality targets for		Limit of determination (μg/l)
	AES (μg/l)	DWS (μg/l)	
Dichloromethane	10	1	1 ... 10
Tetrachloromethane	0.8	1	0.001 ... 1.0
Tetrachloroethylene	40	1	0.001 ... 0.2
Hexachlorobenzene	0.01	0.1	0.0009 ... 0.01
Nitrobenzene	0.1	10	0.02 ... 0.5
2-Nitrotoluene	50	10	0.02 ... 0.5
4-Chloroaniline	0.05	0.1	0.02 ... 1.0

9.3.2.2.2 Sewage treatment. To improve the quality of surface waters there are set strict demands for waste water treatment in Germany according to the precautionary principle: (1) further waste water purification to reduce nutrients such as nitrogen and phosphate; (2) the requirement to avoid and clean waste water according to the most up-to-date technology.

9.3.2.2.3 Targets for water quality. Despite that effluent discharges are only authorized if they are harmless or after passing through a powerful, top-performance effluent purification system (emission principle), an additional test has to ensure that it cannot damage the waters concerning several kinds of usage (pollution effect principle). Furthermore, surface waters are not only adversely affected by industrial and municipal waste waters but also by diffuse sources like run-off from agriculture, traffic, and precipitation. To control the ecological water quality so-called quality targets are set (emission principle). The quality targets are threshold values (concentrations) for hazardous substances relating to the protected goods (e.g. water ecosystems, drinking water supply, fishing) and should not be exceeded (Table 9.4). Based on the recent scientific knowledge an impairment of the according goods can be excluded by observing the quality targets [8].

9.3.2.2.4 Water pollution from agricultural activities. Agricultural activities are one of the major reasons of water pollution from diffuse sources. Among the nutrients entering the waters about 50–55% of all nitrogen and 40–45% of all phosphorus are from agricultural areas. The water pollution from agriculture can be attributed to erosion and to run-off of soil particles with attached phosphate and pesticides in surface waters as well as leaching of nitrate and pesticides to groundwater.

The improvement of the water quality should be one aim of sustainable farming. The main legal stipulations can be found in the Directives on Water Management (Wasserhaushaltsgesetz), on Fertilizing (Düngemittelgesetz), on Plant Protection (Pflanzenschutzgesetz) as well as in EC Directives on Pesticides and on Nitrate.

To reduce the burden of fertilizers and pesticides land has to be used appropriate to local conditions and careful husbandry. Soil erosion has to be prevented, fertilization has to be done in line with the needs of the plants, integrated plant protection schemes and cultivation methods based on good farming practice have to be applied, marginal strips along surface waters have to be installed, and farmers have to be trained and educated [9].

9.4 THE CURRENT SITUATION

9.4.1 Europe

For the June 1998 Ministerial Conference in Aarhus, Denmark, the European Environment Agency (EEA) prepared the report 'Europe's Environment, the Second Assessment'. The data were collected by the EEA through its European Topic Centers and from several organizations and institutions (Eurostat, UN, OECD, WHO, IEA, European Commission). The report gives a general overview on the situation and the development of the environment in Europe including inland waters, marine, and coastal waters from 1990 to 1995.

9.4.1.1 Gross development product

The economic development as a mirror of industrialized societies can be expressed by the gross development product (GDP). GDP grew in Western Europe by about 1.4% per year between 1990 and 1995, in Central and Eastern Europe and the NIS [1] GDP fell by 32% between 1990 and 1994. There are great differences between the countries (from an 11.8% fall in Ukraine to a 9.5% growth in Albania). In the west and the east the services sector is currently showing the fastest growth.

9.4.1.2 Chemicals

Europe is the largest chemical-producing region in the world (37% of global turnover). In Western Europe the production grew by 12% between 1993 and 1996 compared with a 7% growth in GDP. 100,106 chemical compounds are listed in the European Inventory of Existing Chemical Substances, with 20,000 to 70,000 substances on the market and several hundred new substances each year. For around 25,000 substances there are toxicity data available, about 10,000 are included in the IUCLID European Chemicals Bureau database for risk assessment, but only for 10 of these a risk assessment was completed in 1997.

---

[1] Western Europe (EU + EFTA):Austria, Belgium, Denmark, Finland, France, Germany, Greece, Ireland, Italy, Luxembourg, the Netherlands, Portugal, Spain, Sweden, United Kingdom, + Iceland, Liechtenstein, Norway and Switzerland.Central and Eastern Europe (CEE) (all Central European Countries, the Baltic States, Turkey, Cyprus and Malta):Albania, Bosnia–Herzegovina, Bulgaria, Czech Republic, Croatia, Estonia, FYROM, Latvia, Lithuania, Hungary, Poland, Romania, Federal Republic of Yugoslavia, Slovak Republic, Slovenia, + Turkey, Cyprus and Malta.The European Newly Independent States (NIS)Armenia, Azerbaijan, Belarus, Georgia, Moldava, Russian Federation, Ukraine.

9.4.1.3 Solid waste

In Europe about 4 billion tons of solid waste are produced per year (5 tons per person). The contribution of different sectors is: agriculture 37% (+2% between 1990 and 1995), mining 33% (+7%), manufacturing 19% (−8%), municipal 7% (+11%) and energy 3% (−17%). In OECD European countries the annual municipal waste generation increased to 200 million tons (420 kg per person per year) in 1995. Hazardous waste production is about 42 million tons per year for the mid-1990s, 38% of it from Germany and France.

9.4.1.4 Inland waters

More than half of the 28 European countries have low (2000–5000 m^3 per capita per year), 5 have very low (less than 2000 m^3 per capita per year) availability of freshwater. Since 1980 there has been a general reduction in total water abstraction in several countries. In Eastern Europe water abstraction has fallen by 20% since 1990, with only small reductions or no change in the other European countries. Mainly industrial abstraction has been falling since 1980 in most countries. However, in urban areas the demand of water may still exceed availability and water shortages may occur. In Western and Northern Europe households are the largest users of abstracted water (between 20% and 50%). In Southern Europe 60% of the total abstracted water is used for irrigation. In 1994 8.3% of the land area was irrigated in Southern Europe. In some regions, groundwater abstraction is exceeding the recharge rate and therefore water management is beyond sustainability.

Water quality in rivers has improved since the 1970s, especially in the most polluted rivers. However, there has been no overall improvement of river water quality since 1989/1990 despite the introduction of water quality targets in the EU and the consideration of water quality in the Environmental Action Program for Central and Eastern Europe.

Emissions of phosphorus, heavy metals and organic matter (measured as biochemical oxygen demand BOD) are falling in many parts of Europe: phosphorus emissions by 30% to 60% since the mid-1980s, and heavy metal emissions by 60% to 70% since about 1985, 35% of all river measuring stations had an average BOD concentration of clean rivers (less than 2 mg O_2/l), but 11% had yet an average BOD concentration of polluted water (greater than 5 mg O_2/l). Nitrogen is less a problem of rivers than the marine environment. Therefore, emissions have to be further controlled to protect the marine environment.

Groundwater is still affected by increasing concentrations of nitrate and pesticides, mainly from agriculture. Nitrate concentrations are low in Northern Europe, but high in several countries of Western and Eastern Europe. Eight countries report on nitrate concentrations exceeding the EU drinking water ordinance level of 5.6 mg N/l in a quarter of all sampling sites. Pesticide concentrations above the maximum admissible concentration for drinking water (0.1 µg/l) are found in one third of the reporting countries.

The water quality in lakes has improved. However, eutrophication remains a significant problem despite that the proportion of lakes with phosphorus concentrations above 500 µg/l has halved since the 1970s.

References pp. 314–318

9.4.1.5 Marine and coastal waters

High nutrient and pollutant concentrations, over-fishing and pressure of development on the coasts are the most serious threats to the European seas. Eutrophication is still a problem in European seas, particularly in the North Sea, the English Channel, the Atlantic coast of France, the Baltic Sea, the Black Sea and locally the Mediterranean Sea. Since the beginning of the 1990s nutrient concentrations are generally unchanged. On top of that in the Black Sea nutrient concentrations increased between 1960 and 1992 by a factor of 7 for nitrates and a factor of 18 for phosphates.

Sediments and biota especially in coastal regions and estuaries are contaminated with anthropogenic substances. Hot-spots of heavy metal contamination are the mouth of the River Rhine, the Oslofjord and the area near Gothenburg, parts of the Greek coast, the northwestern Mediterranean coast as well as the east Adriatic coast.

9.4.2 Germany

9.4.2.1 General overview

Germany has a relative high population density of 230 inhabitants/km^2 or 82.012 million inhabitants on an area of 356,970 km^2. Despite the population density and a high level of industrialization 54.1% of the area are used for agriculture, 29.4% are forests, 11.8% are areas used for settlement and traffic with an increasing tendency (+4.3% between 1993 and 1997). About 50% of the area for settlement and traffic is covered with buildings or sealed. Low portions can be attributed to 2.2% water surfaces and 2.6% other (mostly natural areas). 10.4% of the area are water conservation zones.

The GDP has increased threefold in the last 35 years and reached 3014 billion DM in 1995. It increased by 5% in the western part and by 39% in the eastern part of Germany between 1991 and 1996. The energy consumption was doubled from 1960 to 1991 and leveled at about 14,300 petajoule between 1991 and 1995. The individual traffic increased by 4.5% and the freight traffic by 14.5% between 1991 and 1996 [12].

The available freshwater is estimated to about 182 billion m^3. From the used 48 billion m^3 of water 43 billion m^3 flow back to the resources and about 5 billion m^3 return to the water cycle by evaporation.

In Germany the water abstraction in 1995 was in the range of 45.2 billion m^3; 27.8 billion m^3 were used as cooling water for power generation, 10 billion m^3 for industrial purposes and 5.8 billion m^3 for the public water supply. For irrigation 1.6 billion m^3 were used by agriculture [8].

Despite sufficient available freshwater for whole Germany there are areas with water deficiency, particularly in urban areas. An appropriate distribution system with long-distance pipelines compensates excess water demand and excess availability. From the abstracted 5.8 billion m^3 for the public water supply 76% was raw water and 23% pristine water, which could be delivered as drinking water without further treatment. 98.6% of the population were served with drinking water from 17,849 water works. Households and trade used 3.9 billion m^3 resulting in a water consumption of 132 l per capita

TABLE 9.5

CHARACTERISTICS OF THE THREE GREAT RIVER BASINS IN GERMANY

	Danube	Rhine	Elbe
River basin area in km^2			
in Germany	56,000	100,000	97,000
total	817,000	185,000	148,000
River length in km			
in Germany	578	695	727
total	2780	1320	1091
Inhabitants in millions			
in Germany	9	34	18.7
total	82	50	24.7
Area used for agriculture in km^2			
in Germany	28,000	43,000	55,000

per day in 1995. The major resource for public drinking water supply is groundwater (72.2%), followed by surface water (22%) and bank filtrated surface water (5.3%).

The number of public sewage treatment plants increased from 9941 (1987) to 10,279 (1995). In 1995 88.6% of the German population was connected to sewage treatment plants, 92% to sewers. In total 9.9 billion m^3 of sewage water was treated in sewage treatment plants; from that 0.3 billion m^3 only mechanically, about 1.5 billion m^3 biologically without special nutrient removal and about 8.1 billion m^3 biologically with special nutrient removal.

9.4.2.2 Rivers

The water quality survey of rivers is conducted within the scope of national and international programs. The monitoring programs of the Federal States (Länder) and the international commissions for river basins (International commission for the protection of the River Rhine IKSR, of the River Elbe IKSE, and of the River Danube IKSD) include biological and chemical parameters like nutrients, heavy metals and organic micropollutants. The current situation of pollution and its development is compiled by the German Länder Working Party on Water (Länderarbeitsgemeinschaft Wasser LAWA) and described for selected water quality parameters.

The area of Germany is divided into the river basins of Danube, Rhine, Ems, Weser, Elbe, and Oder. About 28%, 27% and 16% of the total area of Germany are covered by the river basins of the Rhine, Elbe, and Danube, respectively (Table 9.5).

The biological quality is described by representative investigations of aquatic ecosystems. The inventory of invertebrates, benthic macroorganisms and fishes. The chemical quality is represented by selected physical and chemical parameters (like dissolved organic carbon (DOC), nitrogen, phosphorus, chlorinated hydrocarbons, benzene derivatives, pesticides, PCBs, PAHs, organic tin compounds, and heavy metals).

The water quality of the great rivers in Germany has improved significantly during the last 20 years. This fact is due to several reasons.

References pp. 314–318

- Additional sewage treatment plants and its improved technology.
- Improved sewage treatment in the industry and adaptation of production processes.
- Usage of phosphate-free washing agents.
- Reduced production or shutting down of production sites, particularly in the eastern part of Germany.

The rivers are polluted from municipal (nutrients, organic matter) and industrial discharges, like the chemical industry (e.g. organic micropollutants), metal industry (e.g. complexing agents, heavy metals), pulp and paper industry (e.g. halogenorganic compounds), and mining (e.g. heavy metals, salts). For several rivers diffuse sources of plant nutrients like phosphorus and nitrogen as well as pesticides are of special interest. In the future diffuse sources of organic micropollutants, heavy metals and yet unknown compounds like pharmaceuticals and endocrinic substances may be of increasing interest [13].

In comparison the River Danube is less polluted, the River Rhine is the most intensively used river with a significant recovery during the last 20 years, and the River Elbe was designated as ecologically damaged before 1990 and is now 'only' critically to moderately polluted. Especially particulate matter and sediments show a high to very high pollution with heavy metals and chlorinated hydrocarbons, also mussels and fish.

TABLE 9.6

TEMPORAL DEVELOPMENT OF CHARACTERISTIC PARAMETERS OF GREAT GERMAN RIVERS (MEAN VALUES (ANNUAL MINIMA [a]); MODIFIED FROM REFS. [8,13])

Parameter	River, location	1985	1990	1996
Flow (m^3/s)	Danube, Jochenstein	1330	1240	1297
	Rhine, Kleve-Bimmen	1990	1930	1753
	Elbe, Schnackenburg	558	447	677
Oxygen [a] (mg/l)	Danube, Jochenstein	7.0	8.1	8.8
	Rhine, Kleve-Bimmen	6.0	6.7	7.3
	Elbe, Schnackenburg	<0.1	2.6	8.6
DOC (mg/l)	Danube, Jochenstein	–	–	3.1
	Rhine, Kleve-Bimmen	4.5	4.1	2.9
	Elbe, Schnackenburg	15	10	5.9
Chloride (mg/l)	Danube, Jochenstein	19.9	14.5	15.0
	Rhine, Kleve-Bimmen	193	182	150
	Elbe, Schnackenburg	250	278	119
Ammonium-N (mg/l)	Danube, Jochenstein	0.2	0.2	0.1
	Rhine, Kleve-Bimmen	0.5	0.2	0.2
	Elbe, Schnackenburg	3.6	1.5	0.5
Nitrate-N (mg/l)	Danube, Jochenstein	2.5	2.4	2.3
	Rhine, Kleve-Bimmen	4.2	3.9	3.5
	Elbe, Schnackenburg	3.2	5.1	4.6
Total phosphorus (mg/l)	Danube, Jochenstein	0.21	0.13	0.09
	Rhine, Kleve-Bimmen	0.48	0.22	0.16
	Elbe, Schnackenburg	0.78	0.71	0.24

TABLE 9.7

HEAVY METAL POLLUTION OF SUSPENDED PARTICULATE MATTER (mg/kg DRY MASS); TEMPORAL DEVELOPMENT OF THE 50-PERCENTILE (RHINE [a], ELBE [b]), AND THE MAXIMUM VALUES (DANUBE [c]), RESPECTIVELY [8]

Parameter	River, location	1990	1994	1996
Cadmium	Danube, Jochenstein	0.7	0.4	0.3
	Rhine, Kleve-Bimmen	1.8	1.2	1.3
	Elbe, Schnackenburg	11.5	13.0	8.5
Chromium	Danube, Jochenstein	58.0	46.0	39.0
	Rhine, Kleve-Bimmen	86.0	54.0	71.5
	Elbe, Schnackenburg	304	150	140
Mercury	Danube, Jochenstein	0.40	0.50	0.20
	Rhine, Kleve-Bimmen	0.60	0.39	0.58
	Elbe, Schnackenburg	21.1	7.5	4.1
Zinc	Danube, Jochenstein	280	167	209
	Rhine, Kleve-Bimmen	540	360	435
	Elbe, Schnackenburg	2180	1840	1355

[a] Centrifuge.
[b] Settling tank (mixing sample).
[c] Particulate matter sampling box.

However, there is a tendency of beneficial development. The different water qualities of the Danube, Rhine and Elbe and the temporal development from 1985 to 1996 is seen obviously comparing the characteristic parameters in Table 9.6 like dissolved oxygen or DOC. However, there remains a considerable pollution with nitrogen especially important with respect to eutrophication of the marine environment.

The heavy metal concentrations generally decreased between 1990 and 1996 (Table 9.7), but there is still a very high pollution level of suspended particulate matter in the River Elbe (mercury, cadmium, zinc). The River Danube shows a general low pollution with heavy metals, the River Rhine an elevated loading with copper and zinc.

Organic micropollutants are no more relevant parameters in the River Danube due to the substitution of chlorine by ozone in the bleaching process of the pulp and paper industry (Table 9.8). A significant decrease for dichloromethane is registered in the rivers Rhine and Elbe. In particular the pollution of the suspended particulate matter of the Rhine and Elbe raises ones attention. There are chlorinated benzene derivatives, PCBs, organic tin compounds and PAHs (Table 9.9). A major problem of the River Elbe is the high hexachlorobenzene (HCB) concentration. HCB is accumulated in the fat tissue, e.g. of fish, to a great extent resulting in metabolic disorder and organ damage. HCB is a persistent organic pollutant and remains therefore in the sediments. DDT and DDD belong also to the persistent chemicals and are present in significant concentrations in the River Elbe. The complexing agents EDTA and NTA were found in average concentrations of 14.8 μg/l and 9.6 μg/l in the River Elbe (Seemannhöft) and in average concentrations of 13.9 μg/l and 1.7 μg/l in the River Rhine (Kleve-Bimmen) in 1996.

References pp. 314–318

TABLE 9.8

CONCENTRATIONS OF ORGANIC MICROPOLLUTANTS IN GREAT GERMAN RIVERS (90-PER-CENTILE, μg/l)

Compound	Concentration (μg/l)		
	Danube, Ulm	Rhine, Kleve-Bimmen	Elbe, Seemannshöft
Trichloromethane	<0.03	0.1	0.2
1,1,1-Trichloroethane	<0.03	0.03	<0.01
Tetrachloroethylene	0.2	0.07	0.09
1,4-Dichlorobenzene	<0.01	<0.3	0.02
Hexachlorobenzene	–	–	0.008
Nitrobenzene	<0.05	<0.5	0.17
2-Nitrotoluene	<0.02	<0.5	0.23

TABLE 9.9

OCCURRENCE OF ORGANIC MICROPOLLUTANTS ON SUSPENDED PARTICULATE MATTER IN RHINE AND ELBE

Compound	Mean concentration in	
	River Rhine, Kleve-Bimmen	River Elbe, Schnackenburg
Organohalogen compounds (μg/kg)		
1,2-Dichlorobenzene	39	16
1,4-Dichlorobenzene	36	43
1,2,4-Trichlorobenzene	25	35
Hexachlorobenzene	26	175
PCB-52	6.9	16
PCB-153	19	17
Chlorinated pesticides (μg/kg)		
β-HCH	<2	29
γ-HCH	<2	4.2
4,4′-DDT	<2	137
4,4′-DDD	<5	64
PAH (mg/kg)		
Fluoranthene	0.90	0.94
Benzo[*a*]pyrene	0.39	0.46

The occurrence of herbicides has been reported especially for the River Rhine. The major herbicides are the ubiquitous diuron, and the agricultural herbicides isoproturon, chlortoluron, atrazine, simazine and terbuthylazine. Metamitrone, chloridazone, MCPA, dichlorprop, mecoprop and metolachlor are of more regional importance [14]. During spring and autumn concentration levels of atrazine, diuron and isoproturon exceed periodically 0.1 μg/l, the parameter value for drinking water. The 90-percentiles of herbicide concentrations in the River Rhine at Mainz and Köln reached or exceeded the quality targets of the IAWR (Internationale Arbeitsgemeinschaft der Wasserwerke im Rheineinzugsgebiet) of 0.05 μg/l (Table 9.10).

TABLE 9.10

HERBICIDE CONCENTRATIONS IN THE RIVER RHINE AT MAINZ AND KÖLN IN 1998 (90-PER-CENTILE, μg/l [15])

Herbicide	Mainz	Köln
Atrazine	0.08	0.08
Desethylatrazine	<0.05	0.05
Simazine	<0.05	<0.05
Diuron	<0.1	0.05
Isoproturon	<0.07	0.14
Chloridazone	0.07	<0.05
Bentazone	0.19	0.05

Pharmaceutical compounds are present in the River Rhine below the μg/l range, e.g. carbamazepin 0.41 μg/l, diclofenac 0.14 μg/l, ibuprofen 0.03 μg/l, bezafibrate 0.05 μg/l [14]. Further compound classes of interest are aromatic sulfonic acids (benzene, naphthalene, stilbene derivatives), amines (e.g. alkyl derivatives, urotropin), as well as triphenyl phosphinoxide (TPPO) a by-product of organic synthesis and N-methyl-N-phenylsulfonyl glycine a stabile degradation product of the anticorrosive agent HPS (6-[N-methyl(phenylsulfonyl)-amino] hexanoic acid).

9.4.2.3 Lakes

The ten great lakes are not equally distributed in Germany and show great differences in their geomorphological and hydrological character. The Lake Constance is the biggest one with an area of 470 km^2 and a maximum depth of 255 m. The Steinhuder Meer is extremely shallow and is the only one without a stabile stratification in summer (Table 9.11).

The major problem of the lakes is still the excess loads of nutrients especially of the growth-limiting phosphorus. In addition nutrients can be stored in the sediments. Therefore, a delayed response in a decreasing algae growth is observed to reduction in nutrient loads. The trophic levels can be described as oligotrophic, mesotrophic,

TABLE 9.11

CHARACTERISTICS OF SELECTED GREAT LAKES IN GERMANY

Lake	Area in km^2 (km^2)	Max. depth (m)	Mean depth [a] (m)	Volume (billion m^3)	Surrounding factor [b]
Lake Constance	476	252	90	48	22
Chiemsee	79.9	73	26	2.1	17
Kummerover See	32.9	25	7.9	2.6	35
Müritz	113	31	6	7.1	6.7
Schweriner See	62.1	57	13	8.1	5.7
Starnberger See	56.4	128	53	3.0	5.9
Steinhuder Meer	29.1	2.8	1.4	0.4	1.8

[a] Mean depth: volume/area.
[b] Surrounding factor: area of the region/area of the lake.

References pp. 314–318

eutrophic, polytrophic and hypertrophic based on a classification system of the LAWA. One important parameter is the concentration of phosphorus in the lake. The introduced nutrients are mainly discharges and run-off from agricultural areas. Improved sewage treatment and the usage of phosphate-free washing agents had a significant effect on the eutrophication level.

For instance in Lake Constance the concentration of phosphorus increased fivefold to almost 90 μg/l between 1960 and 1980, whereas the biomass of plankton algae increased fourfold to about 25 g/m^2. During the 1990s the phosphorus concentration could be decreased to about 30 μg/l and as a consequence the biomass of algae decreased to about 10 g/m^2. However, the phosphorus concentration is still high enough that considerable biomass of algae are produced under disadvantageous circumstances.

The Starnberger See was oligotrophic till 1950 and changed to a mesotrophic or slightly eutrophic level in the 1960s. The phosphorus concentration could be decreased from 30 μg/l to about 12 μg/l between 1982 and 1994. An oligotrophic status is not likely to be recovered because of atmospheric nutrient loading. The Chiemsee shows a comparable development.

The recovery of the shallow Steinhuder Meer is difficult despite of reduced phosphorus loading, because of nutrients in the sediments and the reduced ground stabilization due to reduced vegetation.

9.4.2.4 Groundwater

Groundwater is part of the water cycle and therefore an important factor of ecological balance. In Germany groundwater is the major drinking water source with a portion of 70%. The available amount is sufficient, but groundwater quality is jeopardized by several ways. There are point sources like contaminated sites, accidents or leaking sewers as well as diffuse sources from industry, agriculture and traffic. Remediation is a long-term task and of high financial and technical expenditure. Therefore, a consequent application of the precautionary principle is of great importance. Part of this is a systematic and periodical monitoring to recognize adverse effects early, to develop strategies for remediation and avoidance and to judge the efficacy of prevention measures. Part of the precautionary principle are the low parameter values for pesticides in drinking water: 0.1 μg/l per single compound, 0.5 μg/l for all pesticides including their toxic major metabolites.

An investigation of the groundwater quality in the eastern part of Germany between 1992 and 1994 resulted in only 41% not anthropogenically influenced sampling sites. The major problem of groundwater pollution is still diffuse loading from agriculture, especially nitrate. The distribution of nitrate in groundwater shows that 25% of all samples have distinct (15% in the range of 25 mg/l to 50 mg/l) to increased values (11% above 50 mg/l).

The situation of pesticide pollution is improving, however very slowly [16]. A report from the LAWA evaluates the investigations on pesticides of 13,000 sampling sites in shallow groundwater in the 16 Federal States (Länder) of Germany [17]. 80% of all sampling sites are situated in 4 Federal States (Baden-Württemberg, Bavaria, Hesse and North Rhine-Westphalia) covering 45% of the total area of Germany. The number of licensed active substances was about 240 in 1997. From this about 150 active agents

TABLE 9.12

FREQUENTLY DETECTED PESTICIDES IN GROUNDWATER (1990–1995; DATA FROM [17]) [a]

Rank	Detected compound		
	$\leq$0.1 µg/l	>0.1 µg/l to $\leq$1.0 µg/l	>1.0 µg/l
1	Atrazine	Desethylatrazine	Bromacil
2	Desethylatrazine	Atrazine	Atrazine/Desethylatrazine
3	Simazine	Bromacil	Diuron
4	Propazine	Simazine	Bentazone
5	γ-HCH (lindane)	Hexazinone	Hexazinone
6	Terbuthylazine	Propazine	Simazine
7	Bentazone	Desisopropylatrazine	Mecoprop
8	Diuron/Desisopropylatrazine	Diuron	Terbuthylazine
9	Isoproturon	Bentazone	Propazine/Dichlorprop
10	Mecoprop	Mecoprop	Sebuthylazine/Desethylterbuthylazine/ Desisopropylatrazine

[a] Findings of dikegulac, a by-product of vitamin C synthesis, were not considered.

and metabolites are of relevance for groundwater due to their properties and amount of application. In 70.7% of all sampling sites no pesticides could be measured. In 18.6% at least one pesticide was observed but below 0.1 µg/l. In 8.6% at least one pesticide ranged between 0.1 µg/l and 1.0 µg/l, even in 1.1% levels above 1.0 µg/l were detected.

In Table 9.12 the most frequently detected pesticides are listed for three concentration ranges. It reveals that the triazine herbicides with the most popular atrazine and its metabolite desethylatrazine are always on the top ranks. With exception of lindane all substances are herbicides. Besides atrazine and desethylatrazine bromacil, simazine, hexazinone, diuron, propazine and bentazone are among the ones most frequently exceeding the drinking water parameter value of 0.1 µg/l. Additionally γ-HCH, terbuthylazine, isoproturon and mecoprop as well as the metabolites desisopropylatrazine and desethylterbuthylazine were found frequently.

In contrast to their frequent detection atrazine and bromacil were banned several years ago and the application of simazine, hexazinone and propazine is restricted in water conservation areas. Diuron application on railway tracks and sealed surfaces is banned since 1996. The data of the 1988 banned atrazine reveal that the recovery of groundwater quality is not a short-term process. In Baden-Württemberg a slow decrease of the annual mean values of atrazine has been recognized since 1992. Desethylatrazine only decreased since 1994/1995 [16]. Other substances in Table 9.12 are not restricted or banned for application. Therefore, groundwater monitoring should direct special attention to them to recognize adverse effects of these substances from the very beginning.

9.5 PARAMETERS TO ASSESS WATER QUALITY

The parameters to assess water quality are given in the first instance by the defined quality, which itself depends on the usage, legislative regulations, and the actual state of research. Legislative regulations are given for drinking water (drinking water ordinance), groundwater (groundwater ordinance), waste water (waste water act). The parameter values proclaimed in the drinking water ordinance are the minimum requirements which have to be met. Only a negative list of parameters with maximum allowed concentrations is laid down in the regulation. A positive list with for good drinking water essential parameters is not given [18]. General statements on the properties of drinking water quality are given in the German norm DIN 2000 [19].

(i)   Drinking water shall be almost free of germs and free of pathogens.

(ii)   It must not show properties injurious to health.

(iii)   It shall be appetizing, colorless, clear, cool, odorless and of unobjectionable taste.

(iv)   The dissolved substances shall not be excessive.

(v)   It shall not cause corrosion of materials.

9.5.1 Parameters in the drinking water directive

At the European Community level the minimum quality and control standards for water intended for human consumption are stated in the Council Directive 98/83/EC (drinking water directive; cf. Table 9.13) [20]. Parameters for organoleptic, physical, chemical, and microbiological investigations are defined, including undesirable and toxic substances.

Besides the parameters and their corresponding parametric values there are also specified the performance characteristics of the analyses methods with a demand for trueness, precision and limit of detection (LOD). Undoubtedly the water quality can be judged only by data of analyses with the appropriate quality. The requirements on data quality at low concentration level like for pesticides (LOD of 0.025 μg/l) are not easily fulfilled. The situation is even more difficult for single parameters like acrylamide, epichlorohydrin and bromate for which no standardized analytical methods are available [21]. However it has to be emphasized that several parametric values like the one for pesticides are not of toxicological relevance but have been set according to the precautionary principle at concentration levels near the limits of determination of existing analytical methods.

9.5.2 Demands on raw water for drinking water supply

Surface water, especially river water, is subjected to substantial fluctuation in quality. Therefore, drinking water production from surface waters normally involves some natural purification steps like bank filtration or artificial groundwater recharge to improve the stability of water quality. In these steps refractory pollutants like pesticides require additional steps like oxidation or adsorption on activated carbon. For this

TABLE 9.13

PARAMETERS AND PARAMETRIC VALUES OF DRINKING WATER DIRECTIVE 98/83/EC (1998) [20]

Parameter	Parametric value	Unit
Microbiological parameters		
Escherichia coli (E. coli)	0	number/100 ml
Enterococci	0	number/100 ml
Chemical parameters		
Acrylamide	0.10	μg/l
Antimony	5.0	μg/l
Arsenic	10	μg/l
Benzene	1.0	μg/l
Benzo[*a*]pyrene	0.010	μg/l
Boron	1.0	mg/l
Bromate	10	μg/l
Cadmium	5.0	μg/l
Chromium	50	μg/l
Copper	2.0	mg/l
Cyanide	50	μg/l
1,2-Dichloroethane	3.0	μg/l
Epichlorohydrin	0.10	μg/l
Fluoride	1.5	mg/l
Lead	10	μg/l
Mercury	1.0	μg/l
Nickel	20	μg/l
Nitrate	50	mg/l
Nitrite	0.50	mg/l
Pesticides	0.10	μg/l
Pesticides, total	0.50	μg/l
Polycyclic aromatic hydrocarbons	0.10	μg/l
Selenium	10	μg/l
Tetrachloroethene and trichloroethene	10	μg/l
Trichloromethanes, total	100	μg/l
Vinylchloride	0.50	μg/l
Indicator parameters		
Aluminum	200	μg/l
Ammonium	0.50	mg/l
Chloride	250	mg/l
Clostridium perfringens (including spores)	0	number/100 ml
Color	Acceptable to consumers and no abnormal change	
Conductivity	2500	μS cm^{-1} at 20°C
Hydrogen ion concentration	$\geq$6.5 and $\leq$9.5	pH units
Iron	200	μg/l
Manganese	50	μg/l
Odor	Acceptable to consumers and no abnormal change	
Oxidizability	5.0	mg/l O$_2$
Sulfate	250	mg/l
Sodium	200	mg/l
Taste	Acceptable to consumers and no abnormal change	
Colony count 22°C	No abnormal change	number/100 ml
Coloform bacteria	0	
Total organic carbon (TOC)	No abnormal change	

References pp. 314–318

TABLE 9.14

DEMANDS ON RIVER WATER AS RAW MATERIAL FOR DRINKING WATER SUPPLY [22]

Parameter	Unit	Standard demands	Minimum demands
General characteristics			
Electrical conductivity (20°C)	μS/cm	500	1000
Oxygen saturation	%	80	60
Color (SUVA 436 nm)	m^{-1}	0.3	1
Smell threshold		5	–
Temperature	°C	22	25
pH		6.8–8.5	5.5–9.0
Inorganic constituents			
Sum parameters:			
Total dissolved matter	mg/l	400	800
Suspended inorganic matter [a]	mg/l	25	150
Single substances:			
NH_4^+	mg/l	0.2	0.4
Al	mg/l	0.1	0.5
As	mg/l	0.005	0.01
Pb	mg/l	0.01	0.02
B	mg/l	0.5	1.0
Cd	mg/l	0.001	0.002
Ca, dissolved	mg/l	100	–
Cl^-	mg/l	100	200
Cr	mg/l	0.03	0.05
CN^-	mg/l	0.01	0.05
Fe, dissolved	mg/l	0.2	1.0
F^-	mg/l	1.0	1.0
Cu	mg/l	0.02	0.05
Mg, dissolved	mg/l	30	–
Mn, dissolved	mg/l	0.03	0.25
Na	mg/l	60	120
Ni	mg/l	0.03	0.04
NO_3^-	mg/l	25	40
PO_4^{3-}	mg/l	0.15	0.5
Hg	mg/l	0.0005	0.001
Se	mg/l	0.001	0.01
SO_4^{2-}	mg/l	100	150
Zn	mg/l	0.1	0.3
Organic constituents			
Sum parameters:			
DOC	mg/l	4	8
Biochem. oxygen demand	mg/l	4	8
Suspended organic matter	mg/l	5	25
Group parameters of the filtrated sample:			
Hydrocarbons	mg/l	0.05	0.2
Anionic tensides	mg/l	0.1	0.3
Nonionic tensides	mg/l	0.1	0.3
PAH [a]	mg/l	0.0001	0.0002
AOX	mg/l	0.03	0.06

TABLE 9.14 (CONTINUED)

Parameter	Unit	Standard demands	Minimum demands
Single substances:			
Pesticides, dissolved	mg/l	0.0001	0.0001
Sum of organic chloro compounds: 1,1,1-trichloroethane, trichloroethene, tetrachloroethene, dichloromethane	mg/l	0.002	0.005
Tetrachloromethane	mg/l	0.001	0.001
Trihalogenmethanes: trichloromethane, bromodichloromethane, dibromochloromethane, tribromomethane	mg/l	0.002	0.005
NTA	mg/l	0.01	0.02
EDTA	mg/l	0.005	0.01
Microbiological parameters			
Total coliforms[c]	N/100 ml	50	–
Faecal coliforms	N/100 ml	20	–
Faecal streptococci	N/100 ml	20	–

[a] Mean low water.
[b] Sum of fluoranthene, benzo[*b*]fluoranthene, benzo[*k*]fluoranthene, benzo[*a*]pyren, benzo[*ghi*]perylene, indeno[1,2,3-*cd*]pyren.
[c] Classification according to Drinking Water Ordinance (5 December 1990).
– No minimum demands.

purpose standard demands and minimum demands for river water as raw material for drinking water supply are given by the Technical Standard W251 (Table 9.14 [22]). For water submitted too close to natural treatment methods the parameter values of the standard demands should be observed. The standard demands are close to the parameter values given by the drinking water ordinance. Also water which does not meet the standard demands can be used for drinking water supply applying optimized treatment methods and very worthwhile proved physical and chemical methods. In these cases the values of minimum demands should be met by the raw water. Furthermore a decreased safety precaution has to be accepted.

The parametric values of the standard and the minimum demands on raw water are based on daily average values and can be subdivided into the following groups: (1) general characteristics; (2) single substances; (3) sum parameters; (4) group parameters.

A compilation of the parameters and their values are shown in Table 9.14.

9.5.3 Quality targets

Quality targets for the protection of surface waters are worked out by a Federal–National Working Commission on Quality Targets (Bund-Länder-Arbeitskommission Qualitätsziele BLAK-QZ). The quality targets are concentration values for hazardous chemicals relating to protected goods like drinking water production, aquatic ecosystems, and fishing. The quality targets have an advisory character and are applied with respect to the quality control of surface waters and to support the approval procedure for the dis-

TABLE 9.15

EXAMPLES FOR QUALITY TARGETS FOR THE GOODS DRINKING WATER SUPPLY (DWS) AND AQUATIC ECOSYSTEMS (AES) [8]

Compound	Quality targets	
	for AES (μg/l)	for DWS (μg/l)
Lead	100 [a]	50
Cadmium	1.2 [a]	1
Chromium	320 [a]	50
Mercury	0.8 [a]	0.5
Hexachlorobutadiene	0.5	1
1,4-Dichlorobenzene	10	1
1,2,4-Trichlorobenzene	4	1
1,4-Dichloro-2-nitrobenzene	20	1

[a] In mg/kg (suspended particulate matter).

charge of waste water with legal permission. Up to now quality targets for 28 industrial chemicals and 7 heavy metals for separated goods are set (Tables 9.4 and 9.15). Among them are heavy metals like lead, cadmium, chromium and mercury as well as organic chemicals like chlorinated hydrocarbons (e.g. dichloromethane, trichloroethane), chlorinated benzene derivatives, nitrobenzene derivatives, and chloroanilines. Pesticides are the third class of compounds just being in hand of the commission. The quality targets for pesticides in drinking water production are for every single substance 0.1 μg/l. Meeting the quality targets may exclude a risk for the specified good on the basis of the present knowledge [8]. The quality targets are derived from toxicological data for different trophic levels. The values for aquatic living communities are in general higher than those for drinking water supply. However, in the case of hexachlorobenzene and nitrobenzene they are lower compared to the ones for drinking water supply indicating an ecotoxicological effect of these chemicals.

A more simple concept to assess river water with regard to drinking water supply is called the Basler Model which was developed for the River Rhine. It works on the basis of the parameters DOC, ammonium, oxygen deficiency compared to saturation with air, the sum of neutral salts and AOX. Based on this set of sum parameters the improvement of river water can be followed clearly over the last decades [23].

9.5.4 Specific requirements

In addition to the above-mentioned concepts and legislative requirements to assess water quality the specific requirements of the regarded system have to be observed. This means for example in river water to consider specific parameters with regard to specific water usage (e.g. temperature for cooling water, organic pollutants in industrial effluents, salts and heavy metals from mining, pesticides from agricultural run-off). For lakes in particular excessive nutrients, like phosphorus, are the common cause of eutrophication, whereas in a marine environment nitrogen is the major mineral nutrient controlling eutrophication. In groundwater particularly nitrate and pesticides are of interest. In drinking water the formation of disinfection by-products (DBPs) and microorganisms are of great concern. Beside the well-known trihalomethanes (THM) a huge number of possible DBPs like halocarboxylic acids, halodi- and halotricarboxylic acids including MX, haloacetonitriles, haloalcohols, phenols, and ketones have to be considered due to reactions of water constituents with the disinfection chemicals chlorine or ozone [24]. *Escherichia coli* and Enterococci were found to be suitable indicator bacteria. Furthermore the Protozoa *Cryptosporidium* and *Giardia* as well as their cysts and oocysts are important microorganisms in drinking water treatment [25].

A new compound class of interest in surface waters and groundwater are pharmaceuticals. Pharmaceuticals are used in large quantities in human and veterinary medicine. After their application residues may enter the water cycle. For instance human pharmaceuticals and their metabolites are excreted and discharged into the sewage system. Residues can be detected in the effluents of sewage treatment plants and in surface water [26,27]. Furthermore, veterinary medical compounds threaten groundwater and surface water by manure dispersion on fields or application in fish ponds [28].

9.6 SUM PARAMETERS VS. SINGLE SUBSTANCE ANALYSIS

A great number of water pollutants can be analyzed only with special appropriate methods. In particular it is impossible to identify and quantify all in water present organic micropollutants. With modern analytical methods only about 10% of the organic water constituents are quantitatively detected. The analysis of single organic substances or compound classes is mostly time consuming and expensive. Furthermore, the determination of single compounds in complex water matrices is not necessary in all cases.

For that reason, sum parameters have proved good for water quality characterization. Sum parameters as a first step of investigation can indicate the demand for further analyses and help in this way to reduce analysis time and costs. A lot of sum parameters cover the determination of multiple compounds by one measured parameter. Moreover it might not always be necessary to know the contribution of the single constituents. So the characterization of waters with a complex composition is often achieved by simple and cost-effective determination of a sum parameter.

However, for the application of sum parameters their limited meaningfulness and the margins of interpretation have to observed. It has to be proved in every case which compounds are included in the measurement and whether these are the right parameters concerning the aim of the determination. In Table 9.16 the pros and cons of sum parameters are concluded.

9.6.1 Dissolved organic carbon

Depending on the task, specific determination of single substances can be necessary in addition to the analysis by sum parameters. For instance, from the concentration of dissolved organic carbon (DOC) in groundwater no toxicity can be derived. To evaluate

TABLE 9.16

PROS AND CONS OF THE APPLICATION OF SUM PARAMETERS

Pros	Cons
Low costs	Low sensitivity
Time-saving	Limited interpretation
Reliable survey of the whole system	Limited specificity

the toxicity the determination of the concentration of single toxic compounds (single compound analysis for e.g. PAHs) has to be done. On the other hand DOC and spectral UV absorbance (SUVA) serve as excellent parameters to optimize activated carbon adsorption for drinking water treatment and to measure adsorption isotherms [29,30] or to observe even the loading of a river with secondary effluents from sewage treatment plants [31]. In natural waters a good correlation is found between the DOC and the SUVA, which can be measured simply and cost-effectively [32].

An advanced characterization of the DOC is available applying the on-line coupling of size exclusion chromatography with DOC-detection. The detector consists of a thin-film reactor with a low pressure mercury arc [33]. The instrument is capable of measuring original water samples without any pretreatment or preconcentration (Fig. 9.2) [34,35]. A typical gel chromatogram with DOC and UV detection is shown for water of the Lake Genezareth. The tentative classification of six major fractions is shown in Fig. 9.3. One dominant fraction can be attributed to refractory humic substances. The humic substances are normally the major fraction in natural waters which are minor anthropogenically influenced. In the effluents of municipal sewage treatment plants the organic acids are predominant. This can be explained by the uncompleted biological conversion of the DOC. The influence of the effluent of a sewage treatment plant on the character of the DOC in a little river in southwest Germany can be revealed

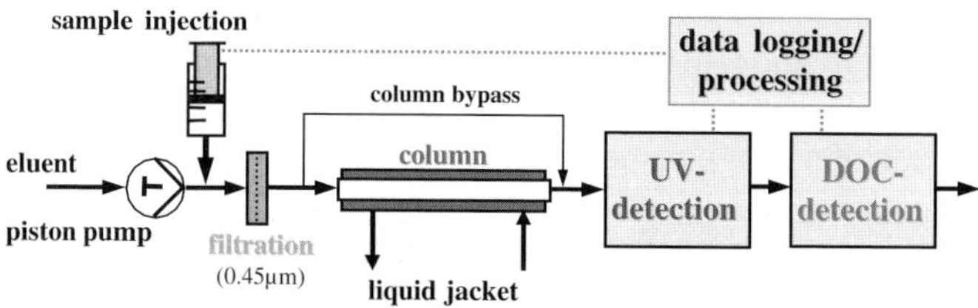

Operating parameters LC-UV/DOC	
Eluent	Phosphate buffer: 28 mmol/L, pH 6.6
	(1.25 g/L Na$_2$HPO$_4$·2H$_2$O and 2.5 g/L KH$_2$PO$_4$)
Flow rate	1.0 mL/min (isocratic)
Column	TSK HW 40/50/55 S (250 x 20 mm, Grom Herrenberg)
Injection volume	2 mL (chromatography), 0.5 mL (column bypass)
Detection limit	20 µg/L DOC

Fig. 9.2. Experimental set-up and operating parameters of the LC/DOC system.

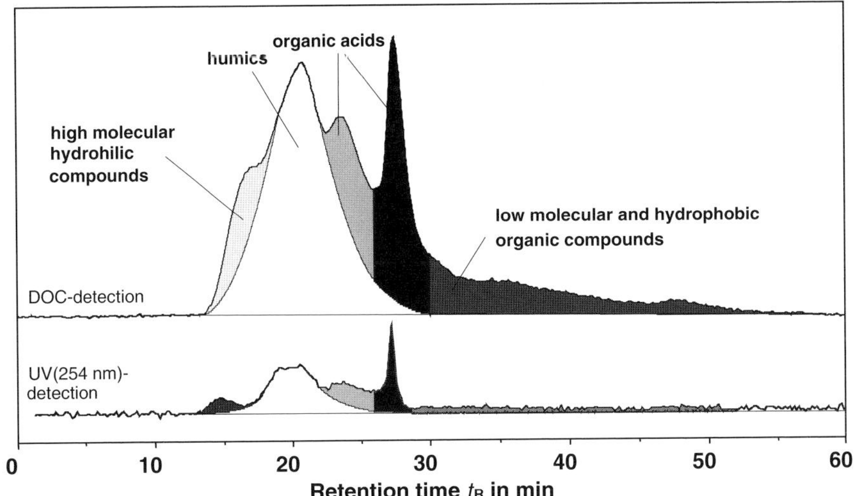

Fig. 9.3. Typical chromatogram of an original water sample from the Lake Genezareth (Israel) and the tentatively attributed fractions.

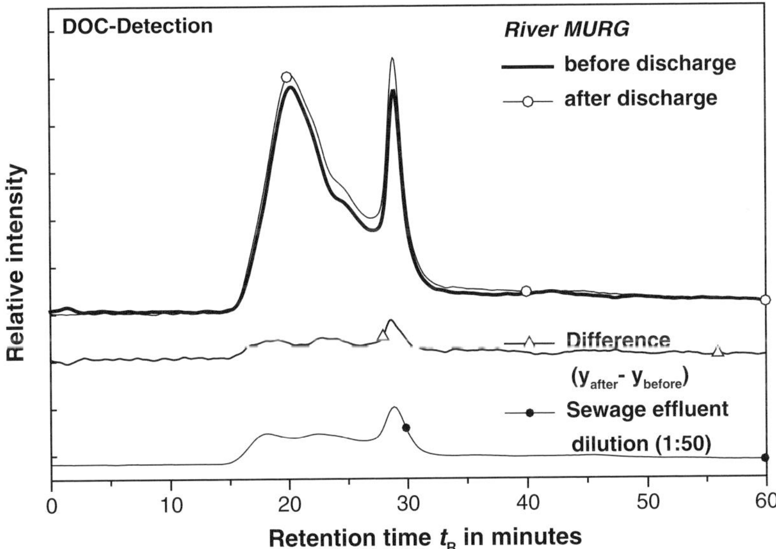

Fig. 9.4. LC/DOC analysis of water samples of the small River Alb (southwest Germany) before and after discharge of sewage effluents.

by LC/DOC analysis (Fig. 9.4) [31]. Whereas near the source of the river the humic substances predominate. After discharge of sewage effluent the fractions of organic acids have been increased. This reveals the difference chromatogram of the samples before and after discharge. In addition the diluted sewage effluent resembles the difference chromatogram. Therefore the changes of the DOC fractions may be attributed to the discharge from sewage treatment. Finally further downstream of the river the character of the DOC equals almost the one at the source due to natural purification of the river [36].

9.6.2 Further parameters

A survey of further available sum parameters can be found in the German Standard Methods for the examination of water, waste water, and sludge [37] and elsewhere [38]. The sum parameters can be subdivided in two groups. The first group of parameters contains mainly the measurement of inorganic substances like pH value, oxygen content, electrical conductivity, oxidation reduction potential (ORP), total solids residue, residue on ignition, filtrate solids residue, settleable matter, turbidity, color, total metal content.

In the second group organic substances measurement is covered. This group itself can be subdivided into the sum parameters, the characterization parameters and the effect parameters as follows.

(1) Sum parameters:

total and dissolved organic carbon (TOC, DOC), assimilable organic carbon (AOC), chemical oxygen demand (COD), biochemical oxygen demand (BOD), permanganate index, adsorbable organically bound halogens (AOX), extractable organically bound halogens (EOX), purgeable organically bound halogens (POX), organic phosphorus, organic nitrogen compounds, organically bound sulfur, total phenols, polycyclic aromatic hydrocarbons (PAH), surfactants.

(2) Characterization parameters:

spectral absorption coefficient, fluorescence, threshold odor number (TON), threshold taste number (TFN).

(3) Effect parameters:

determination of the effect of substances in waters on fish, on microcrustaceans (*Daphnia magna*), on protozoans, on *Pseudomonas putida* (growth inhibition test), on freshwater and marine algae (growth inhibition test with *Scenedesmus subspicatus, Selenastrum capricornutum, Skeletonema, Phaeodactylum tricornutum*), on enzyme activity (e.g. dehydrogenase, cholin esterase), on light emission of *Photobacterium phosphoreum* (luminescent bacteria test).

9.6.3 Parameter measurement

Several of the above-mentioned sum parameters are accessible with on-line analysis systems. Examples are sensors like the pH electrode, and the Clark electrode for oxygen measurement. New systems with miniaturized flow injection analysis, optical sensors and biosensors for single-component analysis are under development. Due to their properties of simple and fast as well as cost-effective measurement they will have the ability to substitute several sum parameters in the future [39,40]. Nevertheless, sum parameters will remain attractive due to their ability to give a reliable survey over whole systems and to perform a simple screening of samples for selection to further analyses.

An important field of the application of toxicity and genotoxicity measurements can be summarized under the titles bioassay- or toxicity-directed chemical analysis, and toxicity identification evaluation (TIE). In these cases the first step is the characterization of a sample or a sample fraction with regard to its toxicity effect. The corresponding positive samples are then subjected to chemical analysis to identify possibly single compounds responsible for the toxic effects. This proceeding is the more reliable way to answer questions on the ecotoxicologic effects of the considered system, since the classic methods of target analysis only show a small section and therefore eventually can miss toxic sample constituents. On the other side instrumental analysis can be applied systematically only for a selection of the samples of interest, which helps to reduce analysis time and costs.

Toxicity-directed water analysis has been employed for surface waters [41,42], drinking water [43], effluents of industry [44–50] and municipal sewage treatment plants [51]. Not in all cases chemical analysis can identify the major part of toxic compounds [44]. This is especially the case for the more polar toxic compounds. Sophisticated analytical procedures and extraction methods are needed. In general the toxicity-directed fractionation of the sample in smaller, less complex mixtures is a suitable analytical approach. A method is described to fractionate mutagens from airborne particulate matter by column chromatography on cyanopropyl-bonded silica, by normal-phase HPLC fractionation and size-exclusion chromatography, respectively [52]. In certain cases a reduction of toxicity caused by H_2S, ammonia, and heavy metals prior to sequential extraction and fractionation is needed. Automated multidimensional development thin layer chromatography (AMD-TLC) was employed for fractionation. For example toxicity testing was performed directly on the TLC plates with a bioluminescence test with *Vibrio fischeri* [50], and the determination of the inhibition of cholinesterase by insecticides was performed after thin layer chromatographic separation on a HPTLC plate [43].

9.7 SAMPLING AND SAMPLING STRATEGIES

In general a very important step of the analytical process is sampling. The quality and accuracy of analytical data are highly dependent on the suitability and accuracy of the sampling procedure. Systematic errors during sampling cannot be corrected in further steps of the analytical determination.

A major task of environmental sampling is to take a representative sample employing a problem oriented proceeding. The procedure is determined by the specific query of the analysis, like parameters to be measured, heterogeneity of the population to be sampled, time and spatial resolution of the data, the accuracy required. Therefore, the sampling process has to be planned under consideration of technical and analytical handling as well as statistical methods. The planning of representative sampling can be based on the methodology of Green [53] and has to take into account the following aspects [54]: place, location, and position; size, quantity, and volume of the sample; number of samples; date, duration, and frequency of sampling; homogeneity of the sample; contamination of the sample; decontamination of the sample; sample conservation and storage.

9.7.1 Standardized methods

Several norms of the European Committee for Standardization (CEN), the German Institute of Standardization (DIN), and the International Organization of Standardization (ISO) are dealing with several aspects of water sampling.
- Water quality; sampling Part 1: Guidance on the design of sampling programs (ISO 5667-1: 1980, EN 25667-1); Part 2: Guidance on sampling techniques (ISO 5667-2: 1991; EN 25 667-2: 1993); Part 3: Guidance on the preservation and handling of samples (ISO 5667-3: 1994; EN ISO 5667-3: 1995); Part 4: Guidance on sampling from lakes, natural and man-made (ISO 5667-4); Part 5: Guidance on sampling of drinking water and water used for food and beverage processing (ISO 5667-5); Part 6: Guidance on sampling of rivers and streams (ISO 5667-6); Part 10: Guidance on sampling of waste waters (ISO 5667-10); Part 11: Guidance on sampling of groundwaters (ISO 5667-11); Part 14: Guidance on quality assurance of environmental water sampling and handling (ISO 5667-14).
- German Standard Methods for the examination of water, waste water and sludge; general information (group A); Part 6: Assessment of the minimum frequency for controlling water constituents in discharges (emission strategy; DIN 38 402-6: 1991); Parts 11–20: Sampling from different aqueous systems: Part 11: Sampling of waste water (A 11; DIN 38 402-11: 1995); Part 12: Sampling from barrages and lakes (A 12; DIN 38 402-12: 1985); Part 13: Sampling from aquifers (A 13; DIN 38 402-13: 1985); Part 14: Sampling of untreated water and drinking water (A 14; DIN 38 402-14: 1986); Part 15: Sampling from rivers and streams (A 15; DIN 38 402-15: 1986); Part 16: Sampling of sea water (A 16; DIN 38 402-16: 1987); Part 17: Sampling of wet precipitation in liquid state (A 17; DIN V 38 402-17 (draft): 1988); Part 18: Sampling of water from mineral and medicinal springs (A 18; DIN 38 402-18: 1991); Part 19: Sampling of water from swimming-pools and baths (A 19; DIN 38 402-19: 1988); Part 20: Sampling from tidal water (A 20; DIN 38 402-20: 1987); Part 22: Sampling of cooling water for industrial use (A 22; DIN 38 402-22: 1991); Part 30: Pretreatment, homogenization and aliquotation of nonhomogeneous water samples (A 30; DIN 38 402-30: 1998).

9.7.2 Sampling procedure

Practical aspects of sampling from different waters are described in [55]. Representative planning of sampling employing statistical methods is an interesting aspect to optimize the number and quantity of samples, as well as the sampling location and frequency [56]. Furthermore, multivariate statistical methods can help to answer questions about sampling times or sampling points necessary to describe e.g. the state of pollution of a river or stream. However, the first step of each planning process is always the formulation of the purpose of sampling, e.g. quality control, prognosis and forecasting, damage or risk assessment. The development of steps for the sampling and analysis plan, the quality assurance program, and instruction and training of the personnel dealing with sampling are the following steps of the sampling program.

To control water quality of rivers and streams often automated sampling is the only cost-effective way to achieve sufficient spatial and temporal resolution. In addition to the liquid phase suspended particulate matter has to be sampled for water quality assessment and control. Sampling and automated analysis for pollutants in aqueous samples was reviewed by Shamas [57]. Stoeppler reviewed the field of sampling for soil and sediment analysis of biological, environmental, and technical materials [58]. Sampling for trace metals with special emphasis on contamination prevention is found in [59]. Sampling for pesticides in ground and surface water covers the subjects of sample containers, sampling techniques, preservation transport, and storage [60].

9.7.3 Semipermeable membrane devices

In recent years more and more publications deal with the use of semipermeable membrane devices (SPMD). SPMDs were introduced by Huckins [61,62] and consist of a low-density polyethylene (LDPE) lay-flat tubing, filled with either natural lipids or the model lipid triolein (1,2,3-tri[*cis*-9-octadecenoyl]glycerol). SPMDs are time-integrating passive samplers and often used for the monitoring of bioavailable hydrophobic pollutants in aquatic environment [63,64]. They may act as surrogates for living organisms. Dialysate preconcentration is controlled by contaminant diffusion rates and molecular size. On the other hand the diffusion rate is controlled by the contaminant activity in the different solvents, which is described by the partition coefficient. Understanding of the accumulation process has employed a three-compartment model. It incorporates both aqueous film (turbulent-diffusive) and polymer film (diffusive) mass transfer and is fit to data from accumulation studies [65]. Suspended solid loads lower than 100 mg/l showed no effect on SPMD uptake efficiency. However, at high solids concentration, e.g. during high river flow or tidal resuspension, reduced SPMD uptake occurred. Back diffusion of contaminants from SPMD dialysate will occur over an extended exposure period following a decline in ambient concentrations [66]. A comparison of chlorophenol uptake by SPMD and by Goldfish (*Crassius auratus*) showed similar overall uptake rate constants and average pollutant sampling rates [67]. The uptake of the four pesticides chlordane, endosulfane and the synthetic pyrethroids fenvalerate and allethrin by triolein-containing SPMDs and the lake mussel *Anodonta piscinalis* was compared. SPMD showed uptake rates 3.5 to 5.5 times higher than the mussel. However, the percent composition (ratios) of analytes in SPMDs and in mussels was similar. This indicates that SPMDs may serve as good surrogates for aquatic organisms with respect to the discriminatory uptake of the pesticides (pyrethroids with lower rates than the organochlorines) [68]. To measure the bioavailable fraction of contaminants in aquatic environments SPMDs were used in direct comparison with the common method of using indigenous or transplanted bivalve mollusks. Factors affecting the use of SPMDs turned out to be membrane thickness, surface area, and lipid substance volume [69].

References pp. 314–318

To measure solute concentrations in pore water Harper et al. discussed the techniques of dialysis and diffusional equilibration in thin films (DET) [70]. The suitability of two more extensively used sampling techniques for sediment pore water, in situ dialysis and centrifugation followed by filtration, was compared in field measurements. In situ dialysis has been expected to measure more accurately pore water concentrations of nutrients and other reactive substances [71].

9.8 NEW SAMPLE PREPARATION METHODS FOR ORGANIC MICROPOLLUTANTS

Sample preparation in environmental analysis normally includes a subdivision of the sample in aliquots, extraction and preconcentration, if necessary a cleanup step to remove interfering sample matrix and finally derivatization to produce suitable analyte derivatives for the subsequent separation (e.g. gas chromatography) and detection. Therefore the analytes in aqueous samples have to be brought to an adequate, matrix-free measuring solution. The sample preparation method is determined by kind and properties of both the matrix and the target analytes.

9.8.1 Liquid–liquid extraction

In general for aqueous samples there exist two main extraction principles. The very traditional liquid–liquid extraction (LLE) and the now dominating solid-phase extraction (SPE). The extraction step normally results in an efficient preconcentration to render a sensitive detection. The LLE is based on the partition of the analytes between two immiscible solvents dependent on the equilibrium constants. For traditional LLE normally is a very laborious method consuming a lot of organic solvents it was thrust into the background. A review of strategies for solvent selection may help to avoid or substitute certain legislative restricted, toxic or environmentally unfriendly solvents (e.g. chlorinated ones) [72]. The strategies are based on solvent classification according to the scheme of Rohrschneider-Snyder [73,74], the solvatochromic scheme [75], and the Hildebrand polarity scheme [76]. To save solvents and to reduce operation time small-scale or miniaturized extraction protocols have been suggested [77–79]. Acoustical levitation of droplets was applied to extract solid and liquid microsamples. Contactless handling of small liquid amounts prevents cross-contamination of the sample [80].

9.8.2 Solid phase extraction

SPE is the method in the foreground with a high potential of automation, the possibility of field application and a solvent-saving working protocol. The aqueous samples are percolated or forced through a column packed with a solid phase material. After drying the loaded cartridge with N_2 gas the analytes are desorbed by small portions of solvents. In this way extraction and preconcentration are performed in one step. SPE is theoretically based on the separation principles of liquid chromatography with the solid phase material as stationary phase and the aqueous sample as the mobile phase. Due to the continuous loading of the solid phase with analytes the plot of analyte concentration against elution volume results in bilogarithmic breakthrough curves. Their maximum equals the analyte concentration in the sample. Aim of the preconcentration is to achieve as high as possible retention volumes (e.g. defined as 1% breakthrough) for the analytes of interest [81]. Retention volumes depend on the physical–chemical properties of the analytes and the solid phase material (e.g. octanol–water partition coefficients) as well as the properties of the solvent (sample). Modified silicas, macroreticular resins and graphitized carbon black are the most common sorbents available for different analytical demands. A compilation of applications can be found in the literature [77,82]. On-line coupling of SPE with analytical determination by GC or HPLC may help to reduce manual work and statistical errors. A major aim is to reduce total analysis time to fulfill the increasing demand on analyses in environmental monitoring programs. A lot of applications can be found in the literature for on-line SPE–HPLC coupling [83] to analyze for instance pesticides [84–89], PAHs [85], explosives [85], polar phenolic compounds [90], an antifouling agent (Irgarol 1051) [91] and further polar compounds [92]. Less applications employ on-line SPE–GC coupling, to analyze for instance butyltin compounds [93], chlorophenols, organophosphorus pesticides or benzene derivatives [94].

9.8.3 Solid phase microextraction

A new alternative to current sample preparation technology is the solid phase microextraction (SPME). SPME technology works without solvents, allows easy automation of the sample preparation step and is amenable to small sample volumes, to on-site analyses and to process monitoring. The analytes are sorbed on a small fused-silica fiber, usually coated with a polymeric phase. For better handling the fiber is mounted on a syringe-like device. The basic principle of SPME is to use small volumes of extracting phase, usually less than 1 µl. The sample volume can be very large. The amount of analyte extracted by the coating phase is determined by magnitude of partition coefficient and concentration. After adsorption the device is transferred to an analytical instrument (e.g. GC, HPLC). Is it inserted into the hot injector of a GC the analytes are thermally desorbed. For HPLC application the analytes are desorbed with solvent in a special device installed in the injection loop. SPME can be applied to all aqueous matrices like drinking water, surface water, groundwater, and waste water. The fiber can be exposed by direct immersion in the sample or to the headspace of the sample. For different demands different coating materials are available, which vary in chemical composition and polarity. The liquid coatings polydimethylsiloxane (PDMS) and polyacrylate (PA) and the solid coatings polydimethylsiloxane/divinylbenzene (PDMS/DVB), polydimethylsiloxane/carboxen (PDMS/CX) and Carbowax/divinylbenzene (CW/DVB) are in use [95]. It has to be emphasized that the distribution coefficients of analytes between the sample and the coating is also strongly dependent on the matrix. Important matrix effects are the pH value for acidic and basic compounds, the salt concentration, organic solvents and compounds in excess due to competitive adsorption.

A lot of SPME applications in environmental analysis can be found in the literature ([96–100], chapters 11–25 in [101]). A validation study of SPME with 20 participating laboratories showed good reproducibility and accuracy, whereas headspace sampling provided better precision compared to direct liquid sampling [102]. SPME followed by GC with a nitrogen phosphorus detector (NPD) and GC–MS compared to U.S. Environmental Protection Agency (U.S. EPA) method 507 for analysis of 46 phosphorus and nitrogen pesticides in water resulted in lower detection limits for 34 and 39 pesticides using SPME–GC–NPD and SPME–GC–MS, respectively [103].

9.9 NEW ANALYTICAL METHODS FOR ORGANIC POLLUTANTS

9.9.1 Instrumental methods

Chromatographic separation methods, in particular gas chromatography (GC) and high performance liquid chromatography (HPLC), were in the last decades and will be also in the near future the most important methods. Thin layer chromatography (TLC), supercritical fluid chromatography (SFC) and capillary electrophoresis (CE) play only a subordinate role and their applications are restricted to special tasks.

Due to improved selectivity and the ability to identify unknown compounds mass spectrometry is of increasing importance and shows a rapid and overwhelming development [104]. The instruments have become smaller and are more powerful with regard to sensitivity.

Recently LC–MS coupling has reached almost GC–MS with regard to performance and sensitivity. New ionization techniques like the atmospheric pressure chemical ionization (APCI) render direct coupling of HPLC and MS possible for a wide polarity range of analysis. However the APCI process yields mass spectra of great simplicity and does not assist in identification of unknown species. The lack of mass spectral information can be overcome applying MS–MS techniques. MS–MS technique improved mass spectral information and is marked by improved selectivity of target analysis. Therefore MS–MS is especially together with soft ionization techniques like APCI of importance in environmental analysis. In particular MS–MS is simply and economically available in the form of ion-trap mass spectrometers with the capabilities to do MS^n.

Above all LC–MS is capable to determine polar organic pollutants in water. For example sulfonic acid degradation products of acetochlor, alachlor, and metolachlor were detected at 0.1 ppb by electrospray LC–MS–MS [105], APCI–MS was used to identify degradation products of organophosphorus pesticides [106], APCI LC–MS was used to determine 0.2–50 ppb of various phthalates, phenols, and tributyl phosphates [107], 1 ppb thiourea [108], pentachlorophenol, phthalates, nonylphenol [109], and interactions of pesticides with dissolved organic matter [110]. SPE followed by LC–electrospray MS has gained wide acceptance for the detection of organic compounds including about 1 ppt of 45 pesticides, about 30 ppt glyphosate and aminomethylphosphonic acid, and in the range of 10 ppt of poly(ethylene glycol)s (#591 [111–113]).

9.9.2 On-line analysis

To comply with the demands of monitoring programs and legislation automated measuring techniques are needed, such as on-line analytical systems working in a continuous or sequential mode or such as sensors for on-line and in situ measurements. The state of the art can be overviewed in a publication on a workshop organized by the Standards Measurements and Testing (SMT) Program of the European Union [114]. High priority for real time monitoring was estimated for the following fields of application.

- *Environmental quality*: monitoring physical, chemical and biological processes.
- *Treatment processes*: feed water alarm and process control.
- *Distribution system and transport*: drinking water and industrial water.

Subdivision for high priority in real time monitoring can also be done according to the following parameters.

- *Physical*: turbidity, particle number concentration.
- *Global*: toxicity, TOC, BDOC.
- *Organic*: DBPs, pesticides.
- *Inorganic*: phosphate, nitrite, ammonium, speciation (i.e. Cr, As, Se), chlorine (free residual), ozone, bromate.
- *Microbiological*: coliforms.

According to their basic working principles in general different types of sensors can be defined, like optical (e.g. measurement of UV absorbance), electrochemical (e.g. ion-selective electrodes, pH electrode) or biochemical sensors (e.g. based on enzymatic or immunochemical reactions). The common feature of sensors is an in most cases miniaturized transducer, which responds qualitatively and quantitatively to a physical or chemical property of the measured medium by means of electrical or other form of information evaluated by data processing.

Optical and microbiological sensors are widespread for water quality monitoring. For instance, applications can be found for nutrients, metal ions, pesticides, total organic carbon (TOC), biological oxygen demand (BOD), suspended solids, algae or bacteria and viruses [115,116].

The field of sensor applications for water quality monitoring is normally distinguished in two sectors, the less polluted waters like surface, ground, and drinking water and the heavily polluted waste water.

Sensor reviews included reports of microbial and enzyme sensor [117] as well as remote electrochemical sensors [118] for monitoring pollutants and nutrients. A review covers remote electrochemical sensors for in situ monitoring of priority organic and inorganic pollutants [118].

An on-line automated titrimetric sensor is described for ammonia–nitrogen determination [119], an automated spectrophotometric field monitor for continuous on-line recording of ammonia [120] and a new sensor module covers the measuring of O_2, pH, NO_3^-, Cl^-, PO_4^{3-}, NH_3, and NH_4^+ [121]. New developments in optical sensor instrumentation and the problems encountered when measuring O_2, pH, CO_2, and ammonia are discussed [122]. Also fiber-optic sensors are described for the measurement of nitrite and nitrate [123], for nitrate and chlorine with detection limits of 22 and 26 μg/l, respectively [124] and for measuring dissolved oxygen and carbon dioxide [125]. Carbon dioxide was also determined with a fiber-optic fluorescence sensor [126]. A microscale sensor is applied for nitrate measurement [127], and an optical sensor for pH [128] and dissolved oxygen measurement [129].

New biosensors are described for rapid measurement of oxygen demand [130], for cyanide [131] as well as for phytotoxic and nutrient substances [132].

Several sensors are described for the determination of organic micropollutants. An evanescent fiber-optic chemical sensor serves as detection system for organic compounds in water [133]. A multi-analyte immunosensor coupled to a total internal reflectance detection is part of a system called river analyzer. It was applied for on-line detection of chlorotriazines and other micropollutants in river water [134]. Fiber-optic biosensors are applied for on-site TNT analysis [135]. Fiber-optic-based biomonitoring of benzene compounds is done by recombinant *Escherichia coli* bearing a luciferase gene fused to a plasmid [136]. The explosives TNT (trinitrotoluene) and RDX (cyclotrimethylenetrinitramine) are determined by continuous-flow immunosensors (CFI) [137]. Time-resolved, laser-induced and fiber-optically guided fluorescence was applied for monitoring of the remediation of a PAH-contaminated site [138]. For the detection of environmental contaminants in water a parallel affinity sensor array was implemented based on chemoluminescence labels and CCD detection [139].

Finally electrochemical biosensors based on bilayer lipid membranes are applied for the automated rapid and sensitive flow monitoring of substrates of hydrolytic enzymes, antigens, and triazine herbicides [140].

9.10 REFERENCES

1   P. Brimblecombe, T.D. Jickells, P.S. Liss and J.E. Andrews, An Introduction to Environmental Chemistry, Blackwell Science, London, 1996.

2   BGW — Bundesverband der Deutschen Gas- und Wasserwirtschaft (Ed.), Entwicklung der öffentlichen Wasserversorgung 1990–1995, Wirtschafts- und Verlagsgesellschaft Gas und Wasser, Bonn, 1996.

3   F.H. Frimmel and B.C. Gordalla, Wasserkreislauf und Wassernutzung, in: F.H. Frimmel (Ed.), Wasser und Gewässer, Spektrum, Heidelberg, 1999, pp. 3–28.

4   European Commission, Internet http://europa.eu.int/scadplus/leg/en/lvb/i28066.htm (1999).

5   European Environment Agency, Internet http://www.eea.eu.int/ (1999).

6   European Commission, Internet http://europa.eu.int./scadplus/leg/en/lbv/ (1999).

7   J. Rechenberg, Rechtliche Grundlagen eines umfassenden Gewässerschutzes, in: F.H. Frimmel (Ed.), Wasser und Gewässer, Spektrum, Heidelberg, 1999.

8   Umweltpolitik — Wasserwirtschaft in Deutschland, Bundesministerium für Umwelt, Naturschutz und Reaktorsicherheit (Ed.), Bonn, 1998.

9   LAWA 2000 — Forderungen der Wasserwirtschaft für eine fortschrittliche Gewässerschutzpolitik, LAWA Länderarbeitsgemeinschaft Wasser (Ed.), Berlin, 1990.

10   National Water Conservation Plan — Current Central Issues, LAWA Länderarbeitsgemeinschaft Wasser (Ed.), Berlin, 1996.

11   UBA-Texte 17/96, UBA — Umweltbundesamt (Ed.), 1996.

12   Umweltdaten Deutschland, Umweltbundesamt und Statistisches Bundesamt (Eds.), Erich Schmidt, Berlin, 1998.

13   Die Beschaffenheit der großen Fließgewässer Deutschlands, LAWA Länderarbeitsgemeinschaft Wasser (Ed.), Berlin, 1997.

14 Gewässergütebericht '97 — Pflanzenbehandlungs- und Schädlingsbekämpfungsmittel in Oberflächengewässern, Landesumweltamt Nordrhein-Westfalen (Ed.), Essen, 1999.

15 ARW Jahresbericht 1998, Arbeitsgemeinschaft Rhein-Wasserwerke (ARW) (Ed.), Köln, 1999.

16 Grundwasserdatenbank Wasserversorgung, Grundwasserdatenbank Wasserversorgung vedewa, Stuttgart, 1999.

17 Bericht zur Grundwasserbeschaffenheit — Pflanzenschutzmittel, LAWA Länderarbeitsgemeinschaft Wasser (Ed.), Kulturbuchverlag, Berlin, 1997.

18 F.H. Frimmel, DVGW-Schriftenr., 201 (1989) 3.1–3.14.

19 DIN 2000, Leitsätze für die zentrale Trinkwasserversorgung, Deutsches Institut für Normung (Ed.), Beuth, Berlin, 1973.

20 Official Journal, L330 (05.12.1998).

21 S. Schmidt, Schriftenr. Ver. Boden-, Wasser- Lufthyg., 102 (1998) 3–9.

22 DVGW Technical Standard W251: Applicability of river water as raw material for drinking water supply, DVGW (Deutscher Verein des Gas- und Wasserfaches) (Ed.), Wirtschafts- und Verlagsgesellschaft Gas und Wasser, Bonn, 1996.

23 25 Jahre AWBR — Jahresbericht 1993, AWBR (Arbeitsgemeinschaft Wasserwerke Bodensee–Rhein) (Ed.), AWBR, St. Gallen, 1993.

24 S. Richardson, Drinking water disinfection by-products, in: R. Allen and R.A. Meyers (Eds.), Encyclopedia of Environmental Analysis and Remediation, Wiley, New York, 1998, pp. 1399–1421.

25 S.D. Richardson, Anal. Chem., 71 (1999) 181R–215R.

26 T. Heberer, U. Dünnbier, C. Reilich and H.-J. Stan, Fresenius' Environ. Bull., 6 (1997) 438–443.

27 H.-J. Stan and T. Heberer, Analusis, 25 (1997) M20–M23.

28 B. Halling-Sørensen, S.N. Nielsen, P.F. Lanzky, F. Ingerslev, H.C.H. Lützhoft and S.E. Jorgensen, Chemosphere, 36 (1998) 357–393.

29 H. Sontheimer, F. Fuchs, B. Haist-Gulde, K. Johannsen and F.H. Frimmel, Vom Wasser, 75 (1990) 183–200.

30 K. Johannsen, M. Assenmacher, G. Kleiser, G. Abbt-Braun, H. Sontheimer and F.H. Frimmel, Vom Wasser, 81 (1993) 185–196.

31 S.A. Huber, A. Balz and F.H. Frimmel, Fresenius' J. Anal. Chem., 350 (1994) 496–503.

32 F.H. Frimmel, Interpretierbarkeit des SAK (254 nm) bei Gewässeruntersuchungen und Trinkwasseraufbereitung, Alte und neue Summenparameter — Einsatz in der Wasser- und Abwassertechnik, Wiener Mitteilungen Wasser Abwasser Gewässer, 1995.

33 S. Huber and F.H. Frimmel, Anal. Chem., 63 (1991) 2122–2130.

34 S.A. Huber and F.H. Frimmel, Liquid chromatography with three dimensional detection of humic substances and oxidation products, 3rd Int. Nordic Symp. Humic Substances, Finnish Humus News, 1991, pp. 127–132.

35 S.A. Huber and F.H. Frimmel, Fresenius' J. Anal. Chem., 342 (1992) 198–200.

36 F.H. Frimmel, Wasser/Abwasser, 137 (1996) 102–109.

37 German Standard Methods for the Examination of Water, Waste Water, and Sludge, Wiley–VCH, Berlin, 1999.

38 L.A. Hütter, Wasser und Wasseruntersuchung, Diesterweg Salle Sauerländer, Frankfurt, 1984.

39 C. Cammann, Kontinuierliche Gewässergüteüberwachung mit Sensoren und Monitoren, in: F.H. Frimmel and B.C. Gordalla, Gewässergütekriterien — Ergebnisse eines Rundgesprächs, Deutsche Forschungsgemeinschaft, VCH, Weinheim, 1996, pp. 91–114.

40 U. Bilitewski, B. Hock and U. Obst, Anwendung immunchemischer Methoden in der Wasseranalytik, in: H. Günzler (Ed.), Analytiker Taschenbuch, Springer, Berlin, 1998, pp. 251–309.

41 A.J. Hendriks, J.L. Mass-Diepeveen, A. Noordsij and M.A. Gaag, Water Res., 28 (1995) 581–598.

42 A. Fernandez, C. Tejedor, F. Cabrera and A. Chordi, Water Res., 29 (1995) 1281–1286.

43 C. Weins and H. Jork, Vom Wasser, 83 (1994) 279–288.

44 L.P. Burkhard, E.J. Durhan and M.T. Lukasewycz, Anal. Chem., 63 (1991) 277–283.

45 K.M. Jop, T.Z. Kendall, A.M. Askew and R.B. Foster, Environ. Toxicol. Chem., 10 (1991) 981–990.

46 S.S. Rao, B.K. Burnison, D.A. Rokosh and C.M. Taylor, Chemosphere, 28 (1994) 1859–1870.

47 O. Fiehn, L. Vigelahn, G. Kalnowski, T. Reemtsma and M. Jekel, Acta Hydrochim. Hydrobiol., 25 (1997) 11–16.

48   T. Reemtsma, O. Fiehn and M. Jekel, Fresenius' J. Anal. Chem., 363 (1999) 771–776.

49   C. Sierdorfer, M. Bungert and H. Kaltwasser, Vom Wasser, 91 (1998) 87–99.

50   T. Reemtsma, A. Putschew and M. Jekel, Vom Wasser, 92 (1999) 243–255.

51   J.R. Amato, D.I. Mount, E.J. Durhan, M.T. Lukasewycz, G.T. Ankley and E.D. Robert, Environ. Toxicol. Chem., 11 (1992) 209–216.

52   M.P. Hannigan, G.R. Cass, B.W. Penman, C.L. Crespi, A.L. Lafleur, W.F. Busby Jr., W.G. Thilly and B.R.T. Simoneit, Environ. Sci. Technol., 32 (1998) 3502–3514.

53   R.H. Grenn, Sampling Design and Statistical Methods for Environmental Biologists, Wiley, New York, 1979.

54   P. Hoffmann, Nachr. Chem. Tech. Lab., 40 (1992) M2.

55   K.-D. Selent and A. Grupe (Eds.), Die Probenahme von Wasser — Ein Handbuch für die Praxis, Oldenbourg, München, 1998.

56   H.W. Zwanziger, S. Geiß and J.W. Einax, Chemometrics in Environmental Analysis, VCH, Weinheim, 1997.

57   J.Y. Shamas, Water Environ. Res., 70 (1998) 418–423.

58   M. Stoeppler, in: M. Stoeppler (Ed.), Sampling and Sample Preparation, Springer, Berlin, 1997, pp. 1–16.

59   E. Helmers, in: M. Stoeppler (Ed.), Sampling and Sample Preparation, Springer, Berlin, 1997, pp. 26–42.

60   R.L. Jones, Sampling, sample preparation and preservation, in: W. Ebing and H.-J. Stan (Eds.), Chemistry of Plant Protection — Analysis of Pesticides in Ground and Surface Water I, Springer, Berlin, 1995, pp. 3–18.

61   J.N. Huckins, M.W. Tubergen and G.K. Manuweera, Chemosphere, 20 (1990) 533–552.

62   H.F. Prest, J.N. Huckins, J.D. Petty, S. Herve, J. Paasivirta and P. Heinonen, Mar. Pollut. Bull., 31 (1995) 306–312.

63   J.D. Petty, C.E. Orazio, J.N. Huckins, R.W. Gale, J.A. Lebo, J.C. Meadows, K.R. Echols and W.L. Cranor, J. Chromatogr. A, 879 (2000) 83–96.

64   K.E. Gustavson and J.M. Harkin, Environ. Sci. Technol., 34 (2000) 4445–4451.

65   R.W. Gale, Environ. Sci. Technol., 32 (1998) 2292–2300.

66   H.R. Rogers, Chemosphere, 35 (1997) 1651–1658.

67   Y. Wang, C. Wang and Z. Wang, Chemosphere, 37 (1998) 327–340.

68   D. Sabaliunas, J. Lazutka, I. Sabaliuniene and A. Södergren, Environ. Toxicol. Chem., 17 (1998) 1815–1824.

69   C.S. Hofelt and D. Shea, Environ. Sci. Technol., 31 (1997) 154.

70   M.P. Harper, W. Davison and W. Tych, Environ. Sci. Technol., 31 (1997) 3110.

71   T.N. Angilidis, Water, Air, Soil Pollut., 99 (1997) 179.

72   V.J. Barwick, Trends Anal. Chem., 16 (1997) 293–309.

73   L.R. Snyder, J. Chromatogr., 92 (1974) 223.

74   L.R. Snyder, J. Chromatogr. Sci., 16 (1978) 223.

75   S.C. Rutan, P.W. Carr, W.J. Cheong, J.H. Park and L.R. Snyder, J. Chromatogr., 463 (1989) 21.

76   L.R. Snyder, Tech. Chem., 12 (1978) 25.

77   M. Akerblom, Extraction and cleanup, in: W. Ebing and H.-J. Stan (Eds.), Chemistry of Plant Protection — Analysis of Pesticides in Ground and Surface Water I, Springer, Berlin, 1995, pp. 19–66.

78   A. Zapf, R. Heyer and H.-J. Stan, J. Chromatogr. A, 694 (1995) 453–461.

79   A. Fernandez-Gutierrez, J.L. Martinez-Vidal, F.J. Arrebola-Liebanas, A. Gonzales-Casado and J.L. Vilchez, Fresenius' J. Anal. Chem., 360 (1998) 568–572.

80   E. Welter and B. Neidhart, Fresenius' J. Anal. Chem., 357 (1997) 345.

81   P. Scribe and M.-C. Hennion, Sample handling for the analysis of organic compounds from environmental water samples, in: D. Barceló (Ed.), Environmental Analysis: Techniques, Applications and Quality Assurance, Elsevier, Amsterdam, 1993, pp. 23–77.

82   M. Mills and E.M. Thurmann, Solid-Phase Extraction: Principles and Practice, Wiley, New York, 1998.

83   R. Wissiack, E. Rosenberg and M. Grasserbauer, J. Chromatogr. A, 896 (2000) 159–170.

84 C. Hidalgo, J.V. Sancho and F. Hernandez, Quim. Anal., 16 (1997) 259.

85 T. Renner, D. Baumgarten and K.K. Unger, Chromatographia, 45 (1997) 199–205.

86 S. Lacorte, J.J. Vreuls, J.S. Salau, F. Ventura and D. Barcelo, J. Chromatogr. A, 795 (1998) 71–82.

87 C. Aguilar, I. Ferrer, F. Borrull, R.M. Marce and D. Barcelo, J. Chromatogr. A, 794 (1998) 147–163.

88 A. Brandt and T.H.M. Noij, HPLC with on-line solid phase extraction for trace analysis of polar pesticides, in: W. Ebing and H.-J. Stan (Eds.), Chemistry of Plant Protection — Analysis of Pesticides in Ground and Surface Water II, Springer, Berlin, 1995, pp. 91–108.

89 C.W. Thorstensen, O. Lode and A.L. Christiansen, J. Agric. Food Chem., 48 (2000) 5829–5833.

90 D. Puig and D. Barcelo, J. Chromatogr., 778 (1997) 313.

91 I. Ferrer, B. Ballestros, M.P. Marco and D. Barcelo, Environ. Sci. Technol., 31 (1997) 3530–3535.

92 A.C. Hogenboom, I. Jagt, R.J.J. Vreuls and U.A.T. Brinkmann, Analyst, 122 (1997) 1371–1377.

93 M.J.-F. Suter, A.C. Alder, M. Berg, C.S. McArdell, S. Riediker and W. Giger, Chimia, 51 (1997) 871–877.

94 A.J.H. Louter, U.A.T. Brinkmann and J.J. Vreuls, Capillary GC with selective detection using on-line solid phase extraction and liquid chromatography techniques, in: W. Ebing and H.-J. Stan (Eds.), Chemistry of Plant Protection — Analysis of Pesticides in Ground and Surface Water II, Springer, Berlin, 1995, pp. 1–31.

95 K. Levsen and C. Grothe, The applications of SPME in water analysis, in: R.M. Smith and J. Pawliszyn (Eds.), Applications of Solid Phase Microextraction, The Royal Society of Chemistry, Cambridge, 1999, pp. 169–187.

96 A. Saraullo, P.A. Martos and J. Pawliszyn, J. Anal. Chem., 69 (1997) 1992–1998.

97 V. Lopez-Avila, R. Young and W.F. Beckert, J. High Resolut. Chromatogr., 20 (1997) 487–492.

98 W.-H. Ho and S.-J. Hsieh, Anal. Chim. Acta, 428 (2001) 111–120.

99 F. Werres, J. Stien, P. Balsaa, A. Schneider, P. Winterhalter and H. Overath, Vom Wasser, 94 (2000) 135–148.

100 S. Aguerre, C. Bancon-Montigny, G. Lespes and M. Potin-Gautier, Analyst, 125 (2000) 263–268.

101 R.M. Smith and J. Pawiszyn (Eds.), Applications of Solid Phase Microextraction, The Royal Society of Chemistry, Cambridge, 1999.

102 T. Nilsson, R. Ferrari and S. Facchetti, Anal. Chim. Acta, 356 (1997) 113–123.

103 T.K. Choudry, K.O. Gerhardt and T.P. Mawhinney, Environ. Sci. Technol., 30 (1996) 3259.

104 H. Engelhardt, Nachr. Chem. Tech. Lab., 47 (1999) 1024–1026.

105 J.D. Vargo, Anal. Chem., 70 (1998) 2699–2703.

106 M. Castillo, R. Domingues, M.F. Alpendurada and D. Barcelo, Anal. Chim. Acta, 353 (1997) 133–142.

107 M. Castillo, M.F. Alpendurada and D. Barcelo, J. Mass Spectrom., 32 (1997) 1100–1110.

108 A. Raffaelli, S. Pucci, R. Lazzaroni and P. Salvadori, Rapid Commun. Mass Spectrom., 11 (1997) 259–264.

109 M. Castillo, A. Oubina and D. Barcelo, Environ. Sci. Technol., 32 (1998) 2180–2184.

110 U. Klaus, T. Pfeifer and M. Spiteller, Environ. Sci. Technol., 34 (2000) 3514–3520.

111 C. Creszenzi, A. Di Corcia and R. Samperi, Environ. Sci. Technol., 31 (1997) 479–488.

112 R.F. Vreeken, P. Speksnijder, I. Bobeldijk-Pastorova and T.H.M. Noij, J. Chromatogr., 794 (1998) 187–199.

113 C. Creszenzi, A. Di Corcia, A. Marcomini and R. Samperi, Environ. Sci. Technol., 31 (1997) 2679–2685.

114 P. Quevauviller and F. Colin (Eds.), Monitoring of Water Quality, Elsevier, Oxford, 1998

115 P. Scully, Optical techniques for water monitoring, in: P. Quevauviller and F. Colin (Eds.), Monitoring of Water Quality, Elsevier, Oxford, 1998, pp. 15–35.

116 P. Vasseur and D. Osbild, Microbiological sensors for the monitoring of water quality, in: P. Quevauviller and F. Colin (Eds.), Monitoring of Water Quality, Elsevier, Oxford, 1998, pp. 37–48.

117 K. Riedel, R.D. Schmid, C. Wittmann and E. Kress-Rogers (Eds.), Handbook Biosensors and Electronic Noses, CRC, Boca Raton, FL, 1997, pp. 299–332.

118 J. Wang, Trends Anal. Chem., 16 (1997) 84–88.

119 K. Gernaey, H. Bogaert, A. Massone, P. Vanrolleghem and W. Verstraete, Environ. Sci. Technol., 31 (1997) 2350.

120   M.J. Bloxham, M.H. Depledge and P.J. Worsfold, Lab. Rob. Autom., 9 (1997) 175.

121   S. Herrmann, W. Vonau, F. Gerlach and H. Kaden, Fresenius' J. Anal. Chem., 362 (1998) 215–217.

122   F. Reininger and W. Trettnak, Optochemical sensors in water monitoring, in: P. Quevauviller and F. Colin (Eds.), Monitoring of Water Quality, Elsevier, Oxford, 1998, pp. 117–136.

123   M. Fiore, M. Brenci and J. Kozlowski, Proc. SPIE-Int. Soc. Opt. Eng., 3105 (1997) 138–143.

124   P. Dress, M. Belz, K.F. Klein, K.T.V. Grattan and H. Franke, Appl. Opt., 37 (1998) 4991–4997.

125   D. Garcia-Fresnadillo and G. Orellana, Fibre-optic chemical sensors: from molecular engineering to environmental analytical chemistry in the field, in: P. Quevauviller and F. Colin (Eds.), Monitoring of Water Quality, Elsevier, Oxford, 1998, pp. 103–115.

126   O.S. Wolfbeis, B. Kovacs, K. Goswami and S.M. Klainer, Microchim. Acta, 129 (1998) 181–188.

127   L.H. Larsen, T. Kjr and N.P. Revsbech, Anal. Chem., 69 (1997) 3527–3531.

128   S. De Marcos and O.S. Wolfbeis, Anal. Chim. Acta, 334 (1996) 149–153.

129   B.-K. Sohn and C.-S. Kim, Sens. Actuators, B, B34 (1996) 435–440.

130   L. An, H. Niu and H. Zeng, Water Environ. Res., 70 (1998) 1070–1074.

131   K. Ikebukuro, A. Miyata, S.J. Cho, Y. Nomura, S.M. Chang, Y. Yamauchi, Y. Hasebe, S. Uchiyama and I. Karube, J. Biotechnol., 48 (1996) 73–80.

132   F. Jacques, Y. Boudey and J.-M. Ory, Conf. Proc.-Jt. Conf.: IEEE Instrum. Meas. Technol. Conf. IMEKO Technol. Comm. 7, Institute of Electrical and Electronics Engineers, New York, 1996, pp. 1354–1359.

133   D.S. Blair and L.W. Burgess, A.M. Brodsky, Anal. Chem., 69 (1997) 2238–2246.

134   E. Mallat, C. Barcen, A. Klotz, A. Brecht, G. Gauglitz and D. Barceló, Environ. Sci. Technol., 33 (1999) 965–971.

135   L.C. Shriver-Lake, B.L. Donner and F.S. Ligler, Environ. Sci. Technol., 31 (1997) 837.

136   Y. Ikariyama, S. Nishiguchi, T. Koyama, E. Kobatake, M. Aizawa, M. Tsuda and T. Nakazawa, Anal. Chem., 69 (1997) 2600–2605.

137   J.C. Bart, L.L. Judd, K.E. Hoffman, A.M. Wilkins and A.W. Kusterbeck, Environ. Sci. Technol., 31 (1997) 1505.

138   R. Kotzick and R. Niessner, Fresenius' J. Anal. Chem., 354 (1996) 72–76.

139   M.G. Weller, A. Schuetz, M. Winklmair and R. Niessner, Anal. Chim. Acta, 393 (1999) 29–42.

140   D.P. Nikolelis and C.G. Siontorou, J. Autom. Chem., 19 (1997) 1.

Subject index